Power Electronic Converters in Advanced Co-Phase Traction Power Supply System

Zeliang Shu

Power Electronic Converters in Advanced Co-Phase Traction Power Supply System

Springer

Zeliang Shu
College of Electrical Engineering
Southwest Jiaotong University
Chengdu, Sichuan, China

ISBN 978-981-95-3275-9 ISBN 978-981-95-3276-6 (eBook)
https://doi.org/10.1007/978-981-95-3276-6
Jointly published with Southwest Jiaotong University Press

The print edition is not for sale in China (Mainland). Customers from China (Mainland) please order the print book from: Southwest Jiaotong University Press.

This Springer imprint is published by the registered company Springer Nature Singapore Pte Ltd.
The registered company address is: 152 Beach Road, #21-01/04 Gateway East, Singapore 189721, Singapore

Preface

The traction power supply system plays a critical role in enabling rapid development of high-speed and heavy-haul railways. However, the conventional systems face serious challenges including neutral section passing, power quality, and limited overload capacity. The advanced co-phase traction power supply system based on power electronics established an interconnected traction network across multiple substations. This system eliminates neutral section, reduces substation capacity requirements and voltage drops, and significantly improves power supply efficiency. It also enhances grid stability by enabling mutual backup support between substations during faults. Additionally, it mitigates traditional power quality issues, including three-phase voltage unbalance, inadequate reactive power compensation, and harmonic distortions.

The advanced co-phase traction power supply system is highly suitable for next-generation high-speed railway and compatible with upgrades to existing traction networks. It also offers an innovative solution for specialized projects such as long-distance traction power supply, weak-grid high-altitude railways, and new energy grid connection. Consequently, it represents an ideal next-generation traction power supply technology. However, realizing this system urgently requires fundamental research on its core component—power electronic converters. Their topologies, operation principles, modulation schemes, and control strategies must be thoroughly analyzed and validated prior to engineering implementation to mitigate technical risks. This book will provide a detailed discussion centered on the aforementioned topics. It is divided into seven chapters:

Chapter 1: Introduction
Chapter 2: Grid-Connected Characteristic Analysis of Advanced Co-phase Traction Power Supply System
Chapter 3: Control and Modulation of Grid-Connected Three-Phase Converter
Chapter 4: Control and Modulation of Single-Phase Converter in Traction Power Supply
Chapter 5: Diode-Clamped Multilevel Three-Phase-to-Single-Phase Converter
Chapter 6: Modular Multilevel Three-Phase-to-Single-Phase Converter
Chapter 7: Analysis and Control of Two-Phase-to-Single-Phase Converter

The author of this book, Professor Zeliang Shu, has participated in the development and implementation of next-generation traction supply system, including the world's first Meishan co-phase power supply system in 2007, heavy-haul railways in Shanxi, and urban lines in Wenzhou. In 2011, he led three projects funded by the National Natural Science Foundation of China (NSFC) on converters and control strategies for ideal co-phase traction systems, three-phase-to-single-phase converters, and two-phase-to-single-phase power electronic transformers of advanced co-phase traction power supply system. This book constitutes a comprehensive summary of these research efforts. It has been made possible through the invaluable support of the following individuals, to whom I extend my deepest gratitude: Prof. Qunzan Li, Prof. Shaofeng Xie, Prof. Xiaoqiong He, Siyang Liu, Biao Yu, Li Zhao, Feng Chu, Han Yan, Yousong Zhou, Linghui Meng, Xufeng Zhang, Yajun Chen, Xiaodong Yin, Ziwei Zhang, Shasha Song, Shitao Wang, Jiangling Nie, Jingci Wu, Jiangpeng Yang, Shun Wang, Yuhao Deng, and others.

Owing to the inherent constraints of this work, certain omissions and inaccuracies may persist. We welcome constructive feedback and corrections from readers.

Chengdu, Sichuan, China
August 2025

Zeliang Shu

Contents

Chapter 1
Introduction

1.1 Brief Introduction to Electrification Development in Railway Transport Field

1.1.1 Development of Electrified Railway and High-Speed Railway

Siemens & Halske AG exhibited an about 550-m section of electrified railway at the Berlin Trade Exhibition held in 1879, marking the first instance of electric traction being used to pull trains in human history. In May 1881, Germany built and put into operation an electrified railway that spanned 2.45 km at Lichterfelde, on the outskirts of Berlin. This railway became the world's first commercially operated electrified railway, ushering in a new era of electric traction for railways. In the mid-twentieth century, societal development created an increasingly strong demand for railway transport, leading to extensive construction of electrified railways (Qian 2003).

On the eve of the 1964 Tokyo Olympic Games, Japan's Tokaido Shinkansen was officially launched into commercial operation as the world's first high-speed railway (HSR). Designed for a maximum speed of 240 km/h, this passenger dedicated line (PDL) has operated at speeds of up to 210 km/h, marking a new chapter in the history of HSRs worldwide. The global development of HSRs can be broadly categorized into three stages (Qian 2003; Li 2012):

Stage I: 1960s to the end of 1980s. The global construction and development of HSRs were primarily driven by Japan, France, Italy, and Germany. During this period, Japan established a nationwide HSR backbone network and achieved significant technical, commercial, financial, and political success through the development and operation of the Shinkansen.

Stage II: the end of 1980s to the mid-1990s. France, Germany, Italy, Spain, Belgium, the Netherlands, Sweden, and the United Kingdom made prominent achieve-

Z. Shu, *Power Electronic Converters in Advanced Co-Phase Traction Power Supply System*, https://doi.org/10.1007/978-981-95-3276-6_1

ments in the construction of high-speed railways. During this period, many countries including Japan, France, Germany, and Italy meticulously planned and expanded their nationwide HSR networks.

Stage III: the mid-1990s to present. With China's remarkable achievements in HSR construction, HSR networks now extend across Asia, Oceania, and all of Europe, driving a global transformation and upgrading of the transport industry.

Currently, the construction and development of HSRs are still on the rise, primarily due to several significant advantages they offer compared to other modes of transport, such as highways and airlines (Li 2012):

Higher speed: HSRs allow trains to operate at average speeds ranging from 200 to 350 km/h over medium and long distances. Chinese-made CRH trains have set a world record of 605 km/h, serving as a benchmark for continuous speed increases in practical operation. Additionally, the typical location of their stations in urban centers provides an advantage over air travel, which often involves extended wait times for flights and lengthy local transport.

Greater capacity: HSRs support trains consisting of coaches that accommodate a large number of passengers and can be configured into various formations for operation on dedicated lines. Furthermore, these trains can operate with a headway as short as 4 min on popular lines. These advantages significantly enhance their transport capacity and efficiency compared to other modes of transport, such as highways and airlines.

Safer and more comfortable: High-speed trains featuring spacious coaches offer a smooth and comfortable riding experience. Additionally, dedicated lines ensure safe and reliable train operation, significantly reducing the likelihood of accidents and casualties.

All-weather operation: HSRs enable trains to operate under all weather conditions with high punctuality rates, remaining largely unaffected by adverse weather such as rainstorms, snowstorms, dense fog, and strong winds.

In addition, HSRs are characterized by lower overall energy consumption, reduced environmental pollution, enhanced social and economic benefits, and decreased transport costs. It is anticipated that high-speed railways will continue to maintain their position as a primary and preferred mode of transport for medium- and long-distance travel for the foreseeable future.

1.1.2 Overview of Electrified Railway Development in China

The Baoji-Chengdu Railway, China's first electrified railway, was completed and launched for operation on August 15, 1961. The Beijing-Tianjin Intercity Railway, China's first high-speed PDL capable of operating at speeds of up to 350 km/h, officially opened on August 1, 2008, heralding a new era for HSR construction in China. As outlined in the *Mid- and Long-Term Railway Network Plan* deliberated and adopted by the Chinese Government in January 2004, by 2020, the nationwide operational railways were projected to exceed 120,000 km, busy trunk corridors were to be expanded for separate freight and passenger traffic, and both national duplication and electrification rates were expected to surpass 50%. The plan included the construction of an additional 41,000 km of new lines, with over 16,000 km of PDLs, aimed to establish the backbone of "four north-south and four east-west" PDLs and to connect provincial capitals and large- to medium-sized cities with rapid passenger transport options. The release of this document signifies the start of a rapid development period for railway construction in China (The People's Republic of China 2008).

According to statistics from China State Railway Group Co., Ltd., by the end of 2020, the nationwide operational railway network totaled 146,300 km, including 38,000 km of HSRs. The density of the railway network reached 152.3 km per 10,000 square kilometers, while the duplication and electrification rates were 59.5% and 72.8%, respectively. In the western regions of China, operational railways extended to 59,000 km. The total national operational railway length was 128,000 km, with duplication and electrification rates of 61.6% and 74.9%. The *National Comprehensive Three-Dimensional Transportation Network Planning Outline*, issued in 2021, sets the development goal of reaching 70,000 km of HSRs and 130,000 km of conventional railways by 2035. These railways are expected to form an HSR network comprising the backbone of "eight north-south and eight east-west" main corridors, connected by regional HSRs, and a conventional railway network featuring interconnected north-south and east-west lines, as well as an intercity railway network linking major cities.

The construction and operation of railways in China, particularly HSRs, will keep long-term development emphasizing independence, reliability, high speed, and efficiency. Traction power supply is considered one of the key technologies in railway transportation, especially for electrified railways. Against the backdrop of the long-term booming development of the railway industry in China, traction power supply technology is expected to evolve gradually based on its currently established mature foundation, fostering innovative development in railways in China, particularly HSRs.

This book focuses on the novel advanced co-phase traction power supply technology for railway traction systems. It thoroughly explores core power electronic conversion techniques that support this technology. The book also discusses associated circuit topologies, application characteristics, control methods, simulation verification, and more. Additionally, it presents preliminary test results from experiments conducted on a prototype verification platform. The aim is to support further research and application of advanced co-phase traction power supply technology.

1.2 Brief Introduction to Research and Application of Traction Power Supply System

Traction power supply in China is configured as a single-phase AC system with a power frequency of 50 Hz and a rated voltage of 27.5 kV or 2 × 27.5 kV (Li 2010). The majority of modern electric traction networks are supplied from the three-phase power grid of their respective electric power systems. Consequently, traction power supply systems need to include traction transformers to convert the three-phase supply from the grid into single-phase traction power supply. The two feeding sections installed at traction substations can be connected to the three-phase power grid alternately. This arrangement allows for the incorporation of specialized balance traction transformers, which can mitigate the impact of single-phase traction supply on the three-phase power grid to some extent. However, the main loads supplied by traction networks are time-varying. Specifically, the position, duration, and current collection level of trains and motor cars supplied by these networks are variable. Such rapidly changing single-phase loads introduce serious impacts, including harmonics and reactive power components, further affecting the electric energy quality of the three-phase power grid and leading to significant challenges such as harmonics, reactive power components, and three-phase unbalance. Additionally, the mode of power supply across neutral sections faces serious problems related to neutral section passing. With the development of high-speed and heavy-haul railways, the aforementioned challenges of electric energy quality and neutral section passing can no longer be effectively addressed within the existing power supply modes. Therefore, upgrading traction power supply systems has become an urgent necessity (Guo et al. 2000; Shi-Lin et al. 2004; Li 2006a).

1.2.1 Overview of Traction Power Supply Technology Abroad

Japan, France, and Germany were early pioneers in the research and application of HSRs, developing a range of mature technologies, including distinctive traction power supply systems, which are briefly discussed below (Li 2006a; Jin 2006):

1. Railway Traction Power Supply System in Japan

   The Japanese railway traction power supply system employs an AC power frequency sectionalized power supply mode. The utility grids connected to traction power supply systems have a low tolerance for deficiencies in electric energy quality originating from electrified railways, due to their limited short-circuit capacity. Japan has extensively implemented electric energy quality compensation devices and balance traction transformers to address this issue and adopted automatic neutral section passing devices with ground control to mitigate problems related to neutral section passing.

The characteristics of the traction power supply system in Japan are as follows: (1) Sectionalized power supply mode is adopted, requiring neutral section passing. (2) The automatic neutral section passing system based on ground control is adopted to prevent power supply interruptions during train operation. Although this system can avoid speed loss when high-speed trains pass through neutral sections, it brings several disadvantages, including huge investment, expensive maintenance, and substantial impacts on onboard equipment. (3) In order to solve electric energy quality deficiencies, a variety of compensation devices are installed, such as static VAR compensators (SVCs) and railway power conditioners (RPCs), considerably increasing investment costs.

2. Railway Traction Power Supply System in France

   French railways are also configured using a single-phase power frequency AC power supply mode. They are connected to large-capacity high-voltage utility grids (either 225 kV or 400 kV) through single-phase traction transformers and a sectionalized power supply mode. Since the energy capacity of these grids is relatively high, they can better support fluctuating railway loads. As a result, electric energy quality deficiencies, such as negative sequences and harmonics caused by traction loads, are less pronounced.

   The characteristics of the traction power supply system in France are as follows: (1) Sectionalized power supply mode is adopted, requiring neutral section passing. (2) The automatic neutral section passing mode based on onboard control is implemented. However, trains passing through neutral sections experience significant speed loss due to brief power interruptions. The complex electrical process involved in neutral section passing contributes to a complicated electrical coupling relationship between the train and the grid. Additionally, this process necessitates frequently opening the onboard main circuit breaker, which generates substantial electrical shocks and affects various devices, including circuit breakers, high-voltage equipment, and signal sensors, to varying degrees. (3) As large-capacity and high-voltage utility grids are connected, the electric energy quality deficiencies are not prominent.

3. Railway Traction Power Supply System in Germany

   Germany implements a railway traction power supply mode completely different from those in Japan and France. German railways are connected to utility grids through private power grids built specifically for railway purposes, using AC-DC-AC converter stations (Li 2010). The characteristics of the traction power supply system in Germany are as follows: (1) The traction power supply system operates at co-phase voltage throughout the network, eliminating the need for neutral section passing devices. This design avoids a series of problems associated with neutral section passing. (2) This configuration also eliminates the need for electric energy quality compensation devices, as deficiencies caused by the railway traction power supply system to the utility grids are avoided. (3) This arrangement reduces the requirements for the connected utility grids, allowing German traction power supply systems to link with utility grid systems at various voltage levels, including medium and high voltage.

The German mode is an ideal power supply mode, offering a valuable reference. However, it is important to consider the following shortcomings: (1) Building private power grids requires a significant investment. (2) The traction power supply system operates at a frequency of 16.7 Hz, which is incompatible with the utility grid. Additionally, low-frequency equipment tends to be heavy and requires substantial space. (3) The German traction power supply system is rated at 15 kV, resulting in a limited power supply capacity for individual substations. This necessitates the installation of more substations at shorter intervals.

1.2.2 Overview of Traction Power Supply System in China

The traction power supply systems currently implemented in China utilize a three-phase-to-two-phase power supply mode across neutral sections. These systems convert the three-phase power from the electric power system into two phases using three-phase-to-two-phase traction transformers installed at the traction substations. This configuration supplies electricity to two single-phase traction networks on both sides of the substations. The notable advantages of this power supply mode include the use of more suitable types of main transformers, which reduces manufacturing requirements and improves economic efficiency. However, these systems also face several significant challenges as follows (Shi-Lin et al. 2004; Li 2006a):

1. **Electric energy quality deficiency**: As single-phase loads within the three-phase electric power systems, electrified railways bring serious problems such as harmonics, reactive power components, and three-phase unbalance.
2. **Line voltage drop**: Due to impedance in the traction wires, the power supply voltage decreases significantly as the distance from the power source increases. In some cases, the terminal voltage may fall below the minimum limit of 20 kV, seriously undermining the traction capacity and operational speed of trains. This issue also restricts the choice of power supply distance between substations.
3. **Neutral section passing**: Traction voltage generally varies in terms of phase, amplitude, and frequency between feeding sections, necessitating the inclusion of neutral sections. When trains pass through these neutral sections, they may enter an operational state lacking tractive force, which can lead to stopping incidents and subsequent rescue operations. The selection of automatic neutral section passing devices based on ground control or onboard control complicates the system and may become a vulnerable component prone to faults within the traction power supply systems (Guo et al. 2000).

1.2.3 Existing Improvement Solutions for Traction Power Supply System

To facilitate the rapid development of railways in China and address the aforementioned challenges, experts and scholars in the country's railway industry have conducted extensive research and innovation efforts focused on traction power supply transformers. They have developed various power supply modes tailored to specific local conditions, including direct feeding, booster transformer (BT) feeding, direct feeding mode with a negative feeder, and autotransformer (AT) feeding. These modes aim to enhance both the traction power supply capacity and quality. The primary solutions to the challenges related to neutral section passing, deficiencies in electric energy quality, and power supply capacity are as follows:

1. **Solutions to neutral section passing issues**

   The relatively mature neutral section passing technologies primarily consist of automatic neutral section passing based on onboard control and those based on ground control (Guo et al. 2000; Shi-Lin et al. 2004; Li 2006a; Jin 2006; Yan 1999). The former tends to result in lower train speeds, a shorter service life for the onboard main circuit breaker, and the generation of strong electrical shocks. As railway design speeds continue to increase, this mode struggles to maintain continuous high-speed train operation, with longer power-off distances leading to more significant speed reductions. For heavy-haul railways, this challenge can adversely affect the carrying and passing capacities of trains. Although the latter has been widely implemented in Japan, issues related to the lifespan and reliability of vacuum switches remain unresolved. Furthermore, overvoltage and overcurrent generated during neutral section passing have negative impacts on both the traction networks and trains.

   Researchers both in China and abroad have also proposed automatic neutral section passing solutions based on frequency conversion and phase shift using power electronic switches (Liu 2012). These systems demonstrate improved performance in preventing shocks during neutral section passing. However, neutral section passing imposes stringent requirements on the converters, particularly regarding their dynamic response and power capacity to handle all load currents, while their utilization rate during other periods remains low.

   In summary, neutral section passing has emerged as a significant bottleneck in the traction power supply systems of HSRs. Reducing or completely eliminating neutral sections is recognized as a critical technical challenge within the HSR field.

2. **Solutions to electric energy quality deficiencies**

   Electric energy quality deficiencies associated with electrified railways mainly include reactive power components, harmonics, negative sequences, as well as voltage fluctuations and flicker. These deficiencies can vary significantly across different historical stages, countries, and regions (Li 2006a, b, 2010; Li et al. 1988; Li and He 2007; Qiu et al. 2010).

In the current stage of transitioning from AC-DC traction drives to AC-DC-AC traction drives in China, the issues related to reactive power components and harmonics have been largely addressed. However, a significant number of existing AC-DC drive locomotives will continue to be used for an extended period, contributing to the persistence of reactive power components and harmonics on electrified railways. Moreover, because electrified railways serve as single-phase loads, the voltage unbalance in three-phase power grids is exacerbated with increasing traction loads. While most traction power supply systems for HSRs in China are connected to utility grids rated over 110 kV, which are better equipped to handle shocks, harmonics, and negative sequences, electric energy quality deficiencies remain a crucial issue for both the electrified railways and electric power systems, particularly when considering the large-scale networking of traction loads and the surge of high-power, long-formation loads.

There are two traditional solutions to address electric energy quality deficiencies in traction power supply systems: passive compensation and active compensation. Active compensation solutions include the inclusion of SVGs, which are based on power electronic technology and exhibit both robust dynamic and static performance. SVGs have been used in many traction substations, including Anding and Sanjie, to control reactive power components and harmonics (Li 2011). Nanxiang Traction Substation in Shanghai serves as another example of active compensation solutions for the compensation of reactive power components and harmonics (Wang 2011). Compensation devices can also be installed on the three-phase electric power system side, and these devices include three-phase thyristor-controlled rectifiers (TCRs), static synchronous compensator (STATCOM), and active power filters (APFs) for electric power system applications (Costa-Castello et al. 2009; Fujii et al. 2005; Tan et al. 2003).

The Railway Technical Research Institute, East Japan Railway, and Meidensha Corporation have collaboratively developed the RPC compensation system, which functions to balance two-phase power supply and compensate for harmonics and reactive power components (Boora et al. 2007; Uzuka and Ikedo 2004; Liu 2010a). This RPC compensation solution has proven effective in the comprehensive control of reactive power components, harmonics, and negative sequences. However, it does not adequately address the issue of neutral sections (Wu 2012).

3. **Solutions to power supply capacity guarantee**

The continuously increasing railway speeds impose higher demands on traction power supply systems, particularly regarding their power supply capacity and continuous reliability. Under existing technical conditions, traction power supply systems with neutral sections face significant challenges related to their power supply capacity (Wu 2008). Additionally, these systems lack a mechanism for mutual support among traction substations, which restricts the power supply distances. Moreover, the capacity of traction substations can only be utilized by the loads within their designated feeding sections, leaving no opportunity for interconnection involving multiple substations or mutual support among them. This results in insufficient capacity utilization.

1.2.4 Characteristics of Ideal Traction Power Supply Systems

Based on the technical characteristics and weaknesses of currently available traction power supply systems, the following four key characteristics are envisioned for the development of ideal traction power supply systems:

1. **Elimination of neutral-section passing**: The elimination of neutral section passing facilitates interconnection throughout the system, enabling mutual support among all substations. This advantage also allows for power supply in feeding sections and across different feeding sections.
2. **High stability**: A high level of system stability ensures bidirectional control controllability of power supply flow, providing redundancy and backup throughout the entire system.
3. **High electric energy quality**: This characteristic ensures high-quality power supply for both three-phase power grids and traction networks.
4. **High flexibility**: This characteristic allows for connection to various power grids, incorporating new energy sources and energy storage, to ensure a flexible layout and high efficiency.

An ideal traction power supply system with the aforementioned characteristics functions as an ideal load to the power source while also acting as an ideal power source for the loads. These notable features emphasize interconnection throughout the entire system and power controllability. With advancements and an increasing level of maturity in the field of power electronic technology, the application of power electronic converters presents a promising opportunity to create ideal traction power supply systems. It is foreseeable that power electronic traction power supply systems will emerge as optimal solutions.

1.3 Overview of Research and Application of Co-phase Traction Power Supply

Co-phase power supply systems represent an innovative mode of traction power supply designed to address challenges related to electric energy quality and neutral section passing. Professor Li Qunzhan from Southwest Jiaotong University, along with other scholars, first proposed a technical solution for constructing co-phase power supply systems in literature (Li et al. 1988). Their research findings established both a theoretical and technical foundation for the implementation of a co-phase power supply system. Unlike traditional feeding methods that rely on neutral sections, co-phase feeding operates without them. In this system, the overhead contact system (OCS) within the feeding section of the traction substations delivers electricity in a co-phase voltage pattern. Co-phase feeding primarily addresses the drawbacks of neutral section passing and the deficiencies in electric energy quality commonly associated with conventional feeding systems as illustrated in Fig. 1.1a.

Figure 1.1b illustrates a co-phase power supply system utilizing power electronic compensation. This system includes three-phase-to-two-phase balance traction transformers that convert three phases from the grid into α and β phases. Phase α is directly connected to the OCS for power supply to the feeding section of substations. The voltage of phase β is connected to phase α through an APC. This device functions as a single-phase back-to-back AC-DC-AC converter, operating in a bidirectional, four-quadrant manner. It transmits half of the active power from phase β to phase α while compensating for all reactive power and filtering all harmonics generated by the loads. In this manner, the active loads served by the traction network are evenly distributed across the two output circuits of the balance transformer in real time. Additionally, reactive power components and harmonics are completely compensated, preventing them from flowing through the transformer and three-phase grid. Consequently, co-phase power supply can effectively address

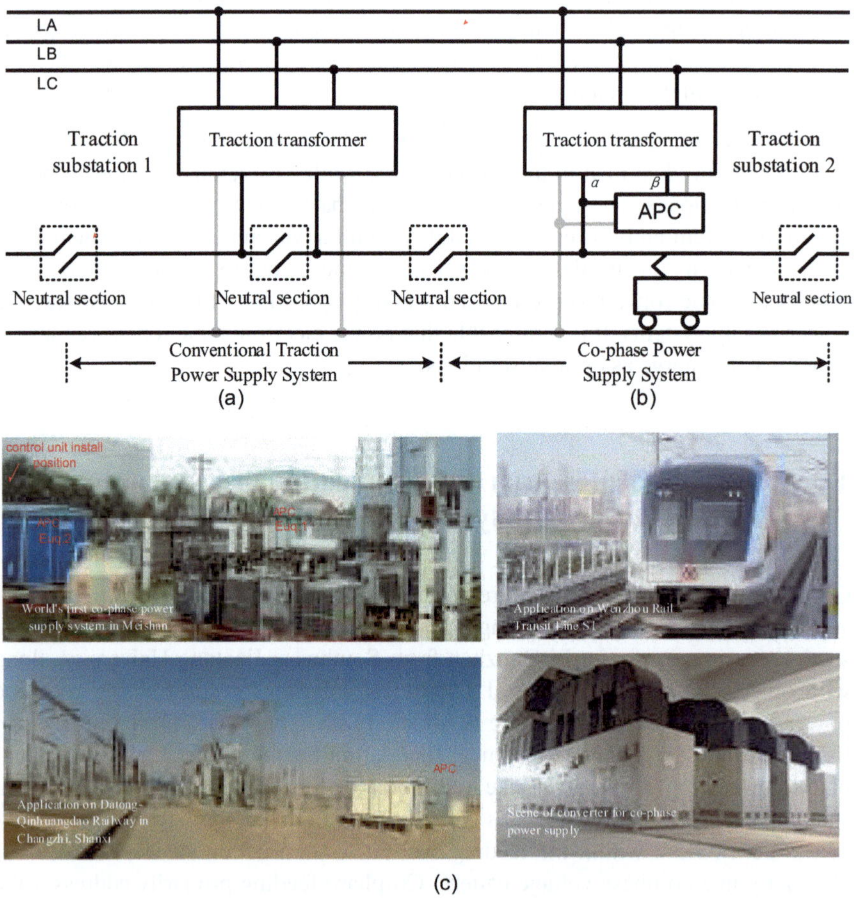

Fig. 1.1 (**a**) Traditional power supply relying on neutral sections; (**b**) sketch of co-phase power supply; (**c**) application scenes of co-phase power supply systems

deficiencies in electric energy quality, such as reactive power components, harmonics, and unbalance in three-phase electric power systems, while eliminating the need for neutral section passing devices at substations. Loads in the original two-phase feeding sections at substations can share the total power supply capacity of the entire traction transformer, indicating an optimized and enhanced load capacity in the feeding sections.

In 2010, the world's first co-phase power supply system, based on a balance transformer and active power compensation device, was successfully put into trial operation at the Meishan Substation of the Chengdu-Kunming Railway (Jin 2006; Liu 2010a; Shu et al. 2013a). This system has successfully achieved the established design objectives, including the elimination of neutral section passing at the substation, interphase compensation, and three-phase equilibrium. It incorporates two sets of parallel APC devices to enhance system reliability. Following this initial implementation, the co-phase power supply system has been effectively applied to various railways, including the Datong-Qinhuangdao Railway in Changzhi, Shanxi, and Wenzhou Rail Transit (Li and Xie 2014; Du 2015). The application scenes are depicted in Fig. 1.1c. The authors participated in the research and engineering construction of these co-phase power supply systems, taking responsibility for their control principles, verification of these principles, coordination of software and hardware for controllers, and other related tasks. This experience has yielded valuable R&D insights, providing a strong foundation for the in-depth study of advanced co-phase traction power supply systems.

1.4 Overview of Research on Advanced Co-phase Traction Power Supply

Co-phase power supply systems can be implemented using various types of balance traction transformers, such as YNvd and Vv (Zhang et al. 2006, 2015; Huang et al. 2016; Zeng and Hao 2003). One of the transformer's bridge arms is connected directly to the traction network, resulting in uncontrollable voltage amplitude, phase, and frequency. As a consequence, adjacent traction substations within a co-phase power supply system cannot be directly interconnected, preventing continuous power supply throughout the system. However, the advanced co-phase traction power supply offers a solution for enabling interconnection across entire railways. This book focuses on exploring different types of power electronic converters for continuous power supply.

Power electronic converters serve as the core components of advanced co-phase traction power supply systems, replacing the transformer distribution network found at traditional substations. The circuit topology options include a three-phase-to-single-phase configuration as depicted in Fig. 1.2a, a two-phase-to-single-phase configuration shown in Fig. 1.2b, and a single-phase-to-single-phase configuration. To match the voltage levels of both the power grid and the traction network, the converter input may still be connected to the three-phase power grid through a traction transformer, while the converter output is linked to the single-phase traction network. Therefore, the power electronic conversion device needs to facilitate both interconnection and energy transmission on both ends of the three-phase power grid and the single-phase traction network.

Advanced co-phase traction power supply systems depend on the functionality of power electronic converters to facilitate active interconnection and electric energy conversion with power grids and traction networks that operate at varying voltage levels, phases, and even different frequencies and phase numbers. When used to supply power to a single substation, these continuous power supply systems are also compatible with traditional traction power supply modes. The ability to control voltage frequency, amplitude, and phase at the ports enables interconnection of multiple substations for continuous feeding, thereby completely eliminating the need for neutral section passing.

Multiple substations interconnected within a continuous power supply system enable mutual support and standby capabilities, allowing any load served by the traction network to draw electric energy from two or more nearby substations. This configuration enhances the efficiency of power transmission and utilization. Additionally, the interconnection of multiple substations offers support between substations, significantly reducing the required installed capacity at each substation. The mutual backup among substations further contributes to improved reliability in power supply.

Moreover, continuous power supply converters based on power electronic technology facilitate active conversion, providing a solution for the electric energy quality deficiencies present in both traction networks and three-phase power grids, such as reactive power components, harmonics, and unbalance. Additionally, these

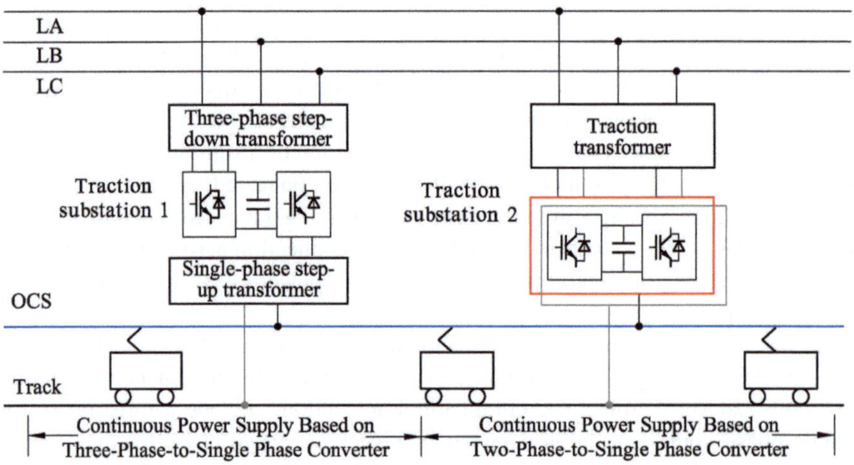

Fig. 1.2 Schematic diagram of continuous power supply system based on (**a**) three-phase-to-single-phase converter and (**b**) two-phase-to-single-phase converter

devices offer interfaces for new energy storage and other AC and DC inputs and outputs.

To sum up, advanced co-phase traction power supply systems can be considered ideal, as they display the four key characteristics discussed in Sect. 1.2.4: the elimination of neutral section passing, high stability, high electric energy quality, and high flexibility (Li et al. 1988; Chen et al. 2016; He 2014; Zhang 2012; Qi et al. 2017; He et al. 2014; Xia 2017). This system design fulfills the requirements of high-speed and heavy-haul railways for efficient and reliable power supply across ultra-long distances without the need for neutral sections. It represents a novel traction power supply solution that is not only suitable for the construction of new railways but also compatible with the upgrading of existing substations. Furthermore, it offers an innovative traction power supply mode for major projects in new railway systems, such as plateau railways and subsea tunnels, which require ultra-long distance power supply or access to new energy sources. Power electronic converters are recognized as the essential components ensuring these system characteristics by connecting three-phase power grids with single-phase traction networks. Various types of power electronic converter systems designed for advanced co-phase traction power supply are analyzed and introduced in detail in the following chapters of this book.

1.5 Balance Principles for Common Types of Traction Transformers

The three-phase side of traction substations is typically connected to electric power systems rated at 110 kV or higher. However, existing power electronic converters face limitations in terms of devices and circuit topologies, leading to an interface voltage of these converters that prevents direct connection to the three-phase high-voltage power grid. Consequently, step-up/step-down transformers are utilized to connect the converters and the grid. Different types of traction transformers possess unique characteristics and impose varying grid connection requirements for the converters. Several typical transformer types are briefly introduced below.

The following types of traction transformers are primarily employed in China: single-phase, impedance matching, Vv, YNd11, Scott, and YNvd (Jin 2006). Among them, single-phase, Vv, and YNd11 transformers are categorized as unbalanced winding connections, whereas Scott connection, impedance matching connection, and YNvd connection as balanced winding connections.

Figure 1.3 shows (a) a single-phase connection, (b) a Vv connection, (c) a YNd11 connection, (d) a Scott connection, (e) an impedance matching connection, and (f) a YNvd connection. Table 1.1 shows the comparative data statistics of the main characteristics for the above common traction transformers.

Based on Fig. 1.3 and Table 1.1, among the six types of traction transformers, only the single-phase traction transformers provide single-circuit output on their secondary side. In contrast, traction transformers with all the other connection modes offer two-phase output on their secondary side. These transformers are used to convert three-phase power into two-phase power for supplying the two feeding sections.

Vv-connected transformers convert three-phase voltage into two components of single-phase voltage, resulting in two-phase outputs with a voltage phase angle difference of 60°. Therefore, to maintain three-phase equilibrium on the primary side, it is necessary to compensate for both the reactive and active power of the two-phase loads until the designated phase angles are achieved.

YNd11-connected transformers are linked to the three-phase power grid on the primary side, while the two single-phase bridge arms for traction power supply are separated on the secondary side of the transformers. Although this connection method offers a simple topology and lower costs, it has limitations in voltage utilization and capacity efficiency. Considering the two-phase outputs with a voltage phase angle difference of 120°, it is also essential to compensate for the reactive power of the two-phase loads and maintain active power balance in order to preserve three-phase equilibrium on the primary side.

Scott-connected transformers feature two windings on the primary side connected in an inverted T shape. In this configuration, one winding is connected to the center point of the bottom winding and to phase A, while the bottom winding is linked between phases B and C. The secondary side outputs two phases with a voltage phase angle difference of 90°. Three-phase current equilibrium is achieved on the primary side when the two-phase loads are in current equilibrium and purely active. Scott-connected transformers exemplify a balanced connection mode.

Impedance matching transformers feature a Y-connection on the primary side, and a Y-connection with an appropriate delay on the secondary side, resulting in

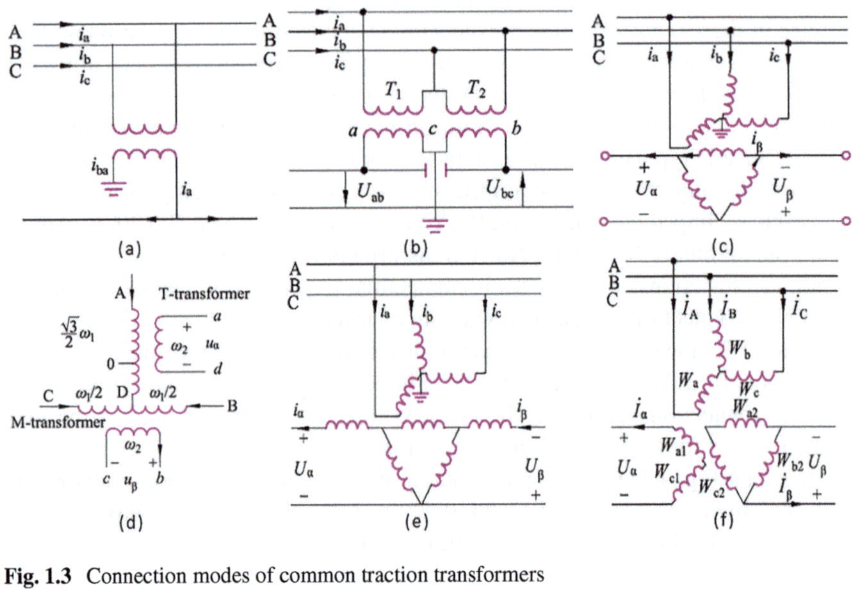

Fig. 1.3 Connection modes of common traction transformers

Table 1.1 Statistics of relevant connection data of common traction transformers

	Single-phase wiring	Vv	YNd11	Scott	Impedance matching	YNvd
Phase number on secondary side	1	2	2	2	2	2
Phase relationship of secondary voltage	–	60°	120°	90°	90°	90°
Whether neutral sections are involved	No	Yes	Yes	Yes	Yes	Yes

two-phase outputs with a voltage phase angle difference of 90°. This transformer arrangement also represents a balanced connection mode. Three-phase current equilibrium is achieved on the primary side when the two-phase loads are in current equilibrium and purely active.

YNvd-connected transformers feature a delta connection on the secondary side, which creates a path for the third harmonic component of the exciting current, thereby improving the voltage waveform. This transformer configuration outputs two independent phases with a voltage phase angle difference of 90°. Three-phase equilibrium is achieved on the primary side when the two-phase loads are in current equilibrium and purely active. YNvd is also a balanced connection configuration.

To sum up, based on the classification of common traction transformers according to phase numbers in inputs and outputs, namely three-phase to single-phase and three-phase to two-phase, as well as the common three-phase-to-three-phase type used in electric power systems, traction transformers in the new-generation continuous power supply systems can be designed to output single-phase, two-phase, and three-phase. These outputs correspond to converter topologies in advanced co-phase traction power supply systems, which are also categorized into three types: single-phase to single-phase, two-phase to single-phase, and three-phase to single-phase.

1.6 Converter Topology in Advanced Co-phase Traction Power Supply System

Based on the analysis in the previous section, if the converter inputs in traction power supply systems are compatible with the three types of transformer outputs—namely single-phase, two-phase, and three-phase—and given that the converter outputs in these systems are connected to the single-phase traction networks, converters for advanced co-phase traction power supply can be categorized into three types according to their input-output topologies: single-phase-to-single-phase, two-phase-to-single-phase, and three-phase-to-single-phase configurations.

1.6.1 Three-Phase-to-Single-Phase Converter Topology

When three-phase-to-three-phase transformers, which are commonly used in electric power systems, are employed as traction transformers, the power electronic converters for continuous power supply are connected to both the three-phase power grid and the single-phase traction network within the electric power system. Consequently, it is evident that converters with a three-phase-to-single-phase configuration present a rational choice.

Voltage reduction at the access point to the operating voltage level of two-level converters within electric power systems creates a two-level three-phase-to-single-phase converter topology, as illustrated in Fig. 1.4 (Liu 2010b). The three-phase grid voltage is stepped down to a level compatible with the two-level converter port using a three-phase-to-three-phase step-down transformer; the single-phase side is boosted by a single-phase-to-single-phase step-up transformer for the interconnection of substations in the traction networks. Given the bidirectional power flow requirements for both traction and braking, four-quadrant operation is necessitated for the converters on both the three-phase and single-phase sides. Although designing such a system can be costly due to the need to achieve significant step-up/step-down transformation ratios between the transformers on the electric power and traction sides, especially with existing power electronic switching devices that have a low withstand voltage level, the transformers provide electrical isolation between the traction network and power grid for enhanced safety, in addition to ensuring matching of voltage levels. Furthermore, the current limitations of existing power electronic devices hinder the ability of single-module three-phase-to-single-phase converters to meet the power demands of traction loads served by substations. As a result, multiple three-phase-to-single-phase converter modules are typically connected in parallel to increase the system's power capacity. However, this approach complicates the system due to the multiple converter modules and the connections required with multi-winding three-phase-to-three-phase traction transformers that are compatible with each respective module.

The configuration consisting of three-phase-to-three-phase transformers and three-phase-to-single-phase converters is both technically mature and cost-effective. However, three-phase-to-three-phase step-down traction transformers are not

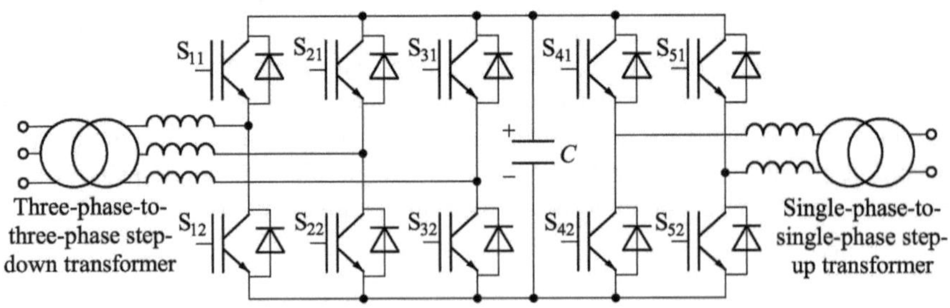

Fig. 1.4 Topology of two-level three-phase-to-single-phase converter

commonly found in existing traction power supply substations. The establishment of continuous power supply systems using three-phase-to-single-phase converters necessitates the addition of these traction transformers. From this standpoint, this solution is better suited for the development of new continuous power supply systems.

1.6.2 Two-Phase-to-Single-Phase Converter Topology

Traction transformers with two-phase outputs, such as impedance matching, Vv, Scott, and YNvd transformers, are widely used in existing traction power supply systems (Guo et al. 2000; Shi-Lin et al. 2004). These highly reliable and mature two-phase output transformers have been implemented in nearly 10,000 traction substations across China. Given the fact that each individual substation is typically equipped with dual transformers in a redundant configuration, the proposed solution involving three-phase-to-single-phase converters would require replacing the existing two traction transformers at each substation. The additional costs incurred from this transformer replacement at each substation amount to approximately RMB 5 million to 10 million. Therefore, the continued utilization of existing two-phase output traction transformers is a critical issue that must be addressed for the research, application, and popularization of continuous power supply systems.

The converter system based on a two-phase-to-single-phase configuration is compatible with existing two-phase traction transformers, making it a viable option for implementing continuous power supply. Therefore, the two-phase-to-single-phase converter topology is more advantageous for transforming existing systems into continuous power supply systems. The substation configuration featuring two-phase-to-single-phase power electronics transformers (PETs) for advanced co-phase traction power supply is illustrated in Fig. 1.5.

In this configuration, two single-phase cascaded rectifiers are, respectively, connected to the two output ports of a three-phase-to-two-phase traction transformer. Each single-phase bridge arm, corresponding to phases A and B, consists of multiple isolated AC-DC-DC-AC modules arranged in series. The output port of the single-phase cascade on the other side is directly linked to the traction network, enabling continuous power supply. On the input side, the two single-phase cascaded converters rectify single-phase AC power into intermediate DC power inputs, with different potentials but generally identical voltage. This intermediate DC power is then transformed by an isolated DC-DC converter into the DC voltage required for the cascaded rectifiers on the single-phase side. Subsequently, the power is converted into single-phase AC power through cascaded inverters on the output side, followed by the ultimate connection to the OCS that supplies power to the traction loads on the railway.

This solution enables the utilization of existing three-phase-to-two-phase traction transformers. As a result, it is not only suitable for upgrading existing substations to continuous power supply systems, but it is also applicable for constructing

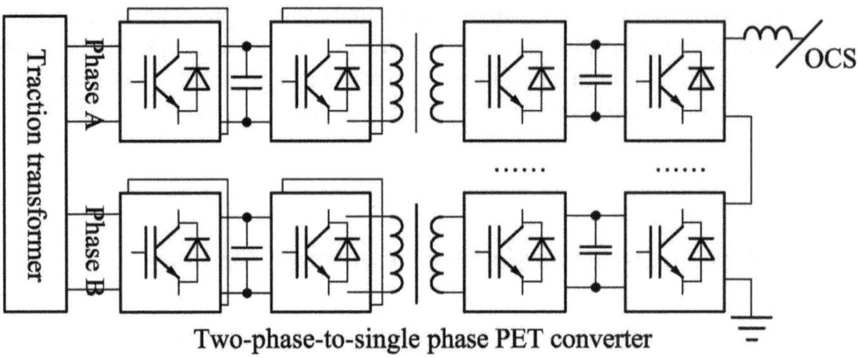

Fig. 1.5 Schematic diagram of substation configuration for advanced co-phase traction power supply based on two-phase-to-single-phase converter

new continuous power supply systems by employing traditional two-phase output traction transformers at new substations.

1.6.3 Single-Phase-to-Single-Phase Converter Topology

Whether using three-phase-to-three-phase or three-phase-to-two-phase transformers, each single-phase output is independently connected to a converter. Traction networks, which also feature a single-phase configuration, and converters connected only with the single-phase outputs from the transformers, constitute a single-phase-to-single-phase topology within continuous power supply systems. An example of this setup, utilizing an H-bridge single-phase-to-single-phase AC-DC-AC converter configuration, is illustrated in Fig. 1.6. In this configuration, a step-down transformer reduces the grid voltage to the level suitable for the converter interface. The H-bridge single-phase-to-single-phase converter facilitates the AC-DC-AC energy conversion, while the converter on the output side is connected to the traction network via a step-up transformer.

Single-phase AC-DC-AC converters can be implemented using various configurations, including two-level H-bridge, two-level half-bridge, and multilevel circuits. However, the power output from a single module often falls short of meeting the load requirements for traction power supply, necessitating the use of multiple modules connected in parallel to enhance the system's capacity. While the transformer and converter systems in this parallel topology are relatively complex in terms of structure and wiring, their operating principles are fundamentally similar to those of a single module, which will not be repeated in this section.

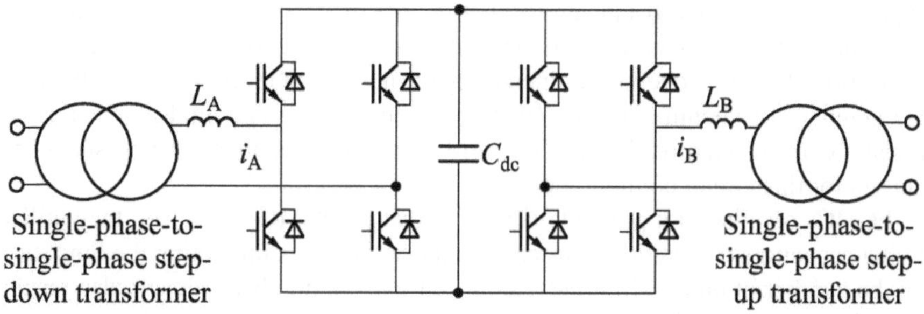

Fig. 1.6 Circuit diagram of two-level H-bridge single-phase-to-single-phase converter

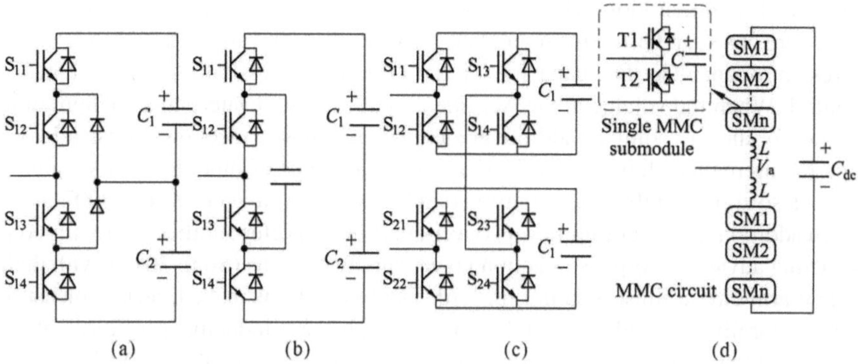

Fig. 1.7 Four multilevel topologies

1.6.4 Multilevel Conversion Circuit Topology

Multilevel converters are topologically categorized into three basic types: the diode-clamped multilevel converter (DCMC), the flying capacitor multilevel converter (FCMC), and the cascaded multicell converter (CMCC). Their specific topologies are illustrated in Fig. 1.7a, b, c respectively (Xu 2013; Xia et al. 2017; Wang et al. 2017; Sheng 2009). Each of these structurally distinct categories has unique features that make them suitable for different applications, as summarized below.

DCMCs are equipped with multiple switching devices connected in series to enhance their withstand voltage levels, along with diodes to clamp the switching tubes, thereby ensuring their respective voltage ranges and creating multilevel output current channels. This circuit configuration offers several advantages, including bidirectional power flow, low total harmonic distortion (THD) of output voltage, and interface voltage multiplication. As a result, it is well suited for applications such as MV/LV large-capacity active rectification, high-power reactive power compensation, and active power filtering (He 2007). However, the presence of inconsistent charging and discharging at the DC-side capacitor can lead to DC voltage excursions, necessitating additional hardware and power electronic circuits for supplementary voltage balancing (Shu et al. 2013b, c). Alternatively, a control or modulation algorithm may be implemented to facilitate voltage balancing (Shu et al. 2007, 2013d). Nonetheless, increasing the number of levels significantly raises the number of switching devices, such as switching tubes and diodes, which complicates control and modulation. Therefore, a moderate number of levels should be chosen for this circuit configuration.

FCMCs operate on a principle similar to that of DCMCs; however, they utilize flying capacitors for clamping instead of diodes, thereby eliminating the limitation on the number of diodes. Similarly, an increase in the number of levels also results in a higher number of clamping capacitors. In cases of voltage unbalance, current surge becomes more pronounced in the voltage balancing paths.

Without clamping elements, CMCCs achieve multilevel output through a modular cascade configuration, which results in a simple topology, easy extensibility, low hardware costs, and effective redundancy. However, each cascaded submodule features an independent DC capacitor on the DC side, necessitating voltage balance control. When the voltage at the AC-side output port is elevated, this configuration allows for the cascading of additional identical modules to meet the withstand voltage requirements. However, an increase in the number of modules and switching tubes also significantly raises control complexity and system costs (Liu 2007).

In addition, modular multilevel converters (MMCs) offer another option for constructing advanced co-phase traction power supply systems (Song 2016), with their single-arm circuit illustrated in Fig. 1.7d. The bridge arms in each phase consist of $2n$ structurally identical submodules and two bridge arm inductors positioned at the upper and lower parts, which are connected in series (Yang et al. 2013). These submodules typically come in two circuit configurations: half-bridge and full-bridge (Kouro et al. 2010). Although the half-bridge structure is more concise than the full-bridge structure, it supports fewer levels and does not enable reverse blocking (Akagi 2011). As the number of levels increases, the system capacity can be easily expanded by cascading more submodules. A greater number of modules results in more output levels, with output voltages that more closely align with the modulation signals. Consequently, MMC circuits composed of multiple modules offer advantages such as low waveform distortion, minimal du/dt, and reduced switching loss (Zhou and Chen 2013). However, certain control challenges also arise, such as the relatively complex voltage balancing within and between modules due to the large number of DC capacitors.

In summary, since traction power supply systems demand high interface voltages and power levels, multilevel converters are regarded as more suitable for providing advanced co-phase traction power supply. However, traditional three-phase multilevel converters need to be modified to achieve single-phase output from these systems. This book analyzes the fundamental circuits and control principles of converters used in traction power supply applications. It also explores special multilevel circuits, including three-phase-to-single-phase, two-phase-to-single-phase, and single-phase-to-single-phase configurations, with a focus on their topologies, modulation, and control strategies. This comprehensive analysis aims to provide a theoretical basis, along with simulations and verifications for converters that are viable for advanced co-phase traction power supply applications.

1.7 Summary

This chapter provides a brief overview of the development history in the field of traction power supply systems. It discusses co-phase power supply technology, as well as the basic components and fundamental principles of continuous power supply systems, with a particular emphasis on the challenges faced by traction power supply systems, including neutral section passage, electric energy quality, and power supply reliability. The chapter further introduces several typical types of traction transformers and multilevel circuits of power electronics transformers. Subsequent chapters delve into a detailed discussion of various converter topologies available for continuous power supply systems, including their circuit structures, control methods, and modulation techniques.

References

Akagi H. Classification, Terminology, and Application of the Modular Multilevel Cascade Converter (MMCC) [J]. IEEE Transactions on Power Electronics, 2011, 26(11): 3119–3130.

Boora A A, Zare F, Ghosh A, et al. Applications of power electronics in railway systems [C]. 2007 Australasian Universities Power Engineering Conference, 2007: 1–9.

Chen M, Li Q, Roberts C, et al. Modelling and performance analysis of advanced combined co-phase traction power supply system in electrified railway [J]. IET Generation, Transmission & Distribution, 2016, 10(4): 906–916.

Costa-Castello R, Grino R, Parpal R C, et al. High-Performance Control of a Single-Phase Shunt Active Filter [J]. IEEE Transactions on Control Systems Technology, 2009, 17(6): 1318–1329.

Du Yunguo. Analysis on Key Technologies of Wenzhou Rail Transit System [J]. *Railway Standard Design*, 2015, 59(12): 27–30+39.

Fujii K, Kunomura K, Yoshida K, et al. STATCOM applying flat-packaged IGBTs connected in series [J]. IEEE Transactions on Power Electronics, 2005, 20(5): 1125–1132.

Guo Yuhua, Lian Jisan, Zhang Kunlun. Influence of Auto-passing of Neutral Section on Electric Locomotive [J]. *Electric Drive for Locomotives*, 2000(02): 13–15.

He Xiangning. *Theory and Application Technology of Multilevel Converter* [M]. China Machine Press, 2007.

He Xiaoqiong. Research on Continuous Co-phase Traction Power Supply System Based on Multi-level Three-Phase-to-Single-Phase Converters [D]. Southwest Jiaotong University, 2014.

He X, Shu Z, Peng X, et al. Advanced Cophase Traction Power Supply System Based on Three-Phase to Single-Phase Converter [J]. IEEE Transactions on Power Electronics, 2014, 29(10): 5323–5333.

Huang Xiaohong, Li Qunzhan, Wu Ping. Co-phase Traction Power Supply System for High-speed Railway Based on Modular H-bridges and Vv-connected Transformers [J]. *Journal of the China Railway Society*, 2016, 38(06): 38–44.

Jin Boquan. Comparative Study on Automation Technology for Traction Power Supply on High-speed Railway [D]. Southwest Jiaotong University, 2006.

Kouro S, Malinowski M, Gopakumar K, et al. Recent Advances and Industrial Applications of Multilevel Converters [J]. IEEE Transactions on Industrial Electronics, 2010, 57(8): 2553–2580.

Li Qunzhan. *Power Supply Analysis and Integrated Compensation Technology for Traction Substation* [M]. Beijing: China Railway Publishing House, 2006a.

Li Qunzhan. Discussion on Electric Energy Quality Deficiencies of Electrified Railway [C]. *Electrified Railway*, 2006b: 1–6.

Li Qunzhan. Several Key Technical Issues in the Development of Traction Power Supply for High-speed Railways in China [J]. Journal of the China Railway Society, 2010, 32(04): 119–124.

Li Rujun. SVC Compensation Optimization Control at Sanjie Traction Substation on Beijing-Shanghai Railway [J]. *Science & Technology Information*, 2011(24): 149–150.

Li Xuewei. *Introduction to High-speed Railway* [M]. Beijing: China Railway Publishing House, 2012.

Li Qunzhan, He Jianmin. Analysis of Traction Power Supply System [M]. Chengdu: Southwest Jiaotong University Press, 2007.

Li Qunzhan, Xie Shaofeng. Development and Application of Co-phase Power Supply Technology for Electrified Railway [A]. Sichuan Association for Science and Technology, Deyang Municipal People's Government. Proceedings of the Seventh Forum for Doctoral Experts in Sichuan [C]. Sichuan Association for Science and Technology, Deyang Municipal People's Government: Sichuan Mechanical Engineering Society, 2014: 6.

Li Qunzhan, Zhang Jinsi, He Weijun. Study on New Power Supply System for Heavy-haul Electric Traction [J]. *Journal of the China Railway Society*, 1988(04): 23–31.

Liu Fengjun. *Multilevel Inverter Technology and Its Application* [M]. China Machine Press, 2007.

Liu Neng. Study on Compensation Optimization Scheme for Co-phase Traction Substation [D]. Southwest Jiaotong University, 2010a.

Liu Xing. Control Strategy of Co-phase Power Supply Device for Continuous Power Supply [D]. Southwest Jiaotong University, 2010b.

Liu Li. Research on Automatic Neutral Section Passing Solution Based on Ground Control Using Power Electronic Switch with Fly-wheel Resistor [D]. Southwest Jiaotong University, 2012.

Qi Weizhi, Liu Quanjing, Xie Yongqiang, et al. Novel Co-phase Power Supply System Solution Based on Three-Phase-to-Single-Phase Conversion. *Power System Protection and Control*, 2017: 93–98.

Qian Lixin. *Technology of the World High-Speed Railways* [M]. Beijing: China Railway Publishing House, 2003.

Qiu Daqiang, Li Qunzhan, Zhou Fulin, et al. Comprehensive Power Quality Control of Electric Railway Based on Back-to-Back SVG [J]. *Electric Power Automation Equipment*, 2010, 30(06): 36–39+44.

Bi Sheng. Research on New Automatic Neutral Section Passing Scheme of AC Locomotive [D]. Southwest Jiaotong University, 2009.

Shi-Lin C, Li R, Pao-Hsiang H. Traction system unbalance problem-analysis methodologies. IEEE Transactions on Power Delivery, 2004, 19(4): 1877–1883.

Shu Z, Tang J, Guo Y, et al. An Efficient SVPWM Algorithm With Low Computational Overhead for Three-Phase Inverters [J]. IEEE Transactions on Power Electronics, 2007, 22(5): 1797–1805.

Shu Z, Xie S, Lu K, et al. Digital Detection, Control, and Distribution System for Co-Phase Traction Power Supply Application [J]. IEEE Transactions on Industrial Electronics, 2013a, 60(5): 1831–1839.

Shu Z, He X, Wang Z, et al. Voltage Balancing Approaches for Diode-Clamped Multilevel Converters Using Auxiliary Capacitor-Based Circuits [J]. IEEE Transactions on Power Electronics, 2013b, 28(5): 2111–2124.

Shu Z, Zhu H, He X, et al. One-inductor-based auxiliary circuit for dc-link capacitor voltage equalisation of diode-clamped multilevel converter [J]. IET Power Electronics, 2013c, 6(7): 1339–1349.

Shu Z, Ding N, Chen J, et al. Multilevel SVPWM With DC-Link Capacitor Voltage Balancing Control for Diode-Clamped Multilevel Converter Based STATCOM [J]. IEEE Transactions on Industrial Electronics, 2013d, 60(5): 1884–1896.

Song Shasha. Research on Three-Phase-to-Single-Phase Converter Based on MMC Structure and Its Control [D]. Southwest Jiaotong University, 2016.

Tan P C, Loh P C, Holmes D G. A robust multilevel hybrid compensation system for 25 kV electrified railway applications [C]. IEEE 34th Annual Conference on Power Electronics Specialist, 2003. PESC'03. 2003: 1020–1025 vol. 3.

The People's Republic of China. *Mid- and Long-Term Railway Network Plan* [R]. 2008.

Uzuka T, Ikedo S. Railway Static Power Conditioner Field Test [J]. Quarterly Report of RTRI, 2004, 45(2): p. 64–67.

Wang Guo. Research on Integrated Active Compensation of Traction Power Supply System on Electrified Railway [D]. Lanzhou Jiaotong University, 2011.

Wang H, Kou L, Liu Y, et al. A Seven-Switch Five-Level Active-Neutral-Point-Clamped Converter and Its Optimal Modulation Strategy [J]. IEEE Transactions on Power Electronics, 2017, 32(7): 5146–5161.

Wu Fengjuan. Analysis of Traction Power Supply Solution and OCS Current-carrying Capacity for Heavy-haul Railway [J]. *Railway Locomotive & Car*, 2008(05): 70–73.

Wu Chuanping. Research on Integrated Compensation Technology for Power Quality of Power Supply System on Electrified Railway [D]. Hunan University, 2012.

Xia Yankun. Design of Direct Power Controller for Co-phase Power Supply System on High-speed Railway [J]. *Journal of Power Supply*, 2017, 15(01): 105–110+118.

Xia C, Zhang G, Yan Y, et al. Discontinuous Space Vector PWM Strategy of Neutral-Point-Clamped Three-Level Inverters for Output Current Ripple Reduction [J]. IEEE Transactions on Power Electronics, 2017, 32(7): 5109–5121.

Xu Min. Design of Multi-level Three-Phase-to-Single-Phase AC-DC-AC Converter for Continuous Power Supply System [D]. Southwest Jiaotong University, 2013.

Yan Yunsheng. Discussion on Automatic Neutral Section Passing Solution for Electric Locomotive [J]. *Electric Drive for Locomotives*, 1999(06): 1–3+7.

Yang Xiaofeng, Lin Zhiqin, Zheng Qionglin, et al. Research Review on Modular Multi-level Converter [J]. Proceedings of the CSEE, 2013, 33(06): 1–15.

Zeng Guohong, Hao Rongtai. Co-phase Traction Power Supply System Based on Active Power Filters and Impedance Matching Balance Transformers [J]. *Journal of the China Railway Society*, 2003(03): 49–54.

Zhang Rui. Research on Protection Solution of Electric Energy Conversion Equipment in Continuous Co-phase Power Supply System [D]. Southwest Jiaotong University, 2012.

Zhang Xiufeng, Gao Shibin, Qian Qingquan, et al. A Novel Co-phase Traction Power Supply System Based on Impedance Matching Balance Transformers and AT Feeding Mode [J]. *Journal of the China Railway Society*, 2006(04): 32–37.

Zhang Chenmeng, Chen Baichao, Yuan Jiaxin, et al. Research on a Hybrid Electric Energy Quality Compensation System for Co-phase Power Supply System Based on V/V Traction Transformers [J]. *Transactions of China Electrotechnical Society*, 2015, 30(12): 496–504.

Zhou Jinghua, Chen Yaai. *Principles and Applications of High-performance Cascaded Multilevel Converter* [M]. China Machine Press, 2013.

References

Chapter 2
Grid-Connected Characteristic Analysis of Advanced Co-phase Traction Power Supply System

2.1 Overview

2.1.1 Basic Architecture of Advanced Co-phase Traction Power Supply System

Advanced co-phase traction power supply systems that incorporate interconnected substations throughout the entire traction network are poised to become a next-generation traction power supply mode. This system fundamentally relies on various types of converters, including three-phase-to-single-phase converters (Shu et al. 2013; Liu 2010; Chen et al. 2016; Xia 2017; Li et al. 1988; He et al. 2013b; Sun et al. 2017; Li 2010). A brief overview of a advanced co-phase traction power supply system, which consists of two substations interconnected and operating in a three-phase-to-single-phase topology, is illustrated in Fig. 2.1. At Substation 1 and Substation 2, high-power power electronic converters are connected to the three-phase power grid and single-phase traction network, on both sides, respectively. These converters are capable of controlling the ports—managing voltage amplitude, frequency, and phase—which facilitates the interconnection on both ends of the three-phase power grid and the single-phase traction network. The two-way flow of energy between the traction network and the three-phase grid, achieved through power control, enables not only power supply to the loads in the traction network from the three-phase power grid but also the feedback of traction braking energy to the three-phase power grid for consumption.

Power electronic converters in advanced co-phase traction power supply systems can be connected to the three-phase power grid and the traction network via input/output matching transformers. They may also be connected directly to the traction network following output voltage step-up through multilevel circuits, such as cascaded multiple modules. Furthermore, these converters can even establish a direct connection with the three-phase power grid through converter topologies like MMCs (He et al. 2014). Subsequent sections provide more detailed discussions on

Z. Shu, *Power Electronic Converters in Advanced Co-Phase Traction Power Supply System*, https://doi.org/10.1007/978-981-95-3276-6_2

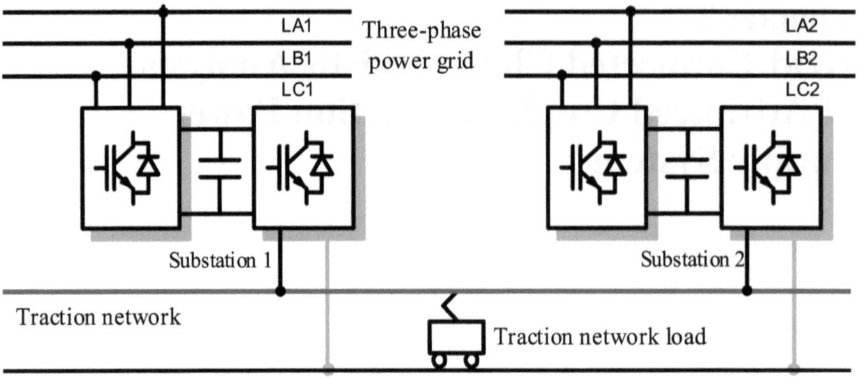

Fig. 2.1 Schematic diagram of advanced co-phase traction power supply system based on three-phase-to-single-phase converter topology

the various types of converters suitable for advanced co-phase traction power supply systems. The analysis of the advanced co-phase traction power supply system architecture shown in Fig. 2.1 reveals several notable characteristics of these systems concerning their traction loads, traction networks, and electric power systems:

First of all, in advanced co-phase traction power supply systems, the traction network is interconnected throughout the system, allowing for operation without the need for neutral sections. This eliminates many issues associated with passing through neutral sections, which are typically required in traditional traction power supply systems (Shu et al. 2013). Additionally, the voltage levels and frequencies of these continuous power supply networks align with those of traditional traction networks, ensuring full compatibility with existing electrical loads and equipment. Therefore, there is no requirement for modifications to current traction drive loads. Consequently, advanced co-phase traction power supply systems provide a power supply solution that is highly compatible with traction loads.

Second, the integration of power electronic converters addresses the electric energy quality deficiencies faced by traditional traction systems. Multiple substations provide mutual power support, significantly enhancing load capacity. These substations interconnected can also serve as backups for one another, eliminating the impact on system operation in the event of disconnection due to maintenance or failure. Traction loads benefit from receiving current from nearby substations, which reduces losses, while braking energy can be utilized by loads throughout the network. As a result, advanced co-phase traction power supply systems can significantly enhance electric energy quality, power supply reliability, and operational efficiency of the traction networks (Liu 2010).

Third, power electronic converters, typically configured in an AC-DC-AC circuit topology, can address electric energy quality deficiencies in traction networks through mechanisms such as dynamic reactive compensation and harmonic active filtering. This approach effectively isolates the deficiencies of single-phase traction networks, preventing them from transferring to and impacting the corresponding three-phase power grid. Additionally, each phase of the converter on the grid side supplies balanced active power to the single-phase traction loads, eliminating issues such as three-phase unbalance on the grid side. Moreover, due to the long power supply distances in continuous traction networks, a significant amount of time-varying traction loads and braking energy with surge characteristics can be absorbed within the traction networks. This results in reduced surges fed back to the three-phase power grid through the AC-DC-AC converter. The potential integration of energy storage systems is anticipated to fully resolve the surge issues associated with traction loads. Therefore, advanced co-phase traction power supply systems provide a solution that is advantageous for the electric power system.

It is evident that advanced co-phase traction power supply systems constitute an ideal mode, particularly advantageous for traction loads, traction networks, and electric power systems. The numerous benefits of this mode stem from its core component—power electronic converters. Unlike conventional converter applications, the power electronic converters utilized in continuous power supply systems face the following challenges:

1. The traction load capacity is typically substantial at individual substations, requiring the power electronic converters used in continuous power supply systems to deliver the full active power of these loads. However, due to limitations imposed by factors such as the rated voltage and current of existing high-power power electronic devices, the unit capacity of the converters tends to be relatively small. Consequently, multiple modules need to be connected in parallel or cascaded to meet the power and voltage requirements of the systems.
2. The ability of converters to bidirectionally control power flow between three-phase and single-phase systems presents several challenges. They need not only transmit active power during traction conditions but also absorb active power during braking conditions for the traction network. Additionally, the converters need to address power quality deficiencies, such as harmonics and reactive power issues on the single-phase traction network side, as well as three-phase unbalance on the three-phase grid side. These requirements pose significant difficulties in maintaining system stability.
3. The interconnection of multiple substations on the traction side needs to maintain voltage alignment in terms of frequency, amplitude, and phase. This requirement places a high demand on the real-time performance in the control of interconnected substations. The considerable distances between the substations and the unique interconnection characteristics further complicate coordination and control.

This chapter concentrates on studying and analyzing the above issues, with particular emphasis on the characteristics of traction networks after interconnection for continuous feeding as discussed in Point 3.

2.1.2 Influence of Traction Network Impedance on Substation Output Voltage

Since the impedance characteristics of traction networks with interconnected substations over long distances influence the interconnection properties and control of the converters, this chapter begins with an analysis of the impedance of traction networks and their distribution characteristics. Based on the impedance droop characteristics of long-distance traction networks, interconnection configurations for continuous power supply involving two substations, three substations, and an arbitrary number of substations were selected, respectively, to study the control strategy based on droop characteristics, as well as the characteristics of the resulting continuous networks.

2.1.3 Analysis of Impedance Characteristics of Traction Network

Continuous power supply systems enable interconnection based on the existing OCS in a sectionalized power supply configuration, without necessitating alterations to the current OCS lines. Consequently, the subsequent analysis includes relevant studies that reference the characteristics of existing single-track OCSs (Li and He 2007; Wen et al. 2011).

Figure 2.2 shows the simplified circuit model of a single-track traction network, which is mainly composed of an OCS, trains, rails, and the ground. In this schematic, $\dot{U}$ represents the traction port voltage, $\dot{U}'$ represents the traction voltage, $\dot{I}$ represents the traction current, $\dot{I}_T$ represents the rail current, l represents the track length, kz is the inductance, z_{12} represents the OCS-ground circuit mutual impedance per unit length (Ω/km) and rail-ground circuit mutual impedance per unit length (Ω/km), and z_1 and z_2 represent the OCS-ground circuit self-impedance per

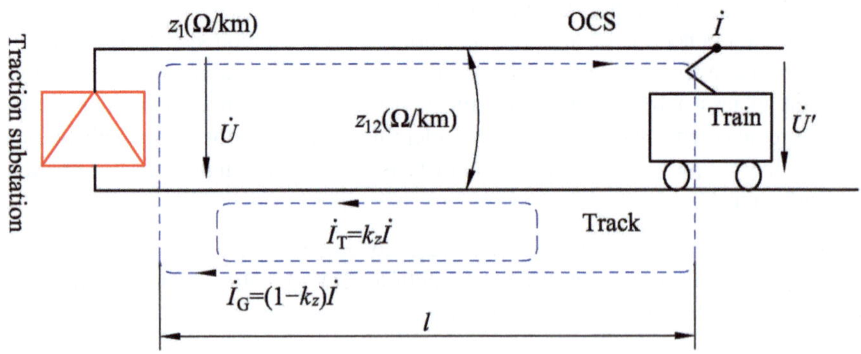

Fig. 2.2 Simplified circuit model of single-track traction network

unit length (Ω/km) and the rail-ground circuit self-impedance per unit length (Ω/km), respectively.

As illustrated, the single-track traction network model primarily comprises two sets of conductor network-ground circuits: (1) the OCS-ground circuit, which constitutes an active network, and (2) the rail-ground circuit, which forms a passive network. The total traction network impedance is the sum of their respective impedances. The voltage circuit equation can be derived from Fig. 2.2 as follows:

$$
\begin{aligned}
\dot{U}-U' &= \left(z_1\dot{I}-z_{12}\dot{I}_{\mathrm{T}}+z_2\dot{I}_{\mathrm{T}}-z_{12}\dot{I}\right)l \\
&= \left(z_1-z_{12}\frac{z_{12}}{z_2}+z_2\frac{z_{12}}{z_2}-z_{12}\right)\dot{I}l \\
&= \left(z_1-\frac{z_{12}^{\ 2}}{z_2}\right)\dot{I}l
\end{aligned}
\tag{2.1}
$$

As a result, the impedance per unit length of the traction network can be expressed as follows:

$$
z=\frac{\dot{U}-U'}{\dot{I}l}=z_1-\frac{z_{12}^{\ 2}}{z_2}
\tag{2.2}
$$

To simplify the analysis and calculations, the conduction processes that result in current changes in the rails at the location of loads and near substations are disregarded.

In this chapter, the impedance value of the traction networks, influenced by protective conductors, is considered as noted in Ma and Chen (2016). Specifically, the impedance per unit length of the traction network is approximately $0.076+j0.167$ Ω/km. It is evident that the resistance component of the traction network impedance is significantly smaller than the reactance component. Taking traction networks with power supply distances ranging from 20 to 100 km as examples, the total resistance of the traction network lines falls between approximately 1.5 and 7.6 Ω, while the total inductance of the lines ranges from about 10.6 to 53.2 mH. To simplify the analysis process, subsequent analyses and simulations are based on a track length of 25 km, total resistance of 2.0 Ω, and total inductance of 15 mH, namely $Z=2+j1.5\pi$.

2.1.4 Influence of Traction Network Impedance on Converter Output Voltage

To achieve voltage stability and power balance, substations in continuous power supply systems are categorized into two types of control: voltage stability control and power control. From this perspective, these two types of substations can be modeled as a voltage source and a power source, respectively. By examining the interconnection model with two substations, which are equivalent to voltage and power sources, the influence of impedance and load positions within continuous traction networks on their voltage, phase, and amplitude distribution was analyzed. The traction load location is explored in two scenarios: between the two substations and at the terminal end of a substation. The positioning of loads, along with the impedance distribution characteristics of the traction networks, impacts the converter output characteristics in different ways. Consequently, modeling analyses were performed for both scenarios, as illustrated in Fig. 2.3.

Scenario I: The traction load is located between the two substations. The two substations jointly provide power support for the load. The position of the traction load and impedance of the traction network affect the output characteristics of the two substations, respectively.

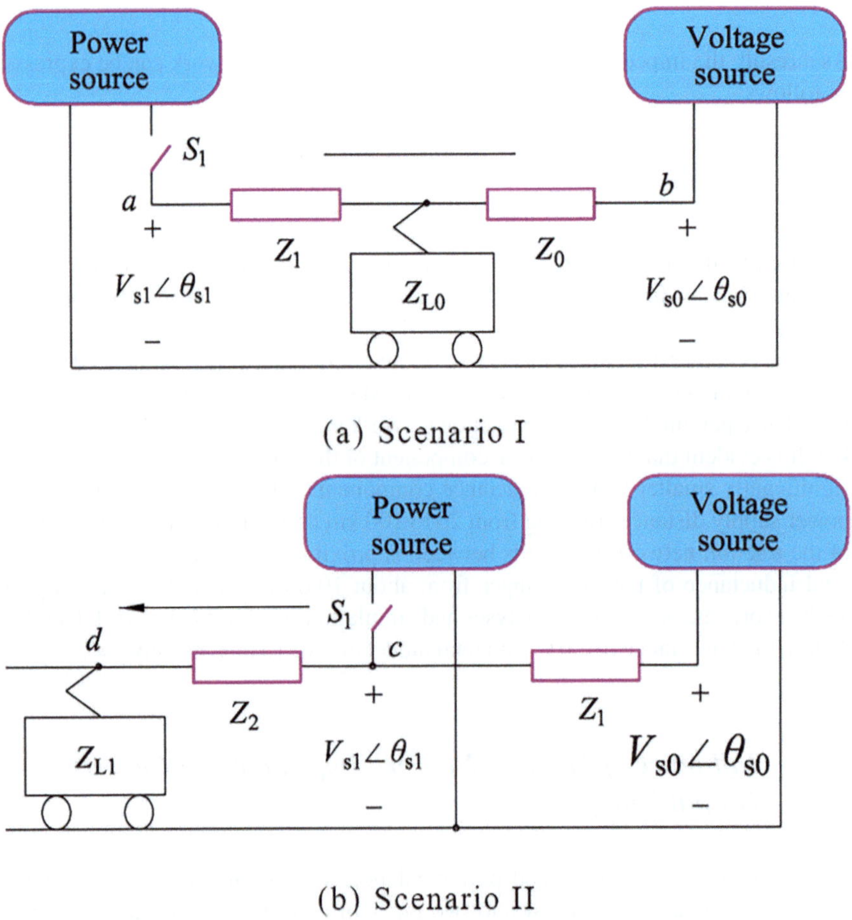

Fig. 2.3 Simplified system model of two traction load positions. (**a**) Scenario I. (**b**) Scenario II

Scenario II: The traction load is located at the terminal end of a substation. When the load does not exceed the capacity of a single substation, the load and the single-ended impedance of the traction network only affect the output characteristics of one substation. When the load exceeds the capacity of a nearby substation, the difference between the load and the capacity of this substation is supplied by remote substations.

Changes in traction network-load impedance due to the movement of the traction load were analyzed under the assumption that the power source substation is not operational (i.e., Switch S1 is open). It is assumed that the output voltage of the voltage source substation is $V_{s0}\angle\theta_{s0} = 27{,}500\sqrt{2}\angle 0^\circ$ V, while the output voltage of the power source is $V_{s1} \angle \theta_{s1}$. The impedances of the traction network in three sections are labeled Z_0, Z_1, and Z_2, where the line resistance is $R_0 = R_1 = R_2 = 2\ \Omega$ and the line inductance is 15 mH. The traction load is represented by impedances Z_{L0} and Z_{L1}. At $V_{s0} = 27.5$ kV, calculations are performed for rated powers of the load at 1 MW, 2 MW, 3 MW, and 8 MW, corresponding to load impedances of 756.25 Ω (rounded to 760 Ω), 378.13 Ω (rounded to 380 Ω), 189.06 Ω (rounded to 190 Ω), and 94.53 Ω (rounded to 95 Ω).

When the traction load Z_{L0} moves in Section I, located between the two substations as shown in Fig. 2.3a, $R = \frac{2}{15}^{*} L_1$ and $Z_0 = j\omega L_0 = 2 + j1.5\pi - \left(j\omega L_1 + \frac{2}{15}^{*} 10^{3*} L_1 \right) = \left(2 - \frac{2}{15}^{*} 10^3 L_1 \right) + j\left(1.5\pi - \omega L_1\right)$.

Based on the voltage circuit, the equation can be expressed as follows:

$$V_{s1}\angle\theta_{s1} = \frac{V_{s0}\angle\theta_{s0}}{Z_0 + Z_{L0}} \times Z_{L0} \tag{2.3}$$

Assuming that the output voltage phase angle of the voltage source (SSO) is 0, the amplitude and the phase angle of the output voltage of the power source (SS1) can be expressed as follows, respectively:

$$\begin{cases} V_{s1} = \dfrac{V_{s0} \times |Z_{L0}|}{|Z_0 + Z_{L0}|} \\ \theta_{s1} = \arctan \dfrac{\omega L_{L0}}{R_{L0}} - \arctan \dfrac{\omega\left(L_{L0} + L_0\right)}{R_{L0}} \end{cases} \tag{2.4}$$

Derived from Eq. (2.4), when the inductance of the traction network impedance, denoted as Z_1, changes from 0 to 15 mH (i.e., load Z_{L0} moves from point a to point b), the trend of the output voltage of the power source is illustrated in Fig. 2.4a, b. These figures demonstrate the impact of different loads Z_{L0} on the amplitude and phase angle of the output voltage of the power source as the traction network impedance varies, based on the calculations performed. It is evident that the load position

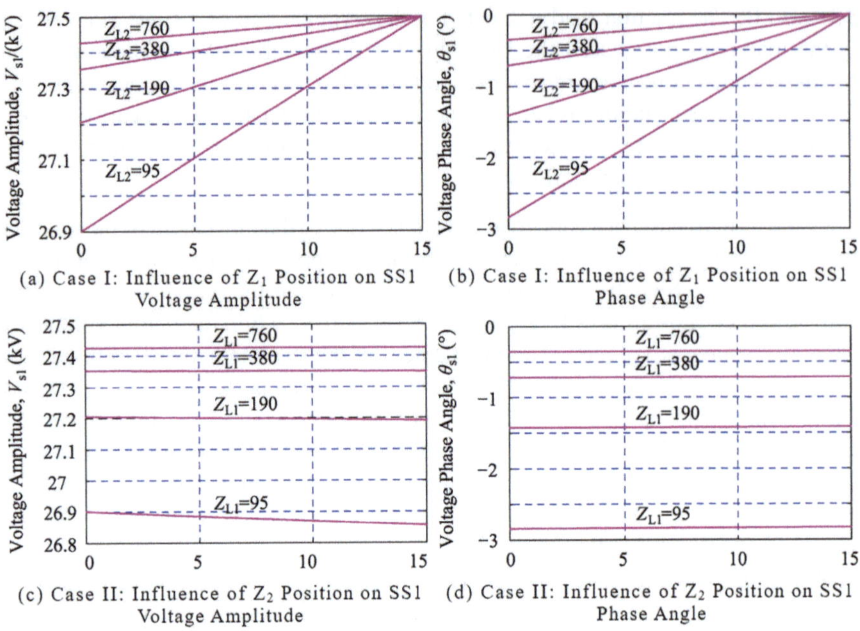

Fig. 2.4 Amplitude and phase angle of output voltage of power source (SS1) under load movement on track in two scenarios. (**a**) Case I: Influence of Z_1 position on SS1 voltage amplitude. (**b**) Case I: Influence of Z_1 position on SS1 phase angle. (**c**) Case II: Influence of Z_2 position on SS1 voltage amplitude. (**d**) Case II: Influence of Z_2 position on SS1 phase angle

significantly influences both the amplitude and phase angle of the output voltage of the power source.

Similarly, when the traction load Z_{L1} moves in the track section scenario II as shown in Fig. 2.3b, based on the voltage circuit, the equation can be expressed as follows:

$$V_{s1}\angle\theta_{s1} = \frac{V_{s0}\angle\theta_{s0}}{Z_1 + Z_2 + Z_{L1}} \times (Z_2 + Z_{L1}) \tag{2.5}$$

Then, the amplitude and phase angle of the output voltage of the power source (SS1) can be expressed as follows:

$$\begin{cases} V_{s1} = \dfrac{V_{s0} \times |Z_2 + Z_{L1}|}{|Z_1 + Z_2 + Z_{L1}|} \\ \theta_{s1} = \arctan\dfrac{\omega(L_{L1} + L_2)}{R_{L1}} - \arctan\dfrac{\omega(L_{L1} + L_1 + L_2)}{R_{L1}} \end{cases} \tag{2.6}$$

Derived from Eq. (2.6), when the inductance of the traction network impedance, denoted as Z_2, changes from 0 to 15 mH (i.e., load Z_{L1} moves from point c to point d), its influence on the output voltage of the power source substation as the distance varies is illustrated in Fig. 2.4c, d. It is evident that in scenario II, changes in load positioning—corresponding to variations in the impedance of the traction network between the load and the substation—have little impact on both the amplitude and phase angle of the output voltage of the power source (SS1).

Based on the above analysis, both the impedance of the traction network and the position of the traction load influence the amplitude and phase distribution of the traction network voltage. Under the operating conditions of scenario I, changes in the traction network impedance have a greater impact on the output voltage of the power source compared to scenario II. Additionally, a larger traction load capacity results in a more pronounced effect. Therefore, in advanced co-phase traction power supply systems, it is essential to rationally select the spacing between traction substations. This ensures that impedance changes in the traction network, resulting from the movement of the traction load, remain within the power range designated for the control of the interconnected system and that the current of the traction network does not exceed the power supply capacity of the conductors.

2.1.5 Analysis of Voltage Step-Down of Continuous Traction Network

Far-end voltage within traction power supply networks decreases significantly as the power supply distance increases. This reduction in catenary voltage limits the output power of the traction drive systems, thereby affecting train operation speeds. In advanced co-phase traction power supply systems, adjacent substations interconnected provide mutual support in maintaining catenary voltage. This collaboration helps sustain the voltage amplitude of the traction networks throughout the entire system.

The analysis examines the changes in catenary voltage under traditional single-end feeding compared to continuous feeding, also considering a distance of 25 km between two substations and a port voltage of 27.5 kV. The comparison of catenary voltage variations as the traction load operates within a 25 km track section is illustrated in Fig. 2.5. It is evident that the voltage amplitude of the traction network in the continuous power supply mode experiences the most significant drop at the midpoint between the two substations. Figure 2.5a shows the maximum voltage drop of approximately 0.5 kV in the traditional traction power supply mode, whereas the drop is only around 0.1 kV in the advanced co-phase traction power supply mode when the load is purely resistive. Figure 2.5b reveals a similar effect for an inductive load, with the maximum voltage drop exceeding 1.1 kV in the traditional traction feeding mode, while under the continuous feeding mode, the maximum drop is only about 0.3 kV.

In continuous feeding systems, adjacent substations connected to the grid and network provide mutual capacity support, allowing the load power demand to be met from nearby substations in both directions. As a result, there are two current collection circuits, which effectively reduce the equivalent line impedance. These systems can significantly suppress voltage drops within the traction network.

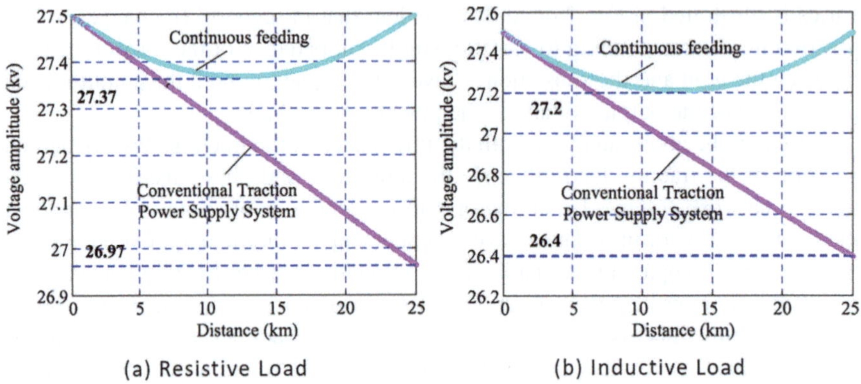

Fig. 2.5 Variation curves of catenary voltage with movement of traction load across continuous feeding and traditional traction feeding systems. (**a**) Resistive load. (**b**) Inductive load

However, it is important to note that inductive loads have a greater impact on catenary voltage compared to resistive loads.

From another perspective, when adhering to the same terminal-end voltage drop amplitude limit without increasing the voltage and power levels of the substations, advanced co-phase traction power supply systems can be designed for longer power supply distances compared to traditional systems. This approach can reduce the number of substations across entire railways by more than half, significantly lowering the design, construction, and operational costs associated with substations.

Hence, advanced co-phase traction power supply systems offer significant advantages in both performance and economic efficiency compared to existing systems. With mutual support along the entire system, this mode has demonstrated enhanced adaptability and greater stability in areas with insufficient power grids, such as the Sichuan-Xizang Railway and the Xinjiang-Xizang Railway. Additionally, the low catenary voltage drops and extended spacing between substations provide further benefits in ultra-long distance power supply applications, such as tunnels crossing mountains and seas.

2.2 Current Distribution Characteristics of Traction Network with Interconnected Substations

2.2.1 Analysis of Output Current Difference Between Substations Connected to Grid and Network

In continuous power supply systems, variations in amplitude, phase, and frequency among the output voltages from converters for the interconnection of multiple traction substations inevitably lead to circulating currents between these substations. This phenomenon significantly increases line losses and impacts the overall system reliability (Liu et al. 2016; Wen et al. 2011; Xu 2014). Unlike the localized parallel connection of multiple converters, traction substations within continuous power supply systems are connected by long wiring with negligible impedance. Consequently, it is essential to incorporate the impedance characteristics of the traction network into the analysis of the current output characteristics for interconnection over long distances in continuous power supply systems.

The equivalent circuit for interconnected two substations was established as shown in Fig. 2.6, based on the interconnection model involving two substations (SS0 and SS1). In this schematic, Z_L represents the traction load, while Z_0 and Z_1 denote the impedances of the traction network. R_0'' and X_0'' are the equivalent output resistance and inductive reactance of Substation SS0, respectively, and R_1'' and X_1'' are the equivalent output resistance and inductive reactance of Substation SS1, respectively. V_{s0} and V_{s1} represent the output voltage amplitudes of the substations (SS0 and SS1), while θ_{s0} and θ_{s1} indicate the phases of the output voltages from these substations. i_{s0} and i_{s1} are the output currents from the two substations, while i_L signifies the load current. Additionally, v_L represents the load voltage, and i_h denotes the current difference between the two substations as they supply energy to the load.

Letting $Z_0' = R_0' + jX_0' = Z_0 + R_0'' + jX_0''$ and $Z_1' = R_1' + jX_1' = Z_1 + R_1'' + jX1''$, it is assumed that these two substations are spaced 25 km apart, and the impedance of traction network per unit length is still taken as $z = (0.076 + j0.167)$ Ω/km. Due to the mobile nature of traction loads, the distance between the substations and traction load constantly varies. Assuming that the distance from Substation SS0 to the traction load is L (km), the following expressions can be established: $Z_0 = z \times L$ and $Z_1 = z \times (25\text{-}L)$. Then, the following expressions can be derived from Fig. 2.6:

$$\begin{cases} v_{s0} - i_{s0} \cdot Z_0' = v_L \\ v_{s1} - i_{s1} \cdot Z_1' = v_L \\ i_L = v_L / Z_L \\ i_L = i_{s0} + i_{s1} \end{cases} \tag{2.7}$$

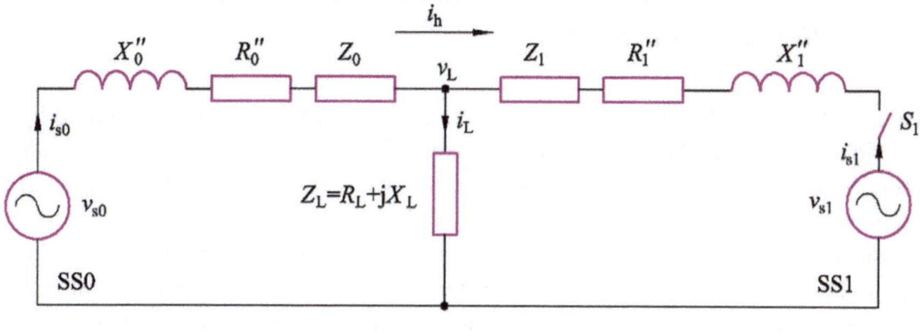

Fig. 2.6 Equivalent circuit of interconnected two traction substations for continuous power supply

According to Eq. (2.7), the output current of each traction substation can be expressed as follows:

$$i_{s0} = \frac{\left(Z_L + Z_1'\right) \cdot v_{s0} - Z_L \cdot v_{s1}}{Z_0' \cdot Z_1' + Z_L \cdot \left(Z_0' + Z_1'\right)} \tag{2.8}$$

$$i_{s1} = \frac{\left(Z_L + Z_0'\right) \cdot v_{s1} - Z_L \cdot v_{s0}}{Z_0' \cdot Z_1' + Z_L \cdot \left(Z_0' + Z_1'\right)} \tag{2.9}$$

Differences denoted as i_h in output currents between the two substations can be expressed as follows:

$$i_h = i_{s0} - i_{s1} = \frac{\left(v_{s0} - v_{s1}\right) + \frac{Z_1'}{Z_L} \cdot v_{s0} - \frac{Z_0'}{Z_L} \cdot v_{s1}}{\frac{Z_0' \cdot Z_1'}{Z_L} + Z_0' + Z_1'} \tag{2.10}$$

As expressed in Eq. (2.10), the differences in output currents between the two substations are primarily associated with their output voltage differences and the impedance of the traction network (Z_1' and Z_0'). Assuming that the distance between the two substations is 25 km, with an output voltage difference of $\triangle V = 100$ V and a load equivalent of $Z_L = 100\ \Omega$, the variation trend of the current difference for the traction load (Z_L) between the two substations is illustrated in Fig. 2.7. It is evident that, due to the influence of traction network impedance, even under purely resistive load conditions, the current differences arising from the voltage differences between the two traction substations interconnected include both active and reactive components. Notably, the reactive component cannot be ignored due to its relatively large amplitude.

2.2.2 *Automatic Current Distribution Characteristics Between Substations*

According to the sketch shown in Fig. 2.5, assuming that the current required by a traction load equals the sum of output currents from power supplies on both sides, there is no active power exchange between the two power supplies. In this scenario, the impedance angles for Z_0' and Z_1' are φ_0 and φ_1, respectively, and the output ports of the two substations are identical in voltage frequency, amplitude, and phase angle, i.e., $V_{s0} = V_{s1}$ and $\theta_{s0} = \theta_{s1}$. Under these conditions, the output currents from the two power supplies depend on the varying impedances of the traction network connected on both sides of the load. The output power of the substations must be automatically adjusted in response to load movements.

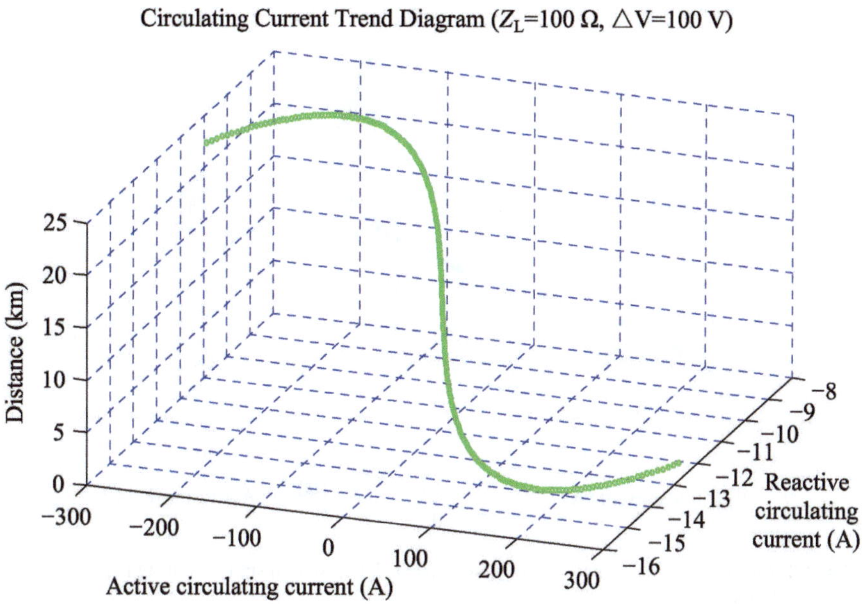

Fig. 2.7 Output current differences between two substations corresponding to movement of traction load ($\triangle V = 100$ V, $Z_L = 100\ \Omega$)

When the total output power of the two substations consistently equals the load demand, the two power supplies interconnected are considered to achieve automatic current distribution, i.e., the load current is equal to the sum of i_{s0} and i_{s1}. The following equation is derived from Eqs. (2.8) and (2.9):

$$i_L = i_{s0} + i_{s1} = \frac{Z_1' \cdot v_{s0} + Z_0' \cdot v_{s1}}{Z_0' \cdot Z_1' + Z_0' \cdot Z_L + Z_1' \cdot Z_L} \tag{2.11}$$

The following equation is also derived from Eq. (2.7), $i_{s1} = \frac{v_{s1} - Z_L \cdot i_{s0}}{2 \cdot Z_L}$. Substituting it into Eq. (2.11) yields:

$$i_{s0} = \frac{Z_1' \cdot v_{s1}}{Z_0' \cdot Z_1' + Z_0' \cdot Z_L + Z_1' \cdot Z_L} \tag{2.12}$$

Similarly, the following equation is established:

$$i_{s1} = \frac{Z_0' \cdot v_{s1}}{Z_0' \cdot Z_1' + Z_0' \cdot Z_L + Z_1' \cdot Z_L} \tag{2.13}$$

Based on the equation above, a three-dimensional curve diagram can be drawn to illustrate the output currents (i_{s0} and i_{s1}) and the load current (i_L) as the load moves from Substation SS0 to Substation SS1, as shown in Fig. 2.8. This indicates that the power demand of the load is jointly supplied by the two adjacent traction

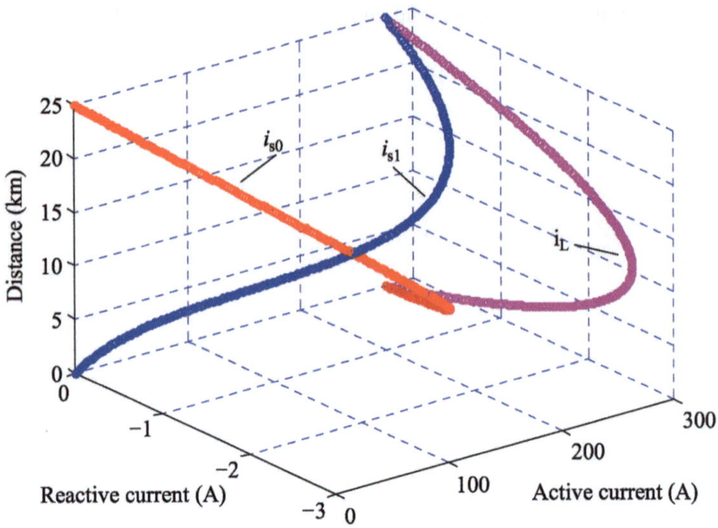

Fig. 2.8 Output currents (i_{s0} and i_{s1}) and load current (i_L) of two substations during load movement from SS0 to SS1

substations. The output current amplitude from each substation is influenced by its distance to the traction load, with output power increasing as this distance decreases. Conversely, a greater distance from the traction load results in a lower output power. Therefore, under the condition that the output voltages from the interconnected traction substations are identical in frequency, amplitude, and phase angle, the output powers of the substations adhere to the principle of nearby automatic current distribution, meaning that the load current is automatically distributed according to different load positions.

According to Sect. 2.2.2, the line impedance of the traction network exerts a certain impact on the active and reactive power outputs from the two substations. Figure 2.8 also indicates that due to the presence of traction network impedance, even when the load is purely resistive, the substations still supply a small amount of reactive current.

However, due to the low line impedance of the traction network between the two substations relative to the high system capacity and voltage level, even very small deviations in phase angle or amplitude of the outputs from the two adjacent substations can lead to significant circulating currents between substations. This phenomenon is also why advanced co-phase traction power supply systems cannot be established by relying solely on direct interconnection through transformers.

In addition, power electronic converters exhibit excellent controllability, offering the possibility for continuous feeding by suppressing circulating currents between substations. However, maintaining complete alignment in voltage frequency, amplitude, and phase angle between the two substations is challenging, especially given that the distance between them is often too great to ensure effective communication synchronization. Therefore, it is essential to study and devise a control strategy for the interconnection of substations within advanced co-phase traction power supply systems.

2.3 Control Strategy of Interconnection for Continuous Feeding

As mentioned in the previous section, substations interconnected for continuous feeding tend to create large circulating currents between them (Liu et al. 2016), which must be eliminated through interconnection control. Currently, the interconnection control of multiple converters is mainly divided into wired active current-sharing control and wireless automatic current distribution control (Xie 2006; Martins et al. 1995).

Wired active current-sharing control can be further categorized into several main types: parallel centralized control, master-slave control, distributed control, and 3C control. This category offers advantages such as straightforward principles, rapid response speeds, easy implementation, and high real-time performance, making it suitable for applications where converters are closely spaced. Given the high voltage and power demands of traction substations, multiple AC-DC-AC converters can be employed at individual substations, utilizing active parallel current-sharing control. This configuration not only enhances the output capacity of individual substations and reduces circulating currents between devices, but also improves the reliability and redundancy of the power supply (Yadong et al. 2003; Lee et al. 2013; Brabandere et al. 2007).

However, this control strategy has several disadvantages, including the limited reliability of long-distance interconnection wiring due to the geographical locations of each unit. In continuous power supply systems, the significant distances between traction substations mean that the line impedance between converters for interconnection at these interconnected substations cannot be overlooked. Additionally, loads in movement can affect the line impedance of different converters. As a result, the inability to achieve complete equivalence in the localized parallel connection of multiple converters leads to shortcomings in the active current-sharing approach, making it challenging to evenly distribute load currents.

As mentioned in the previous chapter, the ideal continuous power supply mode does not necessitate current and power equalization among the outputs of all the interconnected substations. Instead, it requires the automatic distribution of output current or power across all traction substations, based on their distances from the loads. Substations closer to the loads provide higher outputs, while those farther away deliver lower outputs. Therefore, the wireless automatic current distribution strategy is recommended for interconnection control in continuous power supply. This control strategy is primarily characterized by droop control (Xie 2006; Martins et al. 1995; Brabandere et al. 2007; Hu 2013). By eliminating the need for information transmission between converters for interconnection, this method allows for completely independent control, which enhances system adaptability and reliability. As a result, it is suitable for various applications, including parallel converters over long distances, isolated microgrids, and grid-connected systems.

2.3.1 *Droop Control Analysis of Parallel Converters*

Classical control theory in the power system field employs droop control to facilitate multiunit interconnection over long distances without the need for communication. Droop control is also applicable to advanced co-phase traction power supply systems, which function as a specialized form of single-phase long-distance transmission. In power systems, the droop characteristics of generators, represented as *P/f* and *Q/V*, are utilized for frequency and voltage regulation, respectively. In converter-based continuous power supply systems, controlling frequency through phase angle differences offers a more straightforward approach. Therefore, the traditional *P/f* droop control can be effectively replaced by *P/θ* droop control.

The system with two inverters in parallel as shown in Fig. 2.9 is taken as an example to analyze the droop characteristics of parallel converters. As shown, v_1 and v_2 indicate the output voltages of the two converters; i_1 and i_2 indicate the output currents; Z_{11} and Z_{22} indicate the outputs from the two converters and the equivalent impedances of the wiring linking the traction network, respectively; Z_{LL} indicates the load impedance; v_{LL} indicates the load voltage; and i_{LL} indicates the load current. These parameters can be expressed as follows: $v_1 = V_1 \angle \theta_1$, $v_2 = V_2 \angle \theta_2$, $Z_{11} = R_{11} + jX_{11}$, $Z_{22} = R_{22} + jX_{22}$, and $v_{LL} = V_{LL} \angle \theta_{LL}$. According to Fig. 2.9, active power and reactive power outputs from Converter 1 can be calculated using the following equations:

$$\begin{cases} P = \dfrac{V_1}{R_{11}^2 + X_{11}^2}\left[V_1R_{11} - V_{LL}R_{11}\cos(\theta_1 - \theta_{LL}) + V_{LL}X_{11}\sin(\theta_1 - \theta_{LL})\right] \\ Q = \dfrac{V_1}{R_{11}^2 + X_{11}^2}\left[V_1X_{11} - V_{LL}X_{11}\cos(\theta_1 - \theta_{LL}) - V_{LL}R_{11}\sin(\theta_1 - \theta_{LL})\right] \end{cases} \tag{2.14}$$

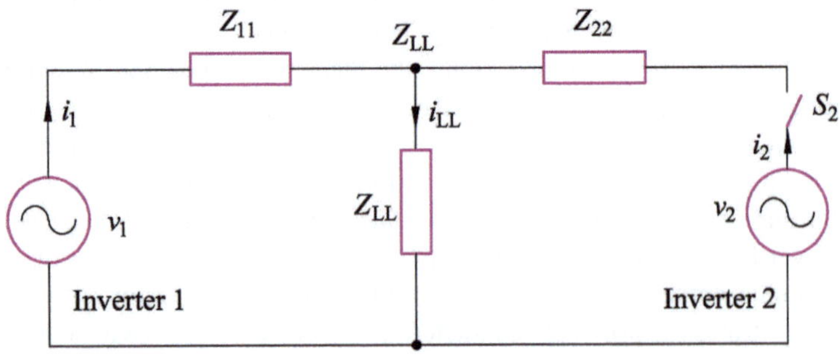

Fig. 2.9 Equivalent circuit diagram of two parallel converters in power system

Based on the analysis in Sect. 2.1.3, it is known that for the traction network transmission wiring, $X_{11} \gg R_{11}$. Therefore, Eq. (2.14) can be approximated as follows:

$$\begin{cases} P = \dfrac{V_1 V_{\mathrm{LL}}}{X_{11}} \sin\left(\theta_1 - \theta_{\mathrm{LL}}\right) \\ Q = \dfrac{V_1}{X_{11}} \left[V_1 - V_{\mathrm{LL}} \cos\left(\theta_1 - \theta_{\mathrm{LL}}\right)\right] \end{cases} \tag{2.15}$$

Considering the small differences between θ_1 and θ_{LL}, the following approximations can be established: $\sin(\theta_1 - \theta_{\mathrm{LL}}) \approx \theta_1 - \theta_{\mathrm{LL}}$ and $\cos(\theta_1 - \theta_{\mathrm{LL}}) \approx 1$. Consequently, Eq. (2.15) can be simplified as follows:

$$\begin{cases} P = \dfrac{V_1 V_{\mathrm{LL}}}{X_{11}} \left(\theta_1 - \theta_{\mathrm{LL}}\right) \\ Q = \dfrac{V_1}{X_{11}} \left(V_1 - V_{\mathrm{LL}}\right) \end{cases} \tag{2.16}$$

Generally, it is expected that the amplitude of load voltage (V_{LL}) remains steady. According to the simplified Eq. (2.16), it can be known that the active power output from converters is primarily influenced by the phase difference ($\theta_1 - \theta_{\mathrm{LL}}$), while the reactive power output is mainly affected by the amplitude difference ($V_1 - V_{\mathrm{LL}}$). If the phase difference ($\theta_1 - \theta_{\mathrm{LL}}$) between the power supply and load changes as a result of variations in the traction load power demand or position, the active power output from the converter will also change. Similarly, if the amplitude difference ($V_1 - V_{\mathrm{LL}}$) between the power supply and load voltages varies as a result of changes in the traction load power demand or position, the reactive power output from the converter will also change.

By analogizing the patterns described above to the parallel applications of multiple converters (Martins et al. 1995; Guerrero et al. 2005), the droop control of the nth converter can be expressed by Eq. (2.17):

$$\begin{cases} \theta_n = \theta_{0n} - m_n P_n \\ E_n = E_{0n} - n_n Q_n \end{cases} \quad (n = 1,2,3\ldots) \tag{2.17}$$

where θ_n and E_n denote the phase angle and amplitude of the output voltage from the nth converter, respectively; θ_{0n} and E_{0n} represent the phase angle and amplitude of the output voltage from the nth converter under no-load conditions; P_n and Q_n indicate the active power and reactive power outputs from the nth converter; and m_n and n_n are the droop control coefficients for phase and amplitude, respectively.

It is evident that droop control facilitates proportional control through the utilization of droop characteristics. While this control method allows for active current equalization among loads, it cannot guarantee a steady state in the output voltage amplitude and phase indices. Some researchers have suggested implementing droop control with a power reference for the parallel control of converters, which can be expressed by Eq. (2.18):

$$\begin{cases} \theta_n = \theta_{0n} - m_n\left(P_n - P_{en}^*\right) \\ E_n = E_{0n} - n_n\left(Q_n - Q_{en}^*\right) \end{cases} \quad (n = 1,2,3\ldots) \tag{2.18}$$

where P_{en}* and Q_{en}* indicate the active power and reactive power references, respectively. If the active current and reactive current outputs from the converter are controlled, instead of active power and reactive power, respectively, the control based on current droop characteristics can be expressed in Eq. (2.19):

$$\begin{cases} \theta_n = \theta_{0n} - m_n\left(I_\mathrm{p} - I_{\mathrm{p}0}\right) \\ E_n = E_{0n} - n_n\left(I_\mathrm{q} - I_{\mathrm{q}0}\right) \end{cases} \quad (n = 1,2,3\cdots) \tag{2.19}$$

where I_p and I_q indicate the active current and reactive current, respectively; $I_{\mathrm{p}0}$ and $I_{\mathrm{q}0}$ indicate the active current and reactive current references.

In localized parallel connection systems involving multiple converters, the output impedance of the converters significantly influences the power-sharing accuracy of these systems. Some studies have introduced the virtual impedance method (He et al. 2013a, b) to mitigate the impact of the output impedance on parallel systems. In continuous power supply systems, the challenges presented by long wiring between substations and the substantial impedance of the traction network require that the interconnection of multiple traction substations not only eliminate circulating currents through active current equalization in localized parallel connections, but also enable dynamic and flexible adjustments of power output. This entails automatically adjusting power distribution based on the varying positions of traction loads, which correspond to changing line impedance in the traction network at load access points. Such requirements can be fulfilled when the voltage amplitude, phase, and frequency at the output port of all substations are synchronized. Therefore, an interconnection control method that maintains alignment of output voltage in amplitude, phase, and frequency across substations can be established for advanced co-phase traction power supply according to the droop characteristics of the system, thereby achieving the goal of automatic power regulation and distribution.

2.3.2 Analysis of Droop Characteristics of Continuous Power Supply System

First, it is essential to determine whether the continuous power supply system exhibits droop characteristics before utilizing these characteristics for interconnection control. Based on the analysis of the equivalent circuit shown in Fig. 2.5, when

Switch S1 is closed, the complex power output by the power source substation (SS1) can be expressed as follows:

$$\begin{aligned}\tilde{S} = v_{s1} i_{s1}^{*} &= v_{s1} \cdot \left(\frac{v_{s1} - v_{L}}{Z_1'}\right)^{*} \\ &= \frac{V_{s1}^{2}}{\left|Z_1'\right|} e^{j\varphi_1} - \frac{V_{s1} \cdot V_{L}}{\left|Z_1'\right|} e^{j(\varphi_1 + \theta_{s1} - \theta_{L})}\end{aligned} \tag{2.20}$$

where φ_1 denotes the impedance angle of Z_1'. According to Euler's formula, the active power (P_1) and reactive power (Q_1) of the power source (SS1) flowing into the traction network can be expressed as follows:

$$\begin{cases} P_1 = \dfrac{V_{s1}}{R_1'^{2} + X_1'^{2}} \left[R_1' \left(V_{s1} - V_{L} \cos(\theta_{s1} - \theta_{L})\right) + V_{L} X_1' \sin(\theta_{s1} - \theta_{L})\right] \\ Q_1 = \dfrac{V_{s1}}{R_1'^{2} + X_1'^{2}} \left[X_1' \left(V_{s1} - V_{L} \cos(\theta_{s1} - \theta_{L})\right) - V_{L} R_1' \sin(\theta_{s1} - \theta_{L})\right] \end{cases} \tag{2.21}$$

Further derivation results in:

$$\begin{cases} V_{s1} - V_{L} \cos(\theta_{s1} - \theta_{L}) = \dfrac{R_1' P_1 + X_1' Q_1}{V_{s1}} \\ \sin(\theta_{s1} - \theta_{L}) = \dfrac{X_1' P_1 - R_1' Q_1}{V_{s1} V_{L}} \end{cases} \tag{2.22}$$

Since the traction load impedance (Z_L) is much smaller than the substation output impedance (Z_1''), the phase difference between the substation output voltage and the load voltage is relatively small, which can be expressed as follows:

$$\begin{cases} \cos(\theta_{s1} - \theta_{L}) \approx 1 \\ \sin(\theta_{s1} - \theta_{L}) \approx \theta_{s1} - \theta_{L} \end{cases} \tag{2.23}$$

By substituting Eq. (2.23) into Eq. (2.22), the latter can be simplified as follows:

$$\begin{cases} V_{s1} - V_{L} = \dfrac{R_1' P_1 + X_1' Q_1}{V_{s1}} \\ \theta_{s1} - \theta_{L} = \dfrac{X_1' P_1 - R_1' Q_1}{V_{s1} V_{L}} \end{cases} \tag{2.24}$$

Based on Eq. (2.24), both voltage amplitude and phase are jointly influenced by active power and reactive power. Given that the relationship between the reactance (X_1) and resistance (R_1) of the traction network impedance is expressed as $X_1 = kR_1$, where k typically ranges from 2 to 4 depending on the type and structure of the traction network, the reactive power output by substations has a greater impact on the amplitude difference between the substation output voltage and the load voltage. Meanwhile, the active power output by substations has a greater influence on the phase difference of the substation output voltage.

Therefore, the traction network impedance in continuous power supply systems creates a coupling between the active and reactive power outputs from the substations, demonstrating the widespread presence of droop characteristics in interconnection involving multiple substations. Consequently, the interconnection control based on droop characteristics offers broad applicability.

2.3.3 Improved Interconnection Control Strategy Based on Droop Characteristics

The initial state of the advanced co-phase traction power supply mode with interconnected multiple substations is defined as follows: the traction network operates under a no-load condition, and the output voltage amplitude, phase position, and frequency are identical across all the substations. When all the substations maintain their outputs in this initial state, any traction load present in the traction network will draw current from nearby sources, thereby achieving automatic power distribution to the load.

To maintain the initial state of the output voltage amplitude, phase, and frequency for converters at substations within continuous power supply systems, the following improved interconnection control strategy based on droop characteristics can be applied. As discussed in the previous section regarding the droop characteristics of traction networks with interconnected substations, the output voltage phase and amplitude by substation nodes are adjusted based on the detection of active and reactive power. Accordingly, Eq. (2.18), which incorporates a power reference, is rewritten as Eq. (2.25):

$$\begin{cases} P^* = m'(\theta_0 - \theta) \\ Q^* = n'(E_0 - E) \end{cases} \tag{2.25}$$

where m' and n' represent the droop coefficients for active power affected by phase and for reactive power affected by amplitude in droop control, respectively; P^* and Q^* are the references of active power and reactive power, respectively. Considering the power regulation control for interconnected substations, the addition of active power and reactive power references for interconnection converters can yield Eq. (2.26):

$$\begin{cases} P^* = m'(\theta_0 - \theta) + P_{em} \\ Q^* = n'(E_0 - E) + Q_{em} \end{cases} \tag{2.26}$$

where P_{em} and Q_{em} denote the active power and reactive power references for the interconnection converter. The output power is controlled by adjusting the phase and amplitude of the output voltage. The droop coefficient is a fixed value, which may cause unreliable results when the traction load power demand fluctuates significantly. To address this issue, various improved control options based on droop characteristics have been developed (Guerrero et al. 2005; He et al. 2013a, b; Sun et al. 2017; Li 2010).

Continuous power supply systems serve time-varying traction loads with changing power and location. To enable effective current collection from nearby substations, the output power of all the substations needs to be adjusted according to the varying power levels and locations of these loads.

The control design mentioned above is depicted in the schematic diagram of interconnection control based on droop characteristics for advanced co-phase traction power supply as shown in Fig. 2.10. Phase and amplitude information for the single-phase output voltage (v_o) is acquired through PLL sampling. After implementing interconnection control based on droop characteristics for automatic current distribution between substations, reference values for active and reactive currents are determined. Real-time values for active, reactive, and harmonic compensation currents are extracted from the single-phase side. The modulated voltage waveform is generated through PQ control at the single-phase converter before PWM signals are produced by the modulation module, driving the single-phase converter into operation.

The interconnection controllers described above ensure that the converter's output current aligns with the reference amplitude, while the single-phase converter's output voltage phase (θ_{s1}) aligns with the no-load output voltage phase (θ_{s10}), which corresponds to the voltage source's output voltage phase. They also ensure that the output voltage amplitude (V_{s1}) aligns with the no-load output voltage amplitude (V_{s10}), which corresponds to the output voltage amplitude of the voltage source. To achieve these objectives, deviation control for both phase and voltage should be established with their respective initial values for these controllers, enabling effective alignment to ensure that the deviation approaches zero. By letting $k_1 = X_1'/V_{s1}V_L$, $k_2 = R_1'/V_{s1}V_L$, $k' = R_1'/V_{s1}$, and $k'' = X_1'/V_{s1}$, Eq. (2.24) can be simplified into Eqs. (2.27) and (2.28):

$$\theta_{s1} - \theta_L = k_1 P_1 - k_2 Q_1 \tag{2.27}$$

$$V_{s1} - V_L = k' P_1 + k'' Q_1 \tag{2.28}$$

Therefore, phase control for interconnection control is as follows:

1. Letting $\theta_{s10} - \theta_{s1} = e_\theta$, substituting it into Eq. (2.27) gives

$$\theta_{s10} - e_\theta - \theta_L = k_1 P_1 - k_2 Q_1 \tag{2.29}$$

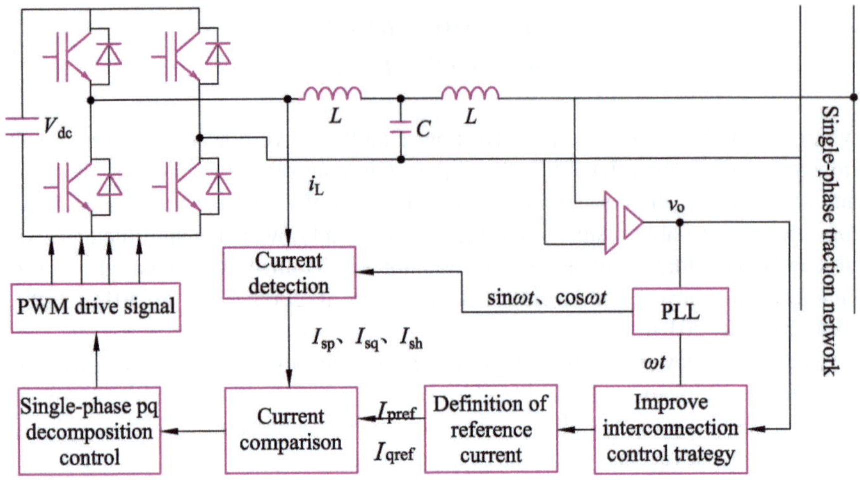

Fig. 2.10 Schematic diagram of interconnection control based on droop characteristics

2. Assuming that the active power (P_1) is a controllable variable, then $\theta_{s1} - \theta_L = k_1P_1 - k_2Q_1$:

$$P_1 = \left(u + k_2Q_1\right)\frac{1}{k_1} \tag{2.30}$$

where u indicates the auxiliary control variable. Substituting Eq. (2.30) into Eq. (2.29) gives

$$\theta_{s10} - e_\theta - \theta_L = u \tag{2.31}$$

Taking u as a controllable variable, the controller is designed as follows:

$$u = \theta_{s10} - \theta_L + k_{p\theta}e_\theta + k_{i\theta}\int e_\theta \mathrm{d}t \tag{2.32}$$

where $k_{p\theta}$ and $k_{i\theta}$ are the proportional and integral constants of the phase control (PI_θ) regulator, respectively. Substituting Eq. (2.32) into Eq. (2.31) gives

$$k_{p\theta}e_\theta + k_{i\theta}\int e_\theta \mathrm{d}t + e = 0 \tag{2.33}$$

In Eq. (2.33), as t approaches infinity, e_θ will converge to 0, indicating that θ_{s1} is aligned with θ_{s10}. Hence, the phase controller can be designed as follows:

$$\begin{aligned} P_1 &= \frac{1}{k_1}\left(\theta_{s10} - \theta_L + k_{p\theta}e_\theta + k_{i\theta}\int e_\theta \mathrm{d}t + k_2Q_1\right) \\ &= \frac{V_{s1}V_L}{X_1^{'}}\left(\theta_{s10} - \theta_L + k_{p\theta}e_\theta + k_{i\theta}\int e_\theta \mathrm{d}t + \frac{R_1^{'}}{V_{s1}V_L}Q_1\right) \end{aligned} \tag{2.34}$$

The derivation of amplitude control is similar to that of phase control. Letting $e_v = V_{s10} - V_{s1}$, voltage amplitude controller can be designed as follows:

$$\begin{aligned} Q_1 &= \frac{1}{k''}\left(-k'P_1 - V_L + V_{s10} + k_{pv}e_v + k_{iv}\int e_v dt\right) \\ &= \frac{V_{s1}}{X_1'}\left(V_{s10} - V_L - P_1\frac{R_1'}{V_{s1}} + k_{pv}e_v + k_{iv}\int e_v dt\right) \end{aligned} \tag{2.35}$$

where k_{pv} and k_{iv} are the proportional and integral constants of the PI regulator for voltage amplitude control, respectively.

The interconnection control for continuous power supply systems, as defined by Eqs. (2.34) and (2.35), can be represented in the schematic diagram illustrated in Fig. 2.11.

Generally, three-phase voltage in electric power systems is relatively steady in terms of both frequency and phase. Ignoring any phase delays in voltage transmission between substations, the phase of the three-phase voltage in an electric power system can serve as a reference for the voltage phase of interconnected multiple traction substations. This means that the voltage phase references of different substations can be synchronized in real time to align with the phase of a selected phase in the three-phase power grid. According to the analysis of droop characteristics presented above, the difference between the phase (θ_t) of the reference phase of the three-phase voltage and the voltage phase (θ_{s1}) of the single-phase traction network at the interconnection point of the converter directly influences the output active power level of the converter. This difference is defined as θ_{dif}, i.e., $\theta_{dif} = \theta_t - \theta_{s1}$. This deviation serves as the active reference output from the PI controller. When the output active power of the converter is insufficient to meet the active power demand

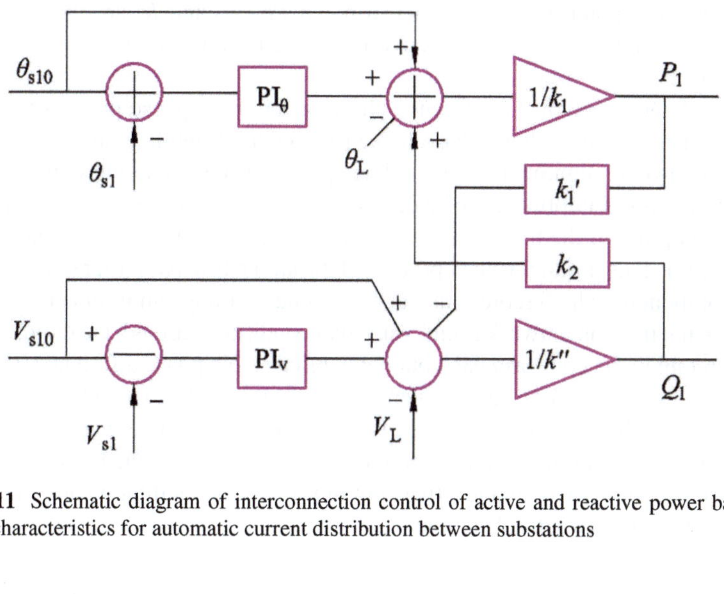

Fig. 2.11 Schematic diagram of interconnection control of active and reactive power based on droop characteristics for automatic current distribution between substations

of a load, the voltage phase at the interconnection point deviates from the reference phase of the three-phase voltage, indicating a change in θ_{dif}. Consequently, the PI controller adjusts the output active reference, prompting the converter to modify its output active power accordingly. A similar adjustment occurs when the output active power of the converter exceeds the active power demand of the load. Through this closed-loop regulation, the output active power of the converter is automatically adjusted in response to changes in the active power of the load, and the voltage phase at the interconnection point of the traction network remains aligned with the voltage phase of the reference phase of the three-phase grid, facilitating nearby current distribution between interconnected substations to meet the active power demands of the traction load.

Assuming that the port voltage amplitude remains unchanged, active power (P_1) and reactive power (Q_1) can be replaced with the reference values of active current (I_{pref}) and reactive current (I_{qref}) by altering the control coefficients given in Fig. 2.11. This approach facilitates the decoupling control of active and reactive currents in the interconnection system, in conjunction with the current detection and active and reactive decoupling techniques discussed in Sect. 2.3.4.1.

Additionally, real-time measurement of both θ_L and V_L is challenging due to the varying positions of traction loads. The terms ($\theta_{s10} - \theta_L$) and ($V_{s10} - V_L$) in Eqs. (2.34) and (2.35) are DC components, which can be obtained through the automatic adjustments of the PI_θ and PI_V controllers if these two elements are omitted. Consequently, the schematic diagram for the simplified calculation of active and reactive reference currents under interconnection control for automatic current distribution between substations is illustrated in Fig. 2.12.

In the said figure, θ_{dif0} and θ_{dif} represent the initial and real-time values of the voltage phase difference between the reference phase on the three-phase grid side and the single-phase interconnection point of the converter, respectively. V_{s10} and V_{s1} denote the initial and real-time values of the voltage amplitude at the single-phase interconnection point of the converter, respectively. K_θ and K_v are the adjustment coefficients for the active and reactive current channels, where $k' = R_1'/V_{s1}$ and $k_2 = R_1'/V_{s1}V_L$.

The voltage phase of the reference phase of the three-phase power grid can be obtained using a three-phase phase-locked loop (PLL), while a single-phase PLL enables real-time alignment for the voltage phase of the single-phase traction network. The voltage amplitude can be detected by multiplying the voltage at the interconnection point of the traction network by the output value from the single-phase PLL ($\sin\theta_{s1}$). This product is then processed through a low-pass filter (LPF) to yield a DC component, which represents the single-phase voltage amplitude (V_{s1}).

When the traction network is under no-load conditions through the entire system, the interconnection converter can obtain the initial stable phase difference (θ_{dif0}) and the initial amplitude (V_{s10}) at the interconnection point. The locked values of θ_{dif0} and V_{s10} are utilized as the references for the interconnection converter. Upon the connection of a load, the interconnection converter keeps aligned with the subsequently calculated active and reactive current references, facilitating automatic current distribution between substations.

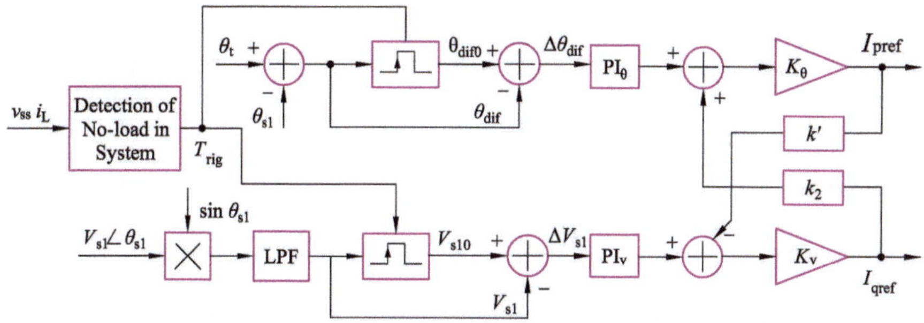

Fig. 2.12 Schematic diagram of calculation of active and reactive reference current under interconnection control for automatic current distribution between substations

2.3.4 Simulation and Analysis of System with Interconnected Two Substations

Figure 2.13 illustrates a simulation system comprising two substations, designed according to the advanced co-phase traction power supply system structure and interconnection control strategy previously mentioned. To simplify simulation and analysis, both the voltage level and capacity of the simulation system were scaled down to conform to conventional single-module converter standards. Specifically, the traction network voltage was reduced to 400 V, akin to the simulation of 380 V converter systems used in industrial applications, while the DC voltage reference is set at 700 V. The converter voltage can be adjusted to match the traction network voltage and the three-phase grid voltage by modifying the transformation ratio of the step-up/step-down transformer, thereby ensuring alignment with the actual traction network voltage.

In the continuous feeding system featuring interconnected two substations as depicted in Fig. 2.13, Substation SS1 is designated as a power source, connecting the three-phase grid to a three-phase-to-single-phase converter via a step-down transformer. The resulting single-phase output is linked to the traction network through a step-up isolation transformer, with $v_{s1} = V_{s1} \angle \theta_{s1}$ indicating the output voltage of SS1. Substation SS0, designated as a voltage source, connects to the traction network via a single-phase inverter, with its output voltage represented as $v_{s0} = V_{s0} \angle \theta_{s0}$. This voltage aligns with the phase of the reference phase of the grid to provide voltage support for the traction network. The DC side of SS0 can be connected to the three-phase power grid through a three-phase converter, similar to SS1, or can be linked to other energy sources to provide active power support.

Traction loads Z_{L0} and Z_{L1} are simplified as a 10 Ω resistor connected in parallel with a 50 mH inductor; the unit impedance of the traction network is represented as $z = 0.076 + j0.167$ Ω/km; the switching frequency of the three-phase-to-single-phase converter is set at 10 kHz.

The system simulation lasts for 20 s. At $t = 0$~1.5 s, the three-phase rectifier is activated to stabilize the intermediate DC voltage. At $t = 1.5$ s, the interconnection

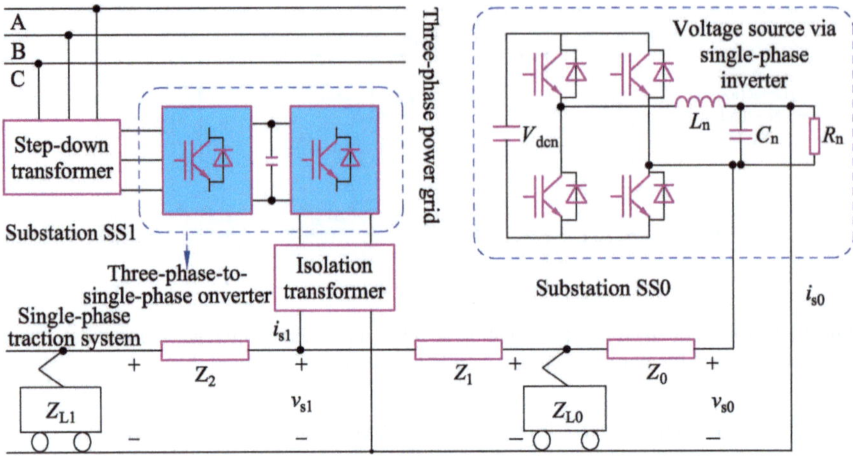

Fig. 2.13 Schematic diagram of simulation system with interconnected two substations

converter on the single-phase side is switched on. Load Z_{L1} is connected at $t = 4$ s, followed by the connection of load Z_{L0} at $t = 9$ s. Load Z_{L1} is disconnected at $t = 14$ s, followed by the disconnection of load Z_{L0} at $t = 17$ s. The simulation waveforms represent per-unit values.

2.3.4.1 Analysis of Control Performance for Three-Phase-to-Single-Phase Converter

Three-phase active and reactive decoupling control is implemented on the three-phase side of the three-phase-to-single-phase converter. Figure 2.14 illustrates the waveforms of the DQ components for active current (I_d), reactive current (I_q), and port control voltage under both no-load and on-load conditions on the three-phase side.

The DQ component of voltage is adjusted in response to variations in load, resulting in corresponding fluctuations in the current components (I_d and I_q) on the DQ axis. Each control variable remains steady after 1 s following a sudden load change. Since the converter on the three-phase side operates at a unity power factor, the reactive current (I_q) of the three-phase converter consistently remains close to 0, regardless of fluctuations in load power on the single-phase side. Despite variations in traction loads, the three-phase grid maintains operation at a unity power factor, while other control variables adapt to changes in single-phase loads and become steady after brief adjustment periods.

Figures 2.15 and 2.16 present the simulation waveforms of the DQ current components of the single-phase controller and the reference voltages (V_{sp}) and (V_{sq}), respectively. In response to the sudden load change, the single-phase reference voltages (V_{sp} and V_{sq}) fluctuate accordingly and return to a steady state after

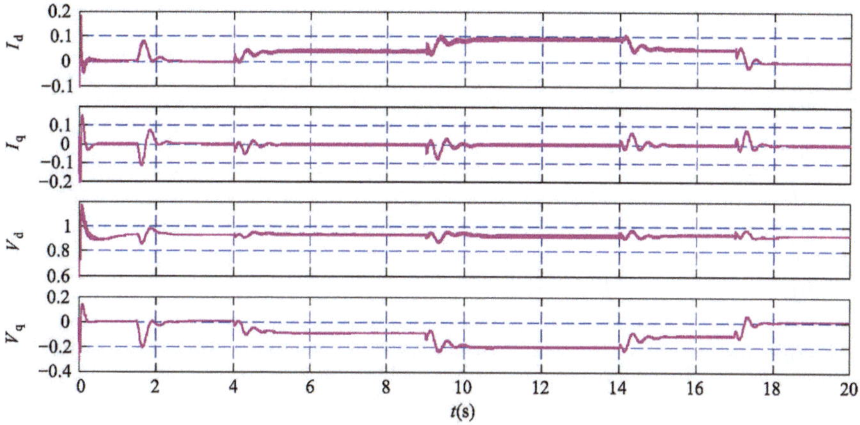

Fig. 2.14 Simulation results of *DQ* decoupling controller on three-phase side (I_d, I_q, V_d, V_q)

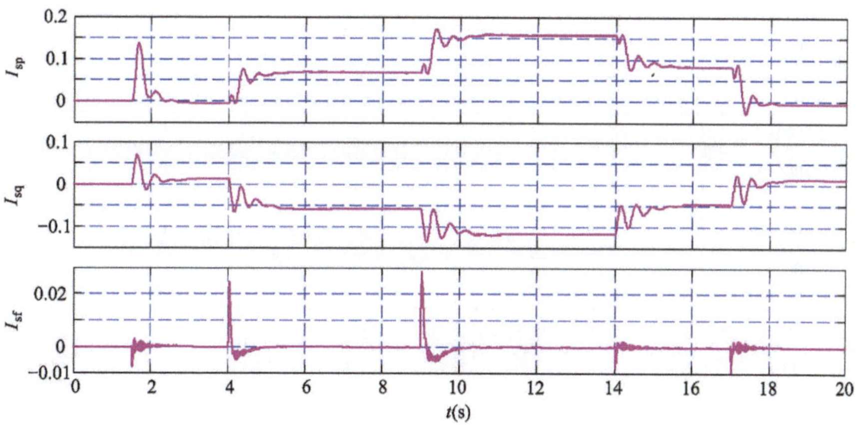

Fig. 2.15 Waveforms of single-phase active current, reactive current, and DC bias current

approximately 1 s of adjustments. The single-phase active current and reactive current also stabilize after adjustments, while the DC bias current consistently remains close to 0 following adjustments. This demonstrates that the controller on the single-phase side can effectively deliver active and reactive outputs as required in response to load changes while controlling the single-phase DC bias current to maintain it near 0.

Figure 2.17 illustrates the simulation waveforms of the total DC voltage, three-phase side current, and single-phase side current after the two substations are interconnected. As shown in the figure, the total DC voltage stabilizes at the reference approximately 1 s after the sudden load change. The AC currents on both the three-phase and single-phase sides, as well as the single-phase port voltage, fluctuate in response to the connection or disconnection of the load and return to a steady state following a brief adjustment period of less than 1 s.

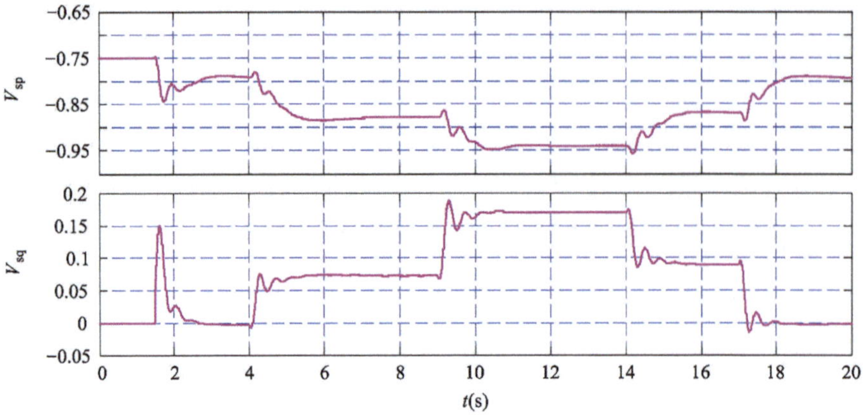

Fig. 2.16 Simulation results of reference voltages of single-phase controller (V_{sp} and V_{sq})

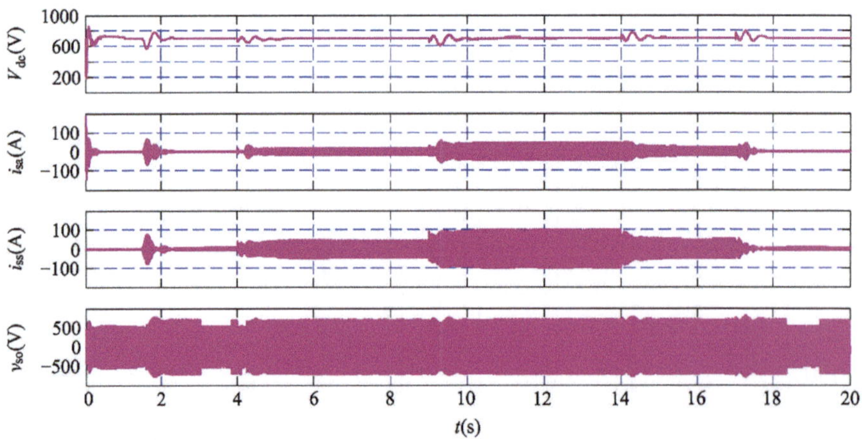

Fig. 2.17 DC voltage (V_{dc}), phase-A AC current (i_{sa}), AC current on single-phase side (i_{ss}), and port voltage on single-phase side (v_{so})

2.3.4.2 Analysis of Active Current and Reactive Current References

Figures 2.18 and 2.19 present the key waveforms from the simulation of the interconnection control strategy for automatic current distribution between the two substations. As illustrated in Fig. 2.18, when sudden load changes occur, the voltage phase (θ_{dif}) aligns with the initial phase (θ_{dif0}) after approximately 1 s of adjustments. Prior to the single-phase interconnection, a phase difference exists due to the non-normalized output of the converter, resulting in a continuous increase in the reference value (I_{pref}) of the active current. Once the converter is enabled for interconnection, the phase difference quickly returns to 0, and I_{pref} also quickly reverts to the no-load value. Furthermore, I_{pref} can be swiftly adjusted in response to sudden load changes occurring at $t = 0.4$, $t = 9$, $t = 14$, and $t = 17$ s.

As illustrated in Fig. 2.19, the output voltage amplitude (V_{s1}) is adjusted in response to changes in the reactive load. These alterations in the reactive load affect the port voltage amplitude, and the substation delivers the reactive currents required by the load. This alignment with the reference value (I_{qref}) of the reactive current facilitates a restoration to the initial amplitude after each load change. This observation validates the effectiveness of the interconnection control strategy for automatic current distribution between substations, as discussed in the previous section, where phase angle differences determine active current and amplitude differences govern reactive current.

2.3.4.3 Simulation Analysis of Distribution of Load Current Between Substations

As illustrated in Fig. 2.20, when loads Z_{L0} and Z_{L1} are connected or disconnected, Substations SS1 and SS0 adjust their output currents (i_{s1} and i_{s0}) in response to these load changes. In the absence of traction load, the minimal active power output from the two substations is due to converter losses.

At $t = 4$ s, load Z_{L1} is connected. Since Z_{L1} is located near Substation SS1, SS1 provides all the load power at this moment. The output current of SS1 increases and approaches 100 A, while the output current of Substation SS0 remains very low, indicating that all the output power is delivered by the power source at Substation SS1.

At $t = 9$ s, load Z_{L0} is connected. Since Z_{L0} is located between Substations SS1 and SS0, the power support is jointly provided by SS1 and SS0. As illustrated in

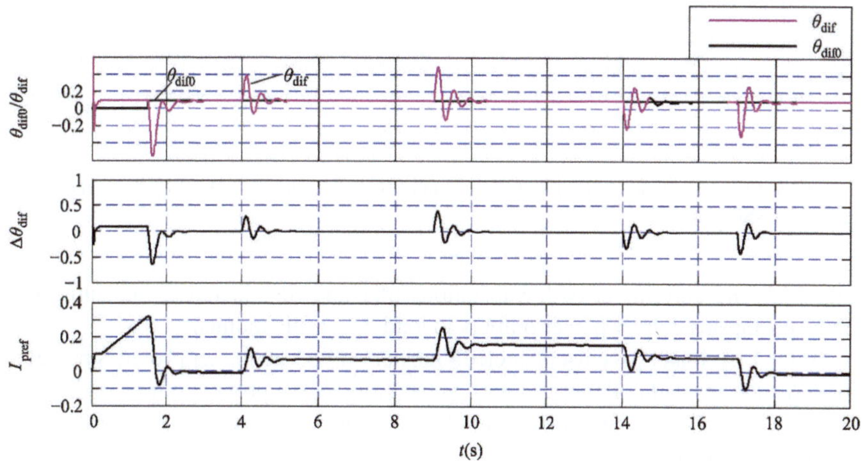

Fig. 2.18 Simulation waveforms of interconnection controller for phase angle: initial value and real-time value of phase angle (θ_{dif0} and θ_{dif}), phase angle increment ($\Delta\theta_{dif}$), and reference value of active current (I_{pref})

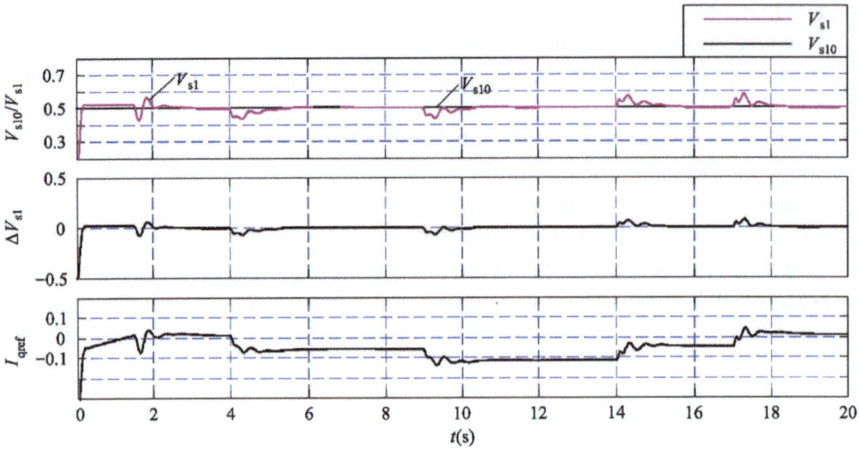

Fig. 2.19 Simulation waveforms of interconnection controller for amplitude: initial value and real-time value of amplitude (V_{s1} and V_{s10}), amplitude increment (V_{s1}), and reference value of reactive current (I_{qref})

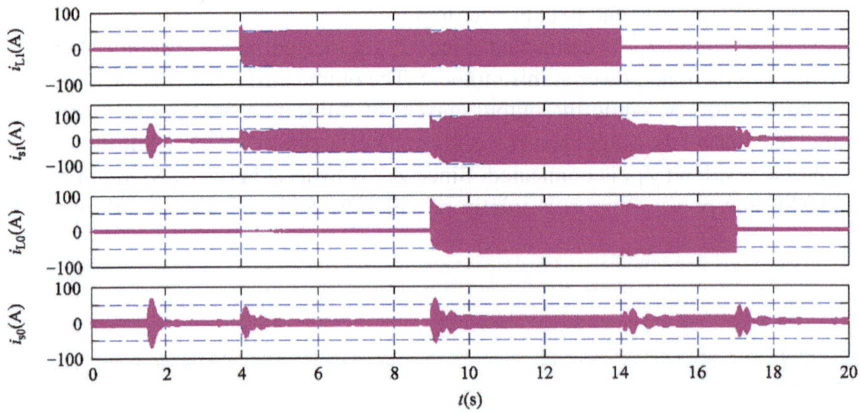

Fig. 2.20 Current of load Z_{L1} (i_{L1}), output current of SS1 (i_{s1}), current of load Z_{L0} (i_{L0}), and output current of SS0 (i_{s0})

Fig. 2.20, the currents at both Substations SS1 and SS0 increase, indicating that both substations deliver power outputs to load Z_{L0} at this moment.

At $t = 14$ s, load Z_{L1} is disconnected. Substation SS1 no longer provides power to load Z_{L1} and instead collaborates with Substation SS0 to provide power support for load Z_{L0}. As a result, the current amplitude at SS1 decreases, while that at SS0 remains relatively unchanged.

At $t = 15$ s, load Z_{L0} is disconnected. Following the complete disconnection of the loads, the currents at both substations return to their no-load values after an adjustment period of approximately 1 s.

The following conclusions can be drawn from the analysis of the simulation results concerning the aforementioned special load combinations:

1. The interconnection control strategy for automatic current distribution between substations allows for the real-time adjustment of power outputs at the substations in response to changes in load position and power demands.
2. The output power level of the substations depends on load demands and distances of substations from the load. The substation that is closer to the load delivers a greater amount of output power, aligning with the optimization control objective of nearby output.

2.3.4.4 Voltage Analysis for Interconnection Point

Figure 2.21 shows the output voltage waveforms of Substations SS0 and SS1. As illustrated, the inverter voltage source substation (SS0) reaches a steady state after 0.5 s following its startup. The voltage source substation (SS0) basically remains unchanged in output voltage following multiple abrupt load changes. The power source substation (SS1) also attains a steady output voltage value after 1 s following the abrupt load change. The details during the intervals from $t = 7$ to $t = 7.1$ s and from $t = 12$ to $t = 12.2$ s show that the output voltages of both substations ultimately align in amplitude, phase, and frequency.

2.3.5 *Simulation and Analysis of System with Interconnected Three Substations*

Based on the simulation of a continuous power supply system with interconnected two substations discussed in the previous section, this section presents the simulation and analysis of a continuous power supply system involving three substations, as illustrated in Fig. 2.22. SS0 is designated as the substation with voltage source characteristics, while SS1 and SS2 are two substations with power source control characteristics. Traction loads Z_{L0}, Z_{L1}, and Z_{L2} are simplified as a 5 Ω resistor in parallel with a 15 mH inductor. The unit impedance of the traction network is represented as $z = 0.076 + j0.167$ Ω/km, corresponding to the traction network impedances of $Z_0 = z \times 2$ Ω, $Z_1 = z \times 1$ Ω, $Z_2 = z \times 2$ Ω, $Z_3 = z \times 1$ Ω, and $Z_4 = z \times 3$ Ω. The DC voltage reference is set at 1000 V, and the simulation duration is maintained at 20 s.

At $t = 0$ s, the inverter voltage source substation is started, and the three-phase side converters of the two power sources are also started to initiate the control for DC voltage stabilization, without being connected to the single-phase traction network. At $t = 0.5$ s, the power source substation (SS1) is connected to the single-phase traction network; at $t = 1.5$ s, the power source substation (SS2) is connected to the single-phase traction network.

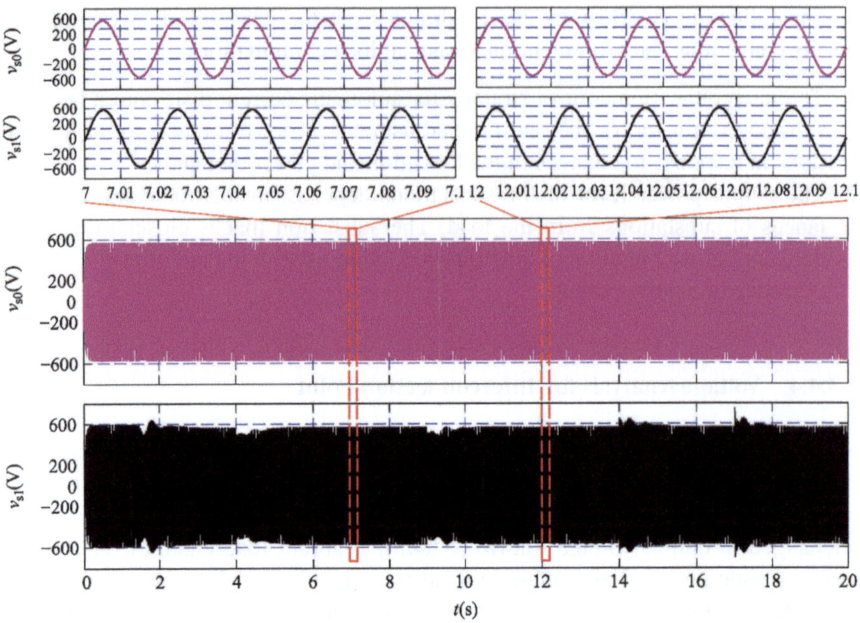

Fig. 2.21 Simulation waveforms of output AC voltage from interconnected two substations

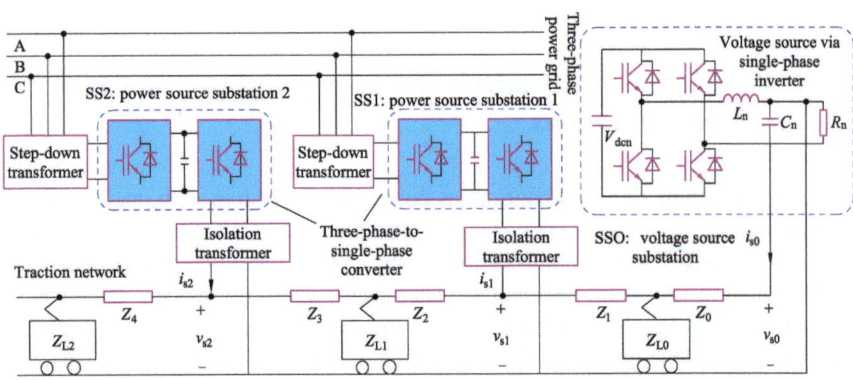

Fig. 2.22 Simulation structure of interconnected three substations

The loads are connected according to the following timing sequence:

Load Z_{L1} is connected at $t = 3$ s, load Z_{L2} at $t = 5$ s, and load Z_{L0} at $t = 8$ s. Load Z_{L1} is disconnected at $t = 11$ s, load Z_{L2} at $t = 14$ s, and load Z_{L0} at $t = 17$ s.

2.3.5.1 Analysis of Simulation Results for Controller

Figure 2.23 illustrates the dynamic simulation waveforms of DC voltage, as well as the three-phase and single-phase AC currents for the three-phase-to-single-phase converter at the power source substation (SS1). As depicted, after the load is connected and subsequently disconnected, the DC voltage exhibits slight fluctuations before returning to a steady state at 1000 V. AC currents on both the three-phase and single-phase sides stabilize within 1 s.

Figure 2.24 displays the active and reactive current references for the interconnection controller at the power source substation (SS1). As shown, these active and reactive current references vary as the three traction loads are connected and subsequently disconnected, each followed by an adjustment period of approximately 1 s. This indicates good dynamic performance.

Figures 2.25 and 2.26 illustrate the simulation waveforms for the reference values of DC voltage, three-phase AC current, single-phase AC current, as well as active and reactive currents in the three-phase-to-single-phase converter at the power source substation (SS2), respectively. Similar to SS1, the DC voltage remains steady at 1000 V. The reference values for AC current and active/reactive current are also adjusted in response to the connection and disconnection of the loads; however, the amplitudes differ from those of SS1 due to the different load positions.

To sum up, the simulation results of the system with interconnected three substations exhibit characteristics similar to those of the system with two substations. This further demonstrates that the interconnection control strategy for automatic current distribution between substations facilitates the automatic distribution of power outputs from nearby substations according to the varying load demands.

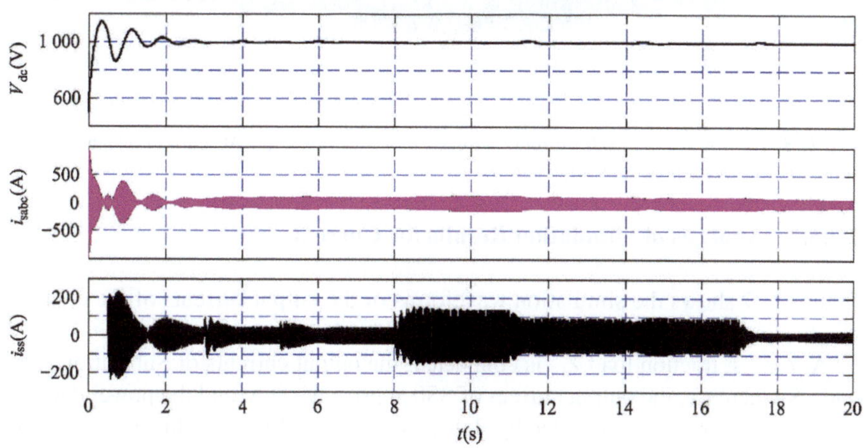

Fig. 2.23 DC voltage (V_{dc}), AC current on three-phase side (i_{sabc}), and AC current on single-phase side (i_{ss}) of power source substation (SS1)

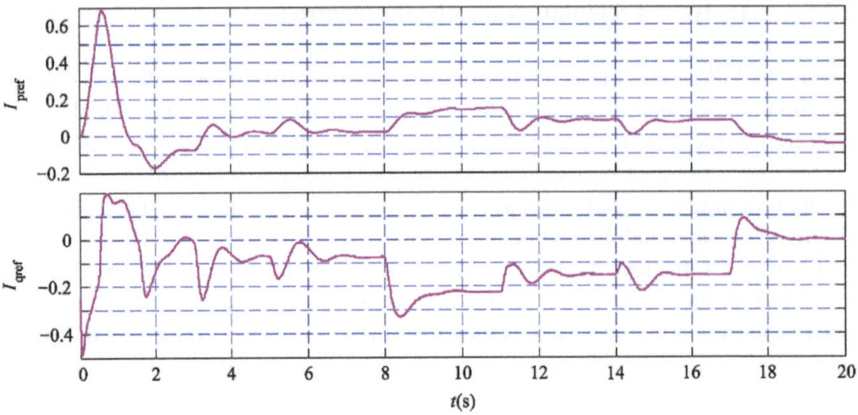

Fig. 2.24 Simulation waveforms of reference values of active and reactive currents of power source substation (SS1)

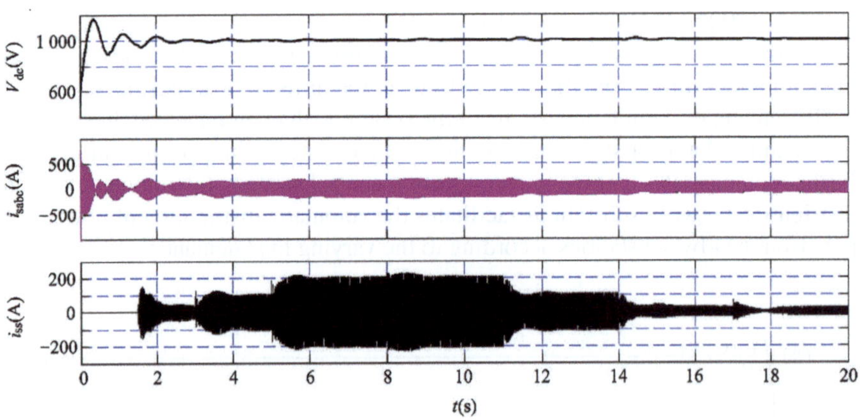

Fig. 2.25 DC voltage (V_{dc}), AC current on three-phase side (i_{sabc}), and AC current on single-phase side (i_{ss}) of power source substation (SS2)

2.3.5.2 Analysis of Simulation Results for Current

Figure 2.27 shows the simulation waveforms of the output currents of the continuous feeding system with three substations and load currents.

At $t = 3$ s, traction load Z_{L1} is connected, the output currents of Substations SS1 and SS2 increase, the output current of SS0 approaches 0 A, and the power demand of load Z_{L1} is primarily provided by Substations SS1 and SS2. At this moment, traction load Z_{L1} is closer to Substation SS2, and the impedance in the traction network (Z_3) is less than Z_2. Consequently, the output current amplitude of SS2 is large, which is in line with the control objective of the automatic distribution of power output from a nearby substation.

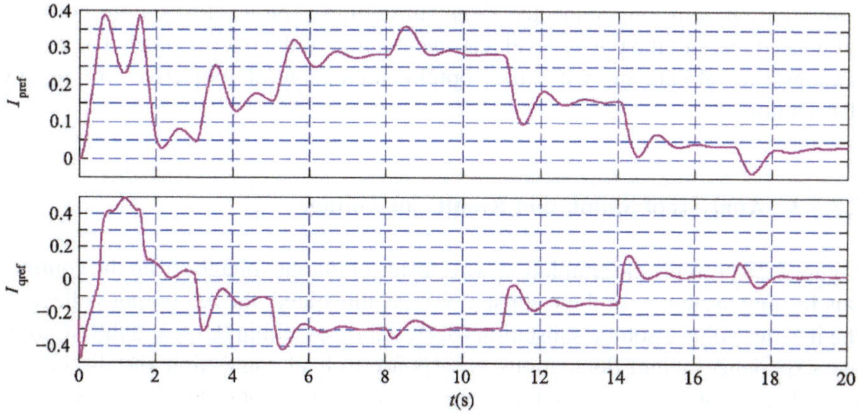

Fig. 2.26 Simulation waveforms of reference values of active and reactive currents of power source substation (SS2)

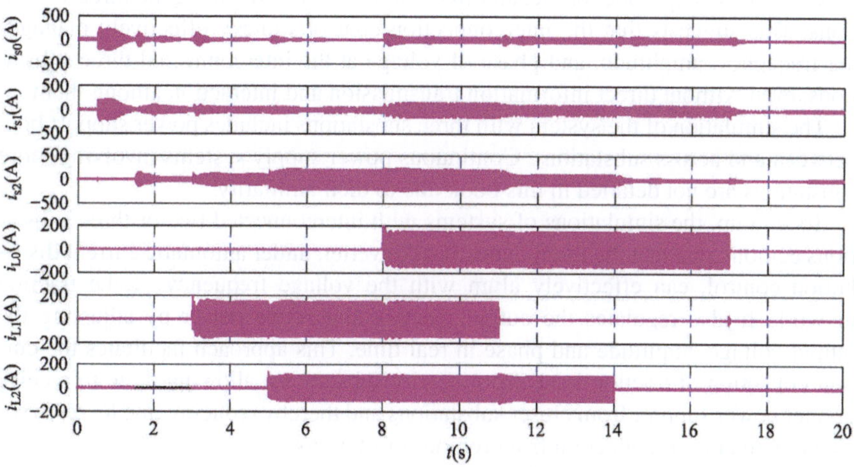

Fig. 2.27 Simulation waveforms of output currents of three traction substations and load currents

At t = 5 s, Lod Z_{L2} is connected, and the output current of Substation SS2 (i_{s2}) stabilizes at 200 A after an adjustment period of approximately 1 s. During this period, i_{s0} and i_{s1} remain almost unchanged. This indicates that the power demand of load Z_{L2} is primarily supplied by Substation SS2.

At t = 8 s, when load Z_{L0} is connected, i_{s2} remains almost unchanged, while i_{s0} and i_{s1} rapidly increase to a new steady value. This indicates that the power demand of load Z_{L0} is primarily provided by Substations SS0 and SS1.

At t = 17 s, when all the loads are disconnected, the output currents of all the substations return to their no-load state.

Therefore, the control strategy is also effective in automatically adjusting the output power of all substations to meet the demands of the traction loads in the configuration with interconnected three substations. The level of output power is influenced by both the power level and the location of the traction load. The results demonstrate alignment with the optimization control objective of nearby distribution.

2.3.5.3 Analysis of Simulation Results for Voltage

Figure 2.28 presents the simulation waveforms of output voltages from the continuous feeding system with three substations. As illustrated, the AC output voltage of the inverter voltage source substation (SS0) stabilizes within 0.5 s, and the AC voltage experiences slight fluctuations in response to load connection and disconnection. The output voltages of the power source substations (SS1 and SS2) also fluctuate slightly at the moments of load connection and disconnection, subsequently returning to their steady state after an adjustment period of 1 s. Details during the intervals from $t = 7$ to $t = 7.1$ s and from $t = 13$ to $t = 13.1$ s indicate that the output voltage amplitude, phase, and frequency are aligned among the three substations. This suggests that the interconnection control strategy effectively manages the frequency, amplitude, and phase of voltage at the interconnected three substations, even without direct information transmission and interaction among them.

The simulation of the system with three substations includes power support both between and across substations. Continuous power supply systems involving more substations are not detailed in this book due to their similarity.

To sum up, the simulations of systems with interconnected two or three substations demonstrate that the interconnection converter, under automatic current distribution control, can effectively align with the voltage frequency of the traction network. It also regulates the output reactive and active power by adjusting the output voltage amplitude and phase in real time. This approach facilitates the current collection of traction loads from nearby substations, allowing them to receive greater power support from closer substations and thereby reducing line losses associated with current collection from remote substations.

In addition, adjacent substations can provide power support for loads throughout the entire system. Ignoring line losses, it is only required for the total power generated by all the substations within the continuous power supply section to exceed the maximum demand of all the loads within that section. This approach can significantly reduce the power requirements at individual substations and ensure the smooth operation of the continuous feeding system, even if one or more substations are disconnected due to faults.

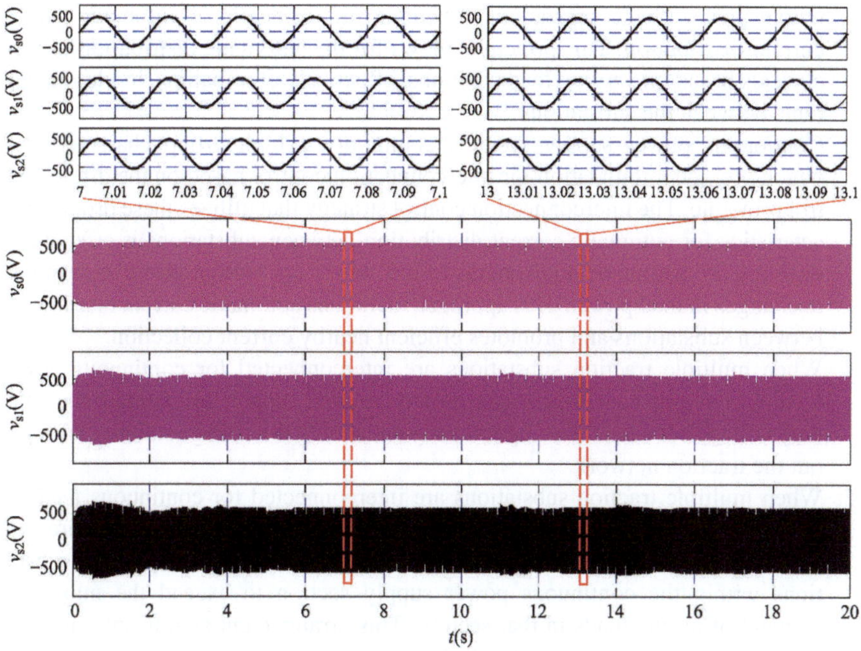

Fig. 2.28 Simulation waveforms of output voltages of three substations

2.4 Summary

In this chapter, the topology structure, mathematical model, and control strategy of three-phase-to-single-phase converters, as well as the control strategy for interconnected multiple substations, are discussed, grounded in the context of advanced co-phase traction power supply across interconnected multiple substations through the three-phase-to-single-phase converter structure. Through theoretical analysis and simulations of continuous feeding configurations with interconnected two or three substations, the following conclusions are drawn:

1. In advanced co-phase traction power supply systems, the impedance characteristics of traction networks with interconnected substations over long distances affect the interconnection control of converters. Both the impedance of the traction network and the positioning of traction loads influence the voltage amplitude and phase distribution in the network. Therefore, a rational control strategy is essential for the interconnection of multiple substations ensuring a continuous power supply across these substations.
2. When there are differences in amplitude, phase, and frequency of the output voltages among interconnection converters, circulating currents inevitably arise between the substations. However, when the output voltage frequency, amplitude, and phase angle are identical across the interconnection converters, the

output power of the substations adheres to the principle of automatic nearby current distribution. This principle allows for the automatic distribution of load currents based on varying load positions, effectively eliminating circulating currents between the substations.

3. Continuous power supply systems commonly exhibit droop characteristics, which provide a foundation for interconnection control based on the droop control principle. The interconnection control strategy that utilizes these droop characteristics for automatic current distribution between substations is effective in enabling the automatic adjustments of the converters' output power in response to changes in load power. This approach facilitates automatic current distribution between substations and promotes efficient nearby current collection.
4. When multiple traction substations are interconnected for continuous power feeding, two adjacent substations provide mutual support to maintain the catenary voltage between them, significantly enhancing the voltage stability throughout the traction network.
5. When multiple traction substations are interconnected for continuous feeding, adjacent substations can provide power support for loads across the entire system. Consequently, it is sufficient for the total power generated by all the substations within the continuous power supply section to exceed the maximum demand of all the loads in that section. This arrangement significantly reduces the power requirement for individual substations.
6. In continuous power supply systems with interconnected multiple substations, the system operation remains unaffected by the exit of one or more substations due to faults, as other functioning substations can continue to provide power support for loads throughout the entire system.

References

Brabandere K D, Bolsens B, Keybus J V D, et al. A Voltage and Frequency Droop Control Method for Parallel Inverters [J]. IEEE Transactions on Power Electronics, 2007, 22(4): 1107–1115.

Chen M, Li Q, Roberts C, et al. Modelling and performance analysis of advanced combined co-phase traction power supply system in electrified railway [J]. IET Generation, Transmission & Distribution, 2016, 10(4): 906–916.

Guerrero J M, Luis Garcia De V, Matas J, et al. Output Impedance Design of Parallel-Connected UPS Inverters With Wireless Load-Sharing Control [J]. IEEE Transactions on Industrial Electronics, 2005, 52(4): 1126–1135.

He J, Li Y W, Guerrero J M, et al. An Islanding Microgrid Power Sharing Approach Using Enhanced Virtual Impedance Control Scheme [J]. IEEE Transactions on Power Electronics, 2013a, 28(11): 5272–5282.

He Xiaoqiong, Zhou Yingying, Peng Xu, et al. Traction Power Supply System Based on Three-Phase-to-Single-Phase Converter and Its Interconnection Control [J]. *Journal of Railway Science and Engineering*, 2013b, (05): 109–115.

He X, Shu Z, Peng X, et al. Advanced Cophase Traction Power Supply System Based on Three-Phase to Single-Phase Converter [J]. IEEE Transactions on Power Electronics, 2014, 29(10): 5323–5333.

Hu Jingyu. Research on Power Flow Control Strategy for Continuous Co-phase Power Supply System [D]. Master's Degree Dissertation of Southwest Jiaotong University, 2013.

Lee C T, Chu C C, Cheng P T. A New Droop Control Method for the Autonomous Operation of Distributed Energy Resource Interface Converters [J]. IEEE Transactions on Power Electronics, 2013, 28(4): 1980–1993.

Li Qunzhan. Several Key Technical Issues in the Development of Traction Power Supply for High-speed Railways in China [J]. *Journal of the China Railway Society*, 2010, 32(4): 119–124.

Li Qunzhan, He Jianmin. *Analysis of Traction Power Supply System* [M]. Chengdu: Southwest Jiaotong University Press, 2007.

Li Qunzhan, Zhang Jinsi, He Weijun. Study on New Power Supply System for Heavy-haul Electric Traction [J]. *Journal of the China Railway Society*, 1988(04): 23–31.

Liu Neng. Study on Compensation Optimization Scheme for Co-phase Traction Substation [D]. Master's Degree Dissertation of Southwest Jiaotong University, 2010.

Liu L, Liu Z, Chen Q. Research and simulation of circulation current control for grid tied inverters connected in parallel [M]. 2016 Chinese Control and Decision Conference (CCDC). 2016: 6036–6041.

Ma L, Chen Q. Research on control strategy for circulation current of parallel inverters [M]. 2016 31st Youth Academic Annual Conference of Chinese Association of Automation (YAC). 2016: 171–174.

Martins A P, Carvalho A S, Araujo A S. Design and implementation of a current controller for the parallel operation of standard UPSs [M]. Industrial Electronics, Control, and Instrumentation, 1995, Proceedings of the 1995 IEEE IECON 21st International Conference on. 1995: 584–589 vol. 1.

Shu Z, Xie S, Lu K, et al. Digital Detection, Control, and Distribution System for Co-Phase Traction Power Supply Application [J]. IEEE Transactions on Industrial Electronics, 2013, 60(5): 1831–1839.

Sun Y, Hou X, Yang J, et al. New Perspectives on Droop control in AC MicroGrid [J]. IEEE Transactions on Industrial Electronics, 2017.

Wen J, Zhou L, Smedley K. Minimizing the Circulation Current in Hexagram Medium-Voltage Variable-Speed Drive by Coupled Inductors [J]. IEEE Transactions on Industrial Electronics, 2011, 58(8): 3421–3430.

Xia Yankun. Design of Direct Power Controller for Co-phase Power Supply System on High-speed Railway [J]. *Journal of Power Supply*, 2017, (01): 105–118.

Xie Meng. Output Control of Single-Phase 400 Hz Medium-Frequency Voltage Source Inverter and Control for Parallel Operation [D]. Doctoral Dissertation of University of Chinese Academy of Sciences, 2006.

Peng Xu. Research on Parallel Connection of Voltage Source Inverters for Continuous Traction Power Supply System [D]. Master's Degree Dissertation of Southwest Jiaotong University, 2014.

Yadong L, Yan X, Lipei H, et al. A novel distributed control scheme for the parallel operation of digital controlled UPS [M]. Industrial Electronics Society, 2003 IECON '03 The 29th Annual Conference of the IEEE. 2003: 668–672 vol. 1.

Chapter 3
Control and Modulation of Grid-Connected Three-Phase Converter

3.1 Three-Phase Converter and Circuit Analysis

A feasible architecture for advanced co-phase traction power supply systems is the three-phase-to-single-phase topology depicted in Fig. 3.1a. This design facilitates energy conversion and transmission between the three-phase power grid and the traction network. In this architecture, the converter that connects to the three-phase power grid should be configured as a three-phase AC-DC structure, utilizing a voltage source inverter (VSI) topology (Liu 2010).

Three-phase AC power supplies are the most common mode in electric power systems, with three-phase voltage source inverters (VSIs) being the most widely used circuit in these applications, as shown in Fig. 3.1b. A three-phase VSI primarily consists of a DC-link capacitor, a three-phase VSI bridge arm, and a three-phase AC connecting inductor, enabling parallel connections to three-phase power supply systems alongside other loads. The three-phase VSI circuit is commonly utilized in various applications, including passive inverters, drive speed regulators, PWM rectifiers, active power filters, dynamic reactive power compensators, and uninterruptible power supplies.

Advanced co-phase traction power supply systems are typically connected to three-phase electric power systems, with three-phase converters facilitating this connection. Consequently, three-phase converters serve as fundamental conversion circuits that support the application of advanced co-phase traction power supply systems. This chapter discusses various common circuit topologies for constructing three-phase converters on the grid side of advanced co-phase traction power supply systems, along with three-phase power decoupling control and essential phase-locked loops and modulation methods used in the control systems.

Z. Shu, *Power Electronic Converters in Advanced Co-Phase Traction Power Supply System*, https://doi.org/10.1007/978-981-95-3276-6_3

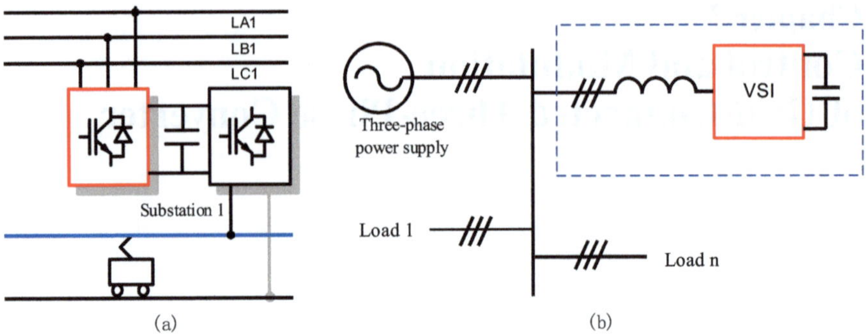

Fig. 3.1 (**a**) Three-phase-to-single-phase converter applied in advanced co-phase traction power supply; (**b**) schematic diagram of application of three-phase converter in three-phase system

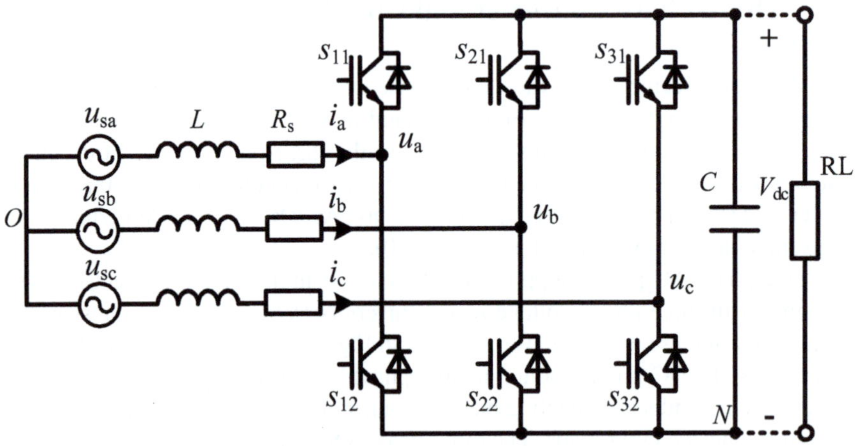

Fig. 3.2 Circuit of three-phase two-level converter

3.1.1 *Three-Phase Two-Level Converter*

In terms of circuit topology, three-phase two-level voltage source converters are the most fundamental type of three-phase AC-DC converters, as illustrated in Fig. 3.2. The AC side features a three-phase symmetrical configuration without a neutral line connection. The filter inductor, denoted as L (with line resistance and loss equivalent represented as R_s), is connected to the power supply and forms a three-phase bridge arm alongside six fully controlled power switching devices and free-wheeling diodes. Point O indicates the AC voltage reference, while point N represents the negative reference of the DC bus (Jie 2006; Zheng 2011; Wang 2007; Shu et al. 2005).

All bridge arms of the three-phase converters are configured in a half-bridge structure, with at most one of the two series-connected switching tubes conducting at any time. The states of six switching tubes (S11–S32) in the three-phase bridge arms a, b, c are represented by s_a, s_b, and s_c, respectively. $s_x(x = a, b, c)$ gives a unipolar two-valued logic switching function.

In the case of $s_x(x = a, b, c) = 1$, the upper switching tube of the phase x bridge arm is turned on and the lower tube is turned off, resulting in high-level output from the bridge arm.

In the case of $s_x(x = a, b, c) = 0$, the upper tube of the phase x is turned off and the lower tube is turned on, resulting in low-level output from the bridge arm.

Therefore, three-phase two-level converters offer a total of $2^3 = 8$ switching modes.

In current circuits under the same switching mode, different instantaneous directions of catenary-side current correspond to different current circuits. In order to simplify the analysis, the study focused on the current circuits of three-phase two-level converters with operating unity power factor rectifiers. In the case of $i_a > 0$, $i_b < 0$, $i_c > 0$, the analysis was conducted based on on-off states of 110 and 101. The conduction path analysis is shown in Fig. 3.3.

When the on-off state is 110, the switching tubes s11, s21, and s32 are turned on, with the circuit flow path shown in Fig. 3.3a. The phases AB of the voltage source on the three-phase AC-side and DC-side capacitor simultaneously provide energy to loads, with $u_{ao} = V_{dc}/3$, $u_{bo} = V_{dc}/3$, $u_{co} = -2V_{dc}/3$, $u_{ab} = 0$, $u_{ac} = V_{dc}$, and $u_{bc} = V_{dc}$.

When the on-off state is 101, the switching tubes s11, s22, s31 are turned on, with the circuit flow path shown in Fig. 3.3b. The phases A and C of the voltage source on the three-phase AC side provide energy to the DC-side capacitor and loads simultaneously, with $u_{ao} = V_{dc}/3$, $u_{bo} = -2V_{dc}/3$, $u_{co} = V_{dc}/3$, $u_{ab} = V_{dc}$, $u_{ac} = 0$, and $u_{bc} = -V_{dc}$.

Similarly, the port levels corresponding to other on-off states can be generated, as shown in Table 3.1.

As shown in the table, at the bridge arm on-off state of 1, regardless of positive or negative current, the upper bridge arm switch is turned on, and the bridge arm midpoint is connected to the DC positive terminal for high-level output. At the

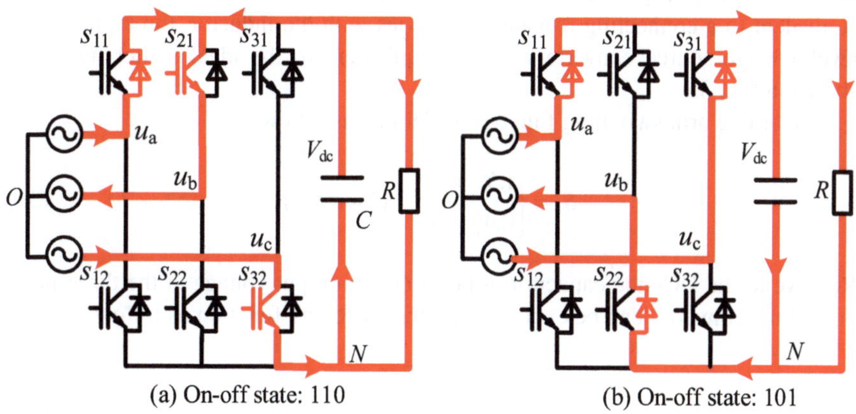

Fig. 3.3 Current flow path analysis of three-phase two-level converter

Table 3.1 Port levels corresponding to on-off states of three-phase two-level converter

On-off state			Port level					
s_a	s_b	s_c	u_{ao}	u_{bo}	u_{co}	u_{ab}	u_{ac}	u_{bc}
1	1	0	$V_{dc}/3$	$V_{dc}/3$	$-2V_{dc}/3$	0	V_{dc}	V_{dc}
1	0	1	$V_{dc}/3$	$-2V_{dc}/3$	$V_{dc}/3$	V_{dc}	0	$-V_{dc}$
1	0	0	$2V_{dc}/3$	$-V_{dc}/3$	$-V_{dc}/3$	V_{dc}	V_{dc}	0
0	1	0	$-V_{dc}/3$	$2V_{dc}/3$	$-V_{dc}/3$	$-V_{dc}$	0	V_{dc}
0	0	1	$-V_{dc}/3$	$-V_{dc}/3$	$2V_{dc}/3$	0	$-V_{dc}$	$-V_{dc}$
0	1	1	$-2V_{dc}/3$	$V_{dc}/3$	$V_{dc}/3$	$-V_{dc}$	$-V_{dc}$	0

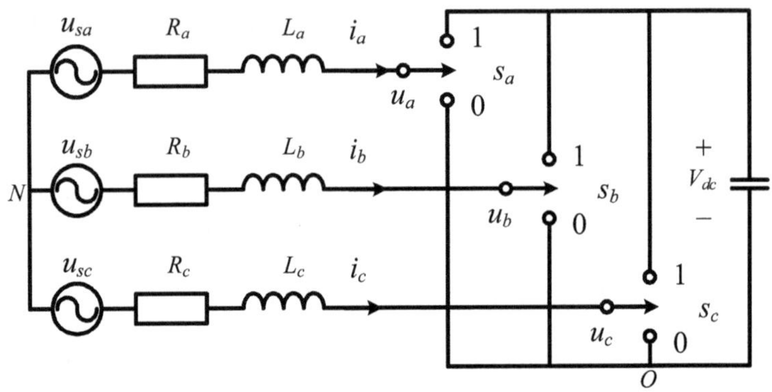

Fig. 3.4 Equivalent circuit of three-phase two-level converter circuit

bridge arm on-off state of 0, regardless of positive or negative current, the lower bridge arm switch is turned on, and the bridge arm midpoint is connected to the DC negative terminal for low-level output. Therefore, one bridge arm in the converters is equivalent to an ideal single-pole double-throw switch. Additionally, ignoring the dead time of switching devices and considering the line resistance of each phase equivalent to a connecting resistance in series with the inductor, three-phase two-level voltage source converters can be equivalent to the ideal switching circuit shown in Fig. 3.4.

The bridge arm switching function is defined as follows:

$$S_i = \begin{cases} 1, V_i = V_{dc} \\ 0, V_i = 0 \end{cases} \quad (i = a,b,c) \tag{3.1}$$

Relative to the DC-side capacitor at point O, the output voltage at the three-phase port of the converter can be expressed by the switching function as follows:

$$\begin{cases} u_{ao} = S_a V_{dc} \\ u_{bo} = S_b V_{dc} \\ u_{co} = S_c V_{dc} \end{cases} \tag{3.2}$$

The three-phase voltage equation derived from Kirchhoff's voltage law is as follows:

$$\begin{cases} u_{aN} = u_{ao} + u_{oN} \\ u_{bN} = u_{bo} + u_{oN} \\ u_{cN} = u_{co} + u_{oN} \end{cases} \tag{3.3}$$

For three-phase symmetrical systems,

$$u_{aN} + u_{bN} + u_{cN} = 0 \tag{3.4}$$

Combining Eqs. (3.3) and (3.4), the following equation is established:

$$u_{oN} = -\frac{u_{ao} + u_{bo} + u_{co}}{3} \tag{3.5}$$

Combining Eqs. (3.3) and (3.5), the following equation is established:

$$\begin{cases} u_{aN} = \dfrac{2u_{ao} - u_{bo} - u_{co}}{3} V_{\text{dc}} \\ u_{bN} = \dfrac{2u_{bo} - u_{ao} - u_{co}}{3} V_{\text{dc}} \\ u_{cN} = \dfrac{2u_{co} - u_{ao} - u_{bo}}{3} V_{\text{dc}} \end{cases} \tag{3.6}$$

According to Fig. 3.3, the phase voltage equation of three-phase voltage can be derived from Kirchhoff's voltage law as follows (Liu 2010):

$$\begin{cases} u_{sa} - L\dfrac{\mathrm{d}i_a}{\mathrm{d}t} - Ri_a - u_{aN} = 0 \\ u_{sb} - L\dfrac{\mathrm{d}i_b}{\mathrm{d}t} - Ri_b - u_{bN} = 0 \\ u_{sc} - L\dfrac{\mathrm{d}i_c}{\mathrm{d}t} - Ri_c - u_{cN} = 0 \end{cases} \tag{3.7}$$

According to Eqs. (3.2) and (3.6), Eq. (3.7) is rewritten as follows:

$$\begin{cases} \dfrac{\mathrm{d}i_a}{\mathrm{d}t} = -\dfrac{R}{L}i_a + \dfrac{1}{L}\left(u_{sa} - \dfrac{2S_a - S_b - S_c}{3} V_{\text{dc}}\right) \\ \dfrac{\mathrm{d}i_b}{\mathrm{d}t} = -\dfrac{R}{L}i_b + \dfrac{1}{L}\left(u_{sb} - \dfrac{2S_b - S_a - S_c}{3} V_{\text{dc}}\right) \\ \dfrac{\mathrm{d}i_c}{\mathrm{d}t} = -\dfrac{R}{L}i_c + \dfrac{1}{L}\left(u_{sc} - \dfrac{2S_c - S_b - S_a}{3} V_{\text{dc}}\right) \end{cases} \tag{3.8}$$

Among them, the phase voltage of the three-phase converter output can be expressed as follows:

$$\begin{cases} u_{aN} = \dfrac{2S_a - S_b - S_c}{3} V_{\mathrm{dc}} \\ u_{bN} = \dfrac{2S_b - S_a - S_c}{3} V_{\mathrm{dc}} \\ u_{cN} = \dfrac{2S_c - S_b - S_a}{3} V_{\mathrm{dc}} \end{cases} \tag{3.9}$$

Substituting Eq. (3.9) into Eq. (3.8) gives

$$\begin{cases} \dfrac{\mathrm{d}i_a}{\mathrm{d}t} = -\dfrac{R}{L} i_a + \dfrac{1}{L}\left(u_{sa} - u_{aN}\right) \\ \dfrac{\mathrm{d}i_b}{\mathrm{d}t} = -\dfrac{R}{L} i_b + \dfrac{1}{L}\left(u_{sb} - u_{bN}\right) \\ \dfrac{\mathrm{d}i_c}{\mathrm{d}t} = -\dfrac{R}{L} i_c + \dfrac{1}{L}\left(u_{sc} - u_{cN}\right) \end{cases} \tag{3.10}$$

Equation (3.10) represents the mathematical model of the three-phase voltage source converter shown in Fig. 3.4, expressed in the *abc* static coordinate system. Variables in this coordinate system are represented in the form of AC, which complicates control and alignment. Typically, it is necessary to transform the mathematical model of the converter from the *abc* coordinate system to the *dq* coordinate system for active and reactive power control. For a detailed analysis, refer to the discussion in Sect. 3.2.

The power and voltage levels of three-phase two-level voltage source converters are constrained by the power and voltage levels during a single switching period. To establish a advanced co-phase traction power supply system, multiple two-level converters should be connected in parallel, along with several step-up and step-down transformers to interface with the high-voltage power grid.

3.1.2 Three-Phase Three-Level Converter

Three-level converters can increase voltage and power levels exponentially, making them more advantageous for high-power and high-voltage applications. Three-level converter circuits typically include two types: diode-clamped and flying capacitor three-level converters. Diode-clamped three-level converters are further divided into two categories: type I and type T. This section focuses on the I-type three-level circuit, while the next section addresses the T-type three-level circuit.

The circuit of type I is illustrated in Fig. 3.5. The bridge arm of each phase consists of four switching tubes and two clamping diodes. Taking bridge arm A as an

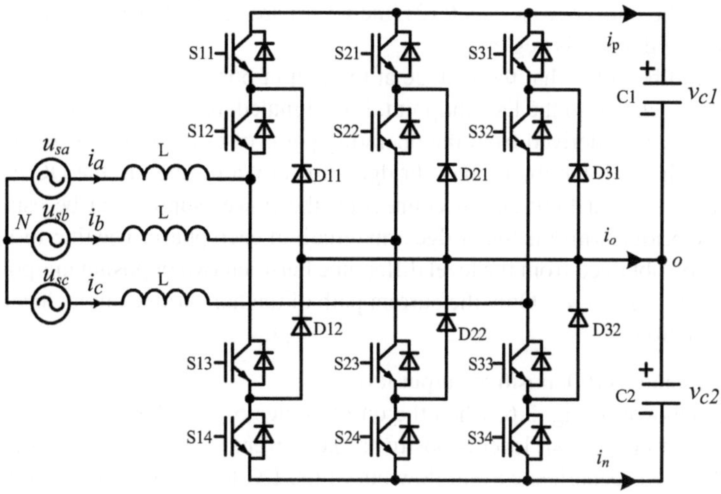

Fig. 3.5 Diode-clamped three-level converter

Table 3.2 Relationships between port output levels and on-off states

Port level	On-off state (phase A)			
	S11	S12	S13	S14
P	1	1	0	0
O	0	1	1	0
N	0	0	1	1

example, the PWM pulse signals for S11 and S13 are complementary, while those for S12 and S14 are also complementary. The DC-side sub-capacitor midpoint is connected to the midpoint of clamping diodes D11 and D12. The total DC-side voltage is divided into three levels: a positive level (P), a zero level (O), and a negative level (N). Assuming balanced capacitor voltages, the positive, zero, and negative voltages can be simplified to +1, 0, and −1 levels, respectively. Consequently, the three-level circuit can produce three output port phase voltage levels, and the line voltage represents the voltage difference between two phases, yielding a total of five output levels: +2, +1, 0, −1, and −2. This increase in the number of levels results in an exponential rise in port voltage. With the switching frequency of each tube remaining constant, the system's equivalent switching frequency increases exponentially, thereby enhancing the output electric energy quality from the converter.

To define switching functions, switch S is represented as 1 when it is turned on and 0 when it is turned off. Using phase A as an example, Table 3.2 illustrates the relationships between the port output levels and the corresponding on-off states, excluding inapplicable combinations.

The current flow paths for each converter on-off state are analyzed below in conjunction with Table 3.2. To simplify the analysis, it is assumed that (1) all switching devices in the converter are ideal; (2) line losses are ignored; and (3) the dead time of switching devices is neglected.

It should be noted that the voltage at the output port (u_a) of the half-bridge arm of phase A, relative to the DC midpoint, is designated as u_{ao}. Similarly, the voltages u_{bo} and u_{co} can be derived from the operating paths of other bridge arms for phases B and C. Since all the three-phase bridge arms are connected to the DC capacitor midpoint, a current flow circuit connecting the power supply can be established across the bridge arms via the bridge arm circuit of each phase. The line-to-line port level can be obtained from the level difference between two opposing midpoints, for example, $u_{ab} = u_{ao} - u_{bo}$. Only the current path of the half-bridge output level (u_{ao}) is analyzed below:

1. On-off state: 1100, positive output level

   As shown in Fig. 3.6, when the on-off states of S11, S12, S13, and S14 are 1100, the port output level is positive. The current flow paths should be considered in two scenarios: current flow out of the DC terminal and current flow into the DC terminal.

   Figure 3.6a illustrates the current flow path out of the DC terminal. With switching devices S11 and S12 turned on, current flows from the positive terminal on the DC side to the grid side through S11 and S12. The voltage of the port relative to the capacitor midpoint is expressed as $u_{ao} = v_{c1}$, and the output level is positive.

   Figure 3.6b shows the current flow path into the DC terminal. With switching devices S13 and S14 turned off, current flows from the grid side to the DC side through switching devices S11 and S12 and through the body diodes of S11 and

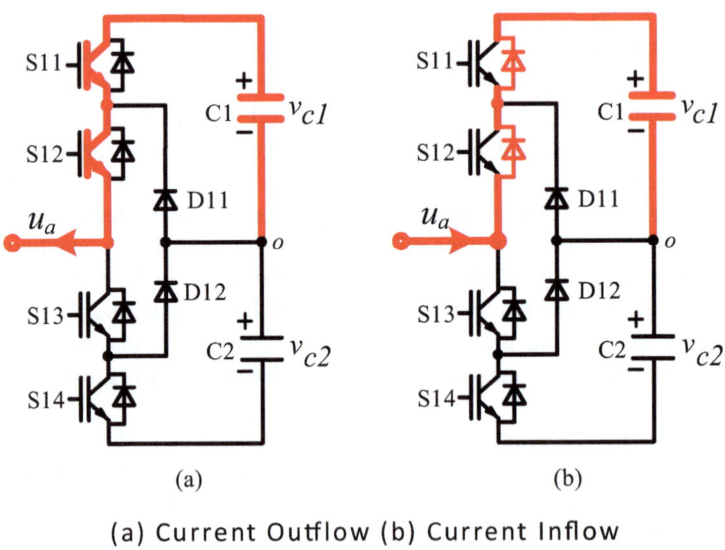

Fig. 3.6 Current flow path (on-off state: 1100, positive output level). (**a**) Current outflow; (**b**) current inflow

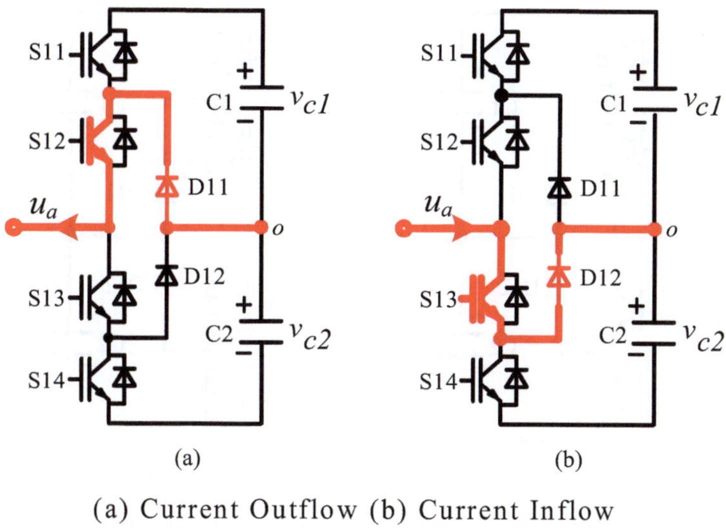

Fig. 3.7 Current flow path (zero output level at port). (**a**) Current outflow; (**b**) current inflow

S12 that are not turned on. The voltage of the port relative to the capacitor midpoint is expressed as $u_{ao} = v_{c1}$, and the output level remains positive.

2. On-off state: 0110, zero output level

   As shown in Fig. 3.7a, with switching devices S11, S13, and S14 turned off, current flows from the DC side to the grid side, through clamping diode D11 and switching device S12. The port output level is zero.

   As shown in Fig. 3.7b, with switching devices S11, S12, and S14 turned off, current flows from the grid side to the DC side through clamping diode D12 and switching device S13. The port output level remains zero.

3. On-off state: 0011, negative output level

   As shown in Fig. 3.8a, with switching devices S11 and S12 turned off, current flows from the DC side to the grid side through the body diodes of switching devices S13 and S14. The port output level is negative.

   As shown in Fig. 3.8b, with switching devices S11 and S12 turned off, current flows from the grid side to the DC side through switching devices S13 and S14. The port output level remains negative. Since the diode-clamped three-level circuit reduces the port voltage to half of the input voltage, Eq. (3.7) can be rewritten as follows, based on Eqs. (3.2) and (3.6):

$$\begin{cases} \dfrac{\mathrm{d}i_a}{\mathrm{d}t} = -\dfrac{R}{L}i_a + \dfrac{1}{L}\left(u_{sa} - \dfrac{2S_a - S_b - S_c}{6}V_{\mathrm{dc}}\right) \\ \dfrac{\mathrm{d}i_b}{\mathrm{d}t} = -\dfrac{R}{L}i_b + \dfrac{1}{L}\left(u_{sb} - \dfrac{2S_b - S_a - S_c}{6}V_{\mathrm{dc}}\right) \\ \dfrac{\mathrm{d}i_c}{\mathrm{d}t} = -\dfrac{R}{L}i_c + \dfrac{1}{L}\left(u_{sa} - \dfrac{2S_c - S_a - S_b}{6}V_{\mathrm{dc}}\right) \end{cases} \tag{3.11}$$

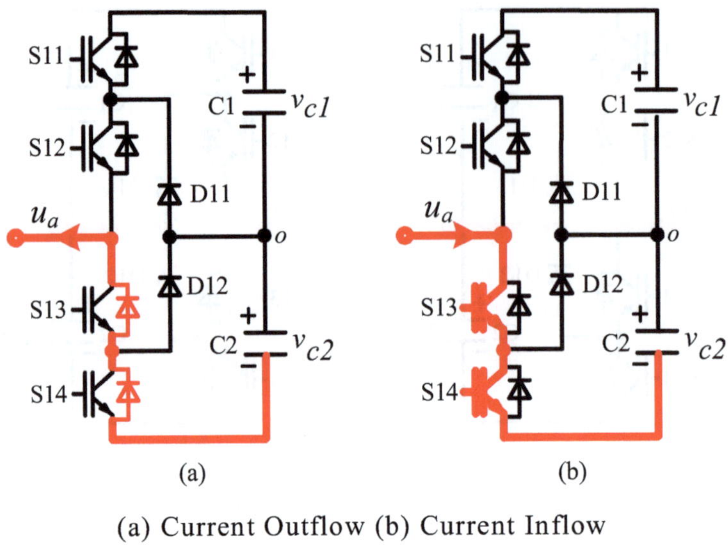

Fig. 3.8 Current flow path (negative output level at port). (**a**) Current outflow; (**b**) current inflow

Based on Eq. (3.11), the phase voltage of the three-phase converter output can be expressed as follows:

$$\begin{cases} u_{aN} = \dfrac{2S_a - S_b - S_c}{6} V_{dc} \\ u_{bN} = \dfrac{2S_b - S_a - S_c}{6} V_{dc} \\ u_{cN} = \dfrac{2S_c - S_a - S_b}{6} V_{dc} \end{cases} \tag{3.12}$$

Substituting Eq. (3.12) into Eq. (3.11) gives

$$\begin{cases} \dfrac{di_a}{dt} = -\dfrac{R}{L} i_a + \dfrac{1}{L}(u_{sa} - u_{aN}) \\ \dfrac{di_b}{dt} = -\dfrac{R}{L} i_b + \dfrac{1}{L}(u_{sb} - u_{bN}) \\ \dfrac{di_c}{dt} = -\dfrac{R}{L} i_c + \dfrac{1}{L}(u_{sc} - u_{cN}) \end{cases} \tag{3.13}$$

Equation 3.13 represents the mathematical model of the diode-clamped three-level circuit in the *abc* static coordinate system.

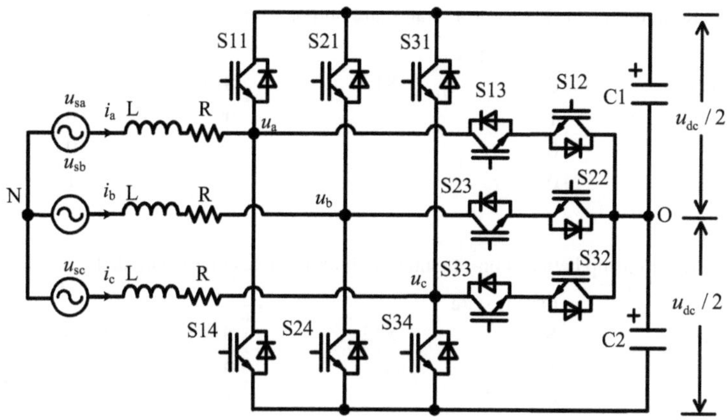

Fig. 3.9 Circuit topology of diode-clamped three-level converter

3.1.3 T-Type Three-Level Converter

The T-type topology for three-phase three-level converter circuits is illustrated in Fig. 3.9. The DC side consists of upper and lower capacitors connected in series, with the midpoint linked to the bridge arm midpoint through two power switching tubes in a series-opposing connection to clamp the midpoint voltage. Each phase of the bridge arm comprises four power switching tubes. The operating states of this circuit are analyzed below, by taking one phase of the converter as an example.

In Fig. 3.9, each bridge arm comprises four switching tubes and can output at three levels. It is assumed that the upper and lower capacitor voltages on the DC side are balanced, meaning that both the upper and lower capacitors have a voltage of $u_{dc}/2$. When S11 is turned on and S14 is turned off, the converter's output level is positive. In this case, S13 must be turned off to prevent a short circuit across the upper capacitor on the DC side. Conversely, when S14 is turned on and S11 is turned off, the converter's output level is negative. In this case, S12 must also be turned off to prevent a short circuit across the lower capacitor on the DC side. When S11 and S14 are both turned off while S12 and S13 are turned on, the voltage output at the DC-side capacitor midpoint from the converter is zero. Thus, the T-type three-level circuit can produce three output levels: $u_{dc}/2$, 0, and $-u_{dc}/2$. The on-off states and their corresponding output states are listed in Table 3.3. There is an ideal drive mode, referred to as mode II. In this drive mode, when the output level is positive, only S11 is turned on while S12, S13, and S14 are turned off. When the output level is negative, S14 is turned on, while S11, S12, and S13 are turned off. When the output level is zero, S12 and S13 are turned on while S11 and S14 are turned off. The relationships between the on-off states and output states in this mode are presented in Table 3.4.

The following analysis examines the operating states of the circuit, respectively, in three output configurations, based on T-type three-level single-phase equivalent circuits, and compares drive modes I and II.

Table 3.3 Relationship between on-off states and output states in drive mode I

Output state	On-off state			
	S11	S12	S13	S14
P	1	1	0	0
0	0	1	1	0
N	0	0	1	1

Table 3.4 Relationship between on-off states and output states in drive mode II

Output state	On-off state			
	S11	S12	S13	S14
P	1	0	0	0
0	0	1	1	0
N	0	0	0	1

1. When the output state is "P," switching tube S1 is turned on or both S1 and S2 are turned on, while S3 and S4 are turned off. Current flows from the positive terminal of capacitor C1 and out of the converter via S1 when flowing from the DC side to the AC side, as shown in Fig. 3.10a. Current flows into the positive terminal of capacitor C1 from the port via the body diode of S1 when flowing from the AC side to the DC side, as shown in Fig. 3.10b. The port voltage in this state can be expressed as $u_o = u_{dc}/2$. In the P-level operation process, there is no difference between drive modes I and II.
2. When the output state is "0," switching tubes S2 and S3 are turned on, while other switching tubes are turned off. Current flows from the capacitor midpoint O and out of the converter via S2 and S3 when flowing from the DC side to the AC side, as shown in Fig. 3.11a. Current flows into the capacitor midpoint O from the port via S2 and S3 when flowing from the AC side to the DC side, as shown in Fig. 3.11b. The port voltage in this state can be expressed as $u_o = 0$. During the system switching from positive-level output to zero-level output, as illustrated in Fig. 3.11a, adopting drive mode II results in the turning-on of both S2 and S3, which is the same as drive mode I. However, in practice, it is not feasible to turn on S1 and S3 simultaneously without a dead zone, nor can it be ensured that S2 and S3 will be turned on at the same time. If S3 is turned on before S2, achieving zero-level output becomes impossible, i.e., $U_o \neq 0$. Similarly, when the system switches from negative-level output to zero-level output, as shown in Fig. 3.11b, the turning-on of S2 first in drive mode II also prevents the attainment of zero-level output.
3. When the output state is "N," switching tube S4 or both S3 and S4 are turned on, while the other switching tubes are turned off. Current flows from the negative terminal of capacitor C2 and out of the converter via the body diode of S4 when flowing from the DC side to the AC side, as shown in Fig. 3.12a. Current flows into the negative terminal of capacitor C2 from the port via S4 when flowing from the AC side to the DC side, as shown in Fig. 3.12b. The port voltage in this

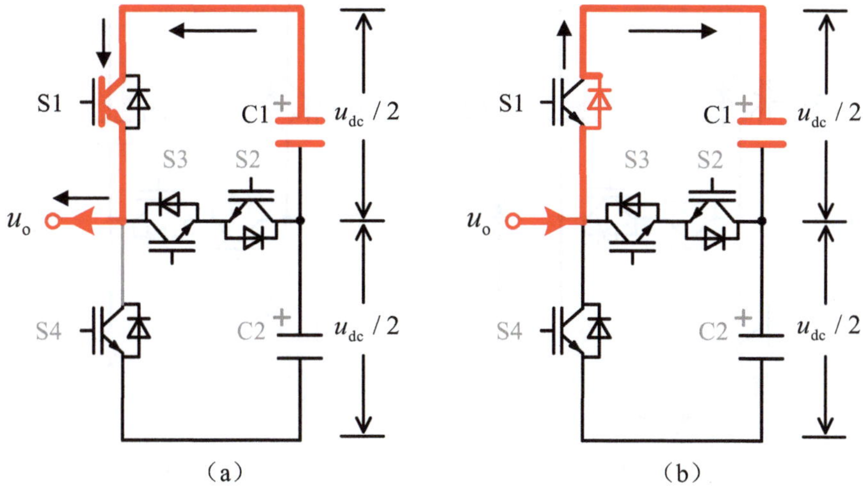

(a) Current Outflow (b) Current Inflow

Fig. 3.10 Current flow path in "P" output state. (**a**) Current outflow; (**b**) current inflow

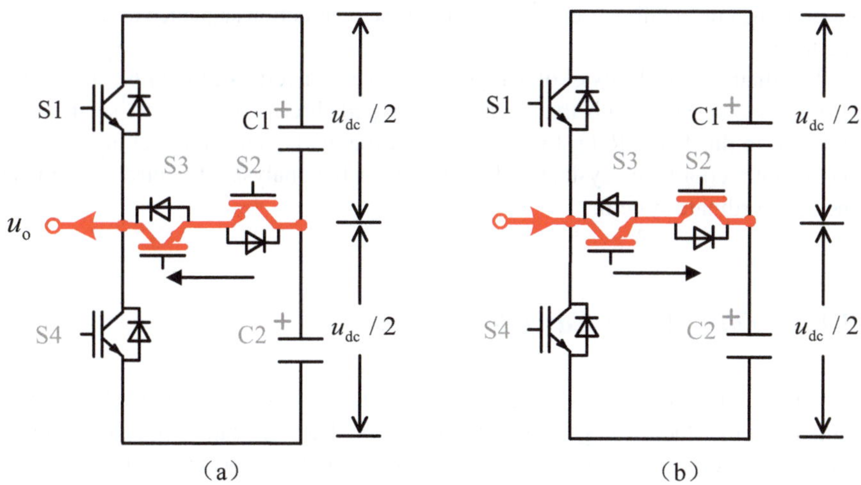

(a) Current Outflow (b) Current Inflow

Fig. 3.11 Current flow path in "0" output state. (**a**) Current outflow; (**b**) current inflow

state can be expressed as $u_o = -u_{dc}/2$. In the N-level operation process, there is no difference between drive modes I and II.

To sum up, while drive modes I and II do not differ in output levels, they are distinct in the commutation process. In mode II, the switching sequence of the internal switches is influenced by the current level state and current direction, whereas in

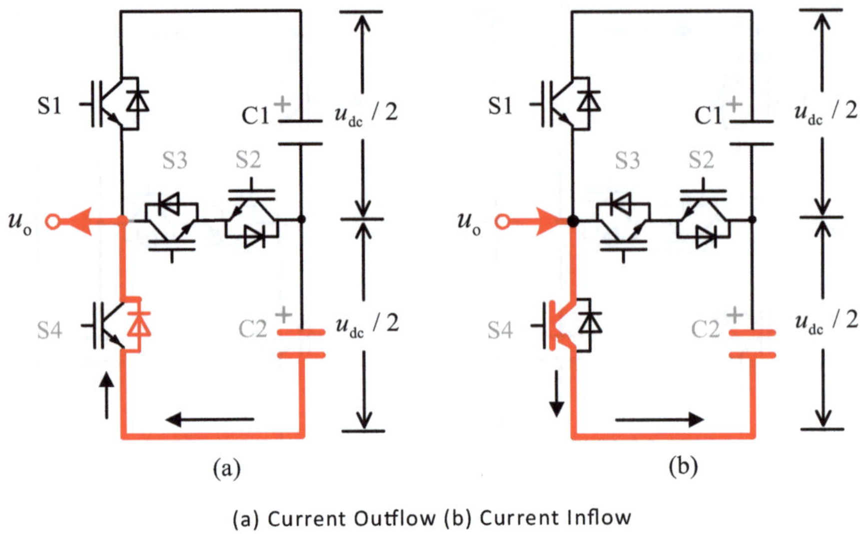

Fig. 3.12 Current flow path in "N" output state. (**a**) Current outflow; (**b**) current inflow

mode I, the current direction does not affect the conduction sequence of the switching tubes.

The mathematical model of T-type three-level converters in the *abc* coordinate system is consistent with that of diode-clamped three-level circuits, so it is not repeated in this book. Refer to Sect. 3.1.2 for the mathematical model in the three-phase static coordinate system and Sect. 3.2 for the analysis of control systems in the *dq* coordinate system.

3.1.4 Three-Phase Cascade Converter

Cascade multilevel circuits undoubtedly provide a suitable topology for further enhancing interface voltage levels in converters. These circuits comprise multiple modular units connected in series, typically utilizing two-level or three-level H-bridge circuits as their unit circuits. The cascaded arrangement of these units allows for voltage multiplication, making the circuits well suited for high-voltage and high-power applications. However, the addition of more modules in cascade converters, along with the relative independence of their DC voltage, complicates various aspects of the multi-module configurations, including driving, voltage balancing, and power control.

The topology of three-phase converters with cascaded two-level units is illustrated in Fig. 3.13, where each cascade unit is composed of two-level H-bridges. A two-level H-bridge circuit consists of four power switching devices, one DC-link capacitor, DC loads, and one AC inductor. Inductor *L* connected in series on the AC

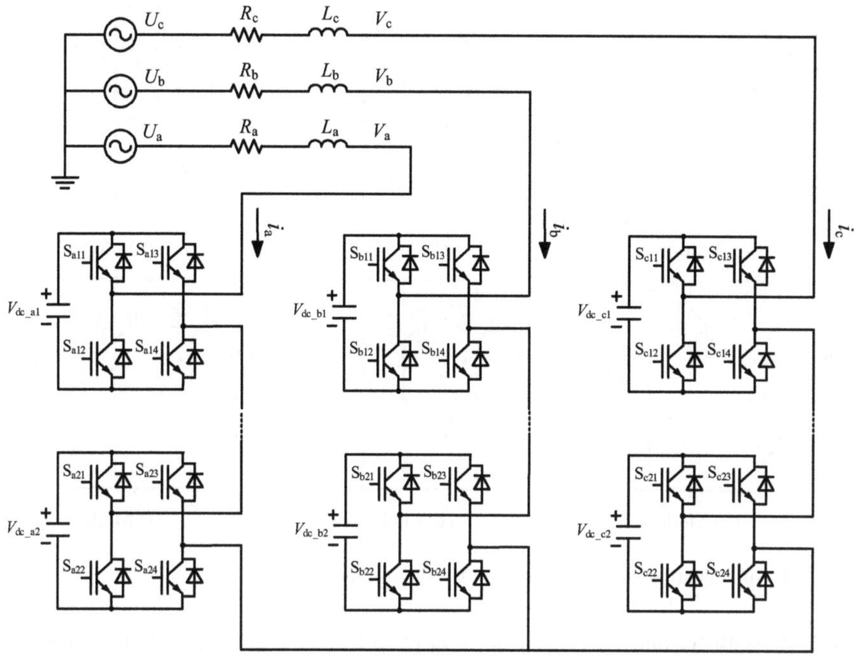

Fig. 3.13 Three-phase cascaded converter composed of two-level H-bridge units

side functions as a current filter, while the DC-side capacitor provides voltage support and serves as temporary energy storage.

3.1.4.1 Analysis of Two-Level H-Bridge Unit Circuit

The sinusoidal pulse width modulation (SPWM) of two-level H-bridge converters is categorized into unipolar modulation and bipolar modulation. In the case of unipolar modulation, the AC-side port voltage alternates between $+V_{dc}$ or $-V_{dc}$ and 0, resulting in lower voltage harmonics. In contrast, the port voltage in bipolar modulation alternates only between $+V_{dc}$ and $-V_{dc}$, leading to higher voltage harmonics. The study in this section focuses on unipolar modulation.

In the unipolar modulation mode, two-level H-bridge converters are classified into six operating modes based on the operating state of power switching devices and the direction of AC flow, as shown in Fig. 3.14.

Mode 1: With S1 and S4 turned on, i_s is greater than 0 and flows via D1 and D4 to charge capacitor C for energy delivery to the load, with increasing output voltage (V_{dc}), as shown in Fig. 3.14a.

Mode 2: With S1 and S4 turned on, i_s is less than 0 and flows via S2 and S4; capacitor C delivers energy to the load and grid side, with decreasing output voltage (V_{dc}), as shown in Fig. 3.14b.

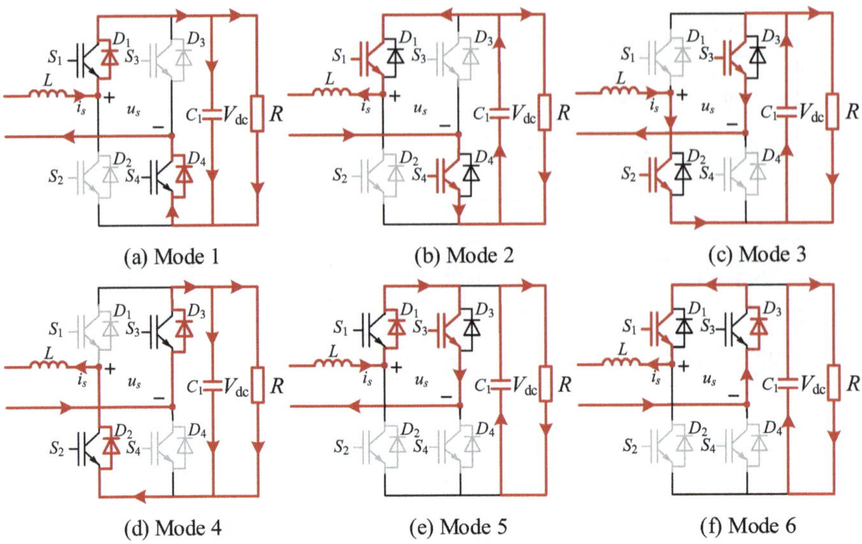

Fig. 3.14 Six operating modes of two-level H-bridge converter

Mode 3: With S2 and S3 turned on, i_s is greater than 0 and flows via S2 and S3; capacitor C delivers energy to the load and grid side, with decreasing output voltage (V_{dc}), as shown in Fig. 3.14c.

Mode 4: With S2 and S3 turned on, i_s is less than 0 and flows via D2 and D3 to charge the capacitor and deliver energy to the load, with increasing output voltage (V_{dc}), as shown in Fig. 3.14d.

Mode 5: With S1 and S3 turned on, i_s is greater than 0 and flows via D1 and S3 to charge inductor L; capacitor C delivers energy to the load, with decreasing output voltage (V_{dc}), as shown in Fig. 3.14e.

Mode 6: With S1 and S3 turned on, i_s is less than 0 and flows via S1 and D3 to charge inductor L; capacitor C delivers energy to the load, with decreasing output voltage (V_{dc}), as shown in Fig. 3.14f.

According to the analysis of the six operating modes presented in Fig. 3.14, the port voltage (u_s) can be classified into three types: $+V_{dc}$, 0, and $-V_{dc}$. The same type of us across different modes corresponds to either the AC current charging or discharging of the DC-link capacitor. Therefore, stability control of the DC-link capacitor voltage can be achieved in different operating modes by maintaining the same type of port voltage output. For instance, in both mode 1 and mode 2, the port voltage is $+V_{dc}$. In mode 1, the DC capacitor is charged, resulting in an increase in DC voltage, whereas in mode 2, the DC capacitor is discharged, leading to a decrease in DC voltage, regardless of the DC load loss.

The relationships between the six switching modes in Fig. 3.14 and the output port voltage (u_s) are summarized in Table 3.5. The table presents three operating states: +1, 0, and −1, which correspond, respectively, to the port voltages of $+V_{dc}$, 0, and $-V_{dc}$.

Table 3.5 Operating state of H-bridge converter

S(State)	S1	S2	S3	S4	u_s
+1	1	0	0	1	$+V_{dc}$
0	1	0	1	0	0
−1	0	1	1	0	$-V_{dc}$
0	0	1	0	1	0

Assuming that all devices in the H-bridge converter are ideal, according to Table 3.5, the switching function $S(t)$ for the port voltage of the H-bridge converter can be expressed as follows:

$$S(t) = S_1(t) - S_3(t) = \begin{cases} +1, S_1, \text{and } S_4 \text{ turned on} \\ 0, S_1, \text{and } S_3 \text{ turned on or } S_2 \text{ and } S_4 \text{ turned on} \\ -1, S_2, \text{and } S_3 \text{ turned on} \end{cases} \tag{3.14}$$

where $S_1(t)$ and $S_3(t)$ indicate the operating states of switching tubes S1 and S3, respectively.

The cascade circuits discussed above are composed of two-level H-bridge circuits, and these unit circuits may also be multilevel ones. Multilevel unit circuits offer higher voltage and power levels and can output at more levels. The analysis of these unit modules is not repeated in this book, as it is similar to that of three-phase diode-clamped multilevel circuits mentioned earlier.

3.2 Active and Reactive Decoupling Control of Three-Phase Interconnection Converter

The application of three-phase converters for interconnection necessitates the stabilization of DC voltage, as this stability reflects the active power stability control of the converters. Consequently, addressing the issue of active/reactive power control is essential for these three-phase converters. The active and reactive decoupling control in the *dq* coordinate system is commonly employed as a solution to ensure stable and reliable three-phase power control. Thus, this section focuses on discussing this control method.

3.2.1 *Model of Three-Phase Converter in Two-Phase Static Coordinate System*

The mathematical model of three-phase rectifier circuits in the *abc* coordinate system can be established according to Kirchhoff's voltage law (KVL) and Kirchhoff's current law (KCL), which is rewritten as Eq. (3.15):

$$\begin{cases} U_a - Ri_a - L\dfrac{\mathrm{d}i_a}{\mathrm{d}t} - V_a - \left(-V_b - L\dfrac{\mathrm{d}i_b}{\mathrm{d}t} - Ri_b + U_b\right) = 0 \\ U_b - Ri_b - L\dfrac{\mathrm{d}i_b}{\mathrm{d}t} - V_b - \left(-V_c - L\dfrac{\mathrm{d}i_c}{\mathrm{d}t} - Ri_c + U_c\right) = 0 \\ U_a - Ri_a - L\dfrac{\mathrm{d}i_a}{\mathrm{d}t} - V_a - \left(-V_c - L\dfrac{\mathrm{d}i_c}{\mathrm{d}t} - Ri_c + U_c\right) = 0 \end{cases} \tag{3.15}$$

To simplify the analysis, it is assumed that the voltage on the grid side is symmetrically balanced among the three phases, meaning that the vector sum of the three-phase voltages is zero and that the current flowing through the midpoint is also zero:

$$U_a + U_b + U_c = 0 \tag{3.16}$$

According to Eqs. (3.15) and (3.16), the mathematical equation for the rectifier port voltage in the three-phase static coordinate system can be expressed as Eq. (3.17):

$$\begin{cases} V_a = -L\dfrac{\mathrm{d}i_a}{\mathrm{d}t} - Ri_a + U_a \\ V_b = -L\dfrac{\mathrm{d}i_b}{\mathrm{d}t} - Ri_b + U_b \\ V_c = -L\dfrac{\mathrm{d}i_c}{\mathrm{d}t} - Ri_c + U_c \end{cases} \tag{3.17}$$

Although the physical meaning of the mathematical model in the three-phase static *abc* coordinate system is clear, designing and implementing controllers capable of achieving fast alignment free from steady-state errors and more are challenging due to all variables being represented in AC form. Therefore, it is necessary to convert AC components into DC components. The three-phase AC components can be effectively expressed as DC components in the two-phase synchronous rotating *dq* frame. The transformation matrix (P) that converts the three-phase static *abc* frame to the two-phase synchronous rotating *dq* frame can be represented as follows:

$$P = \frac{2}{3}\begin{bmatrix} \sin\omega t & \sin(\omega t - 2\pi/3) & \sin(\omega t + 2\pi/3) \\ \cos\omega t & \cos(\omega t - 2\pi/3) & \cos(\omega t + 2\pi/3) \end{bmatrix} \tag{3.18}$$

Based on Eqs. (3.17) and (3.18), the mathematical expression of the rectifier port voltage in the *dq* coordinate system is as follows:

$$\begin{cases} V_d = -L\dfrac{\mathrm{d}i_d}{\mathrm{d}t} - Ri_d + \omega Li_q + u_d \\ V_q = -L\dfrac{\mathrm{d}i_q}{\mathrm{d}t} - Ri_q + \omega Li_d + u_q \end{cases} \tag{3.19}$$

The *dq* component of the port voltage after simplification can be expressed as follows:

$$\begin{bmatrix} V_d & V_q \end{bmatrix}^{\mathrm{T}} = P\begin{bmatrix} V_a & V_b & V_c \end{bmatrix}^{\mathrm{T}} \tag{3.20}$$

The *dq* component of the grid-side voltage can be expressed as follows:

$$\begin{bmatrix} u_d & u_q \end{bmatrix}^{\mathrm{T}} = P\begin{bmatrix} u_a & u_b & u_c \end{bmatrix}^{\mathrm{T}} = \begin{bmatrix} \sqrt{\frac{3}{2}}u_{\max} & 0 \end{bmatrix}^{\mathrm{T}} \tag{3.21}$$

The *dq* component of the output current can be expressed as follows:

$$\begin{bmatrix} i_d & i_q \end{bmatrix}^{\mathrm{T}} = P\begin{bmatrix} i_a & i_b & i_c \end{bmatrix}^{\mathrm{T}} \tag{3.22}$$

After *dq* transformation, all expressions representing three-phase AC voltage and current in the *abc* coordinate system are converted into their corresponding *dq* components in the *dq* coordinate system. In stable systems, these *dq* components remain stable and unchanged, facilitating the design and implementation of simple and reliable feedback control systems that are free from steady-state errors.

3.2.2 *Active and Reactive Decoupling Control in* DQ *Coordinate System*

As indicated in Eq. (3.19), there is a coupling relationship between the *dq* component of port voltage and the *dq* component of current. Following the feedback control of the *dq* components of current, the *dq* component of port voltage can be derived from Eq. (3.20). This allows for the decoupling control of both the *dq* components and the active/reactive power components of the port voltage.

According to the differential definition of the *d*-axis component of current, the following expression is established:

$$\frac{\mathrm{d}i_d}{\mathrm{d}t} = \frac{\Delta i_d}{\Delta t} = \frac{i_{d\mathrm{ref}} - i_d}{\Delta t} \tag{3.23}$$

Letting the mathematical expression of PI input and output as Eq. (3.23):

$$\mathrm{PI}(s) = \left(K_{\mathrm{P}} + \frac{K_{\mathrm{I}}}{s} \right)\Delta\varepsilon \tag{3.24}$$

where K_{P} is the proportional gain coefficient, K_{I} is the integral gain coefficient, and $\Delta\varepsilon$ is the deviation between the controlled variable and the reference variable.

Hence, the proportional calculation of the current d-axis deviation can be expressed as follows:

$$P\left(i_{dref}-i_{d}\right)=K_{P}\Delta t\frac{di_{d}}{dt} \tag{3.25}$$

The integral calculation of the current d-axis deviation can be expressed as follows:

$$\mathrm{Int}\left(i_{dref}-i_{d}\right)=K_{I}\int\Delta t\frac{di_{d}}{dt}dt=K_{I}\Delta ti_{d} \tag{3.26}$$

As indicated in Eqs. (3.25) and (3.26), the control requirements for the d-axis active power channel can be satisfied through appropriate parameter design of K_P and K_I. Similarly, the PI controller parameters for the q-axis reactive power channel can also be designed. Therefore, with the incorporation of the PI controller, the voltage dq coordinate equation represented by the decoupling of the current dq components can be expressed as follows:

$$\begin{cases} V_{d}=-\left(K_{P}+\dfrac{K_{I}}{s}\right)\left(i_{dref}-i_{d}\right)+\omega Li_{q}+u_{d} \\ V_{q}=-\left(K_{P}+\dfrac{K_{I}}{s}\right)\left(i_{qref}-i_{q}\right)+\omega Li_{d}+u_{q} \end{cases} \tag{3.27}$$

Based on Eq. (3.27), the three-phase voltage source inverter (VSI) can be controlled to operate as a three-phase PWM rectifier. The active and reactive decoupling control of the three-phase PWM rectifier is shown in Fig. 3.15.

As illustrated, the active component of the rectifier port voltage is generated by superimposing the cross-term ωLi_{df} component onto the output of the current d-axis active component PI regulator. Similarly, the reactive component of the port voltage is produced by superimposing the cross-term ωLi_{df} component onto the output of the current q-axis reactive component PI regulator. The active reference i_{dref} is determined by the difference between the three-phase rectifier DC voltage and the voltage reference, as outputted by the PI regulator of the voltage outer loop. The input or output of the active component influences the rise or fall of the DC capacitor voltage. Therefore, the control objective of the active power channel reflects the stability control of the DC voltage. When the active component remains stable under control, the stability of the DC voltage can be maintained regardless of changes in load active demand and circuit losses.

Additionally, the PWM rectifier typically operates at a unity power factor, which means that the reference i_{qref} for the reactive q-axis is set to 0. If the i_{qref} reference responds to the reactive power of the AC load, the interconnection converter can be controlled to function as a dynamic reactive compensator.

The d-axis component (V_d) and q-axis component (V_q) of the port output voltage under three-phase decoupling control in the dq coordinate system are calculated according to Eq. (3.27). The modulation waveform reference in the three-phase

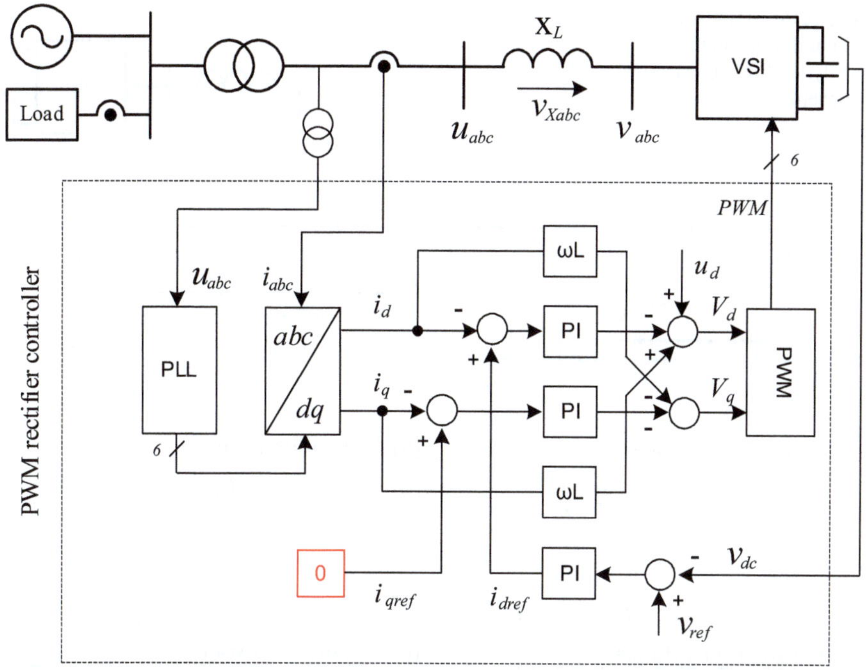

Fig. 3.15 Block diagram of active and reactive power decoupling control of three-phase PWM rectifier

static frame can be obtained through the inverse *dq* transformation. Subsequently, the three-phase switching signals can be generated using the PWM module, allowing the voltage source converter to operate as a PWM rectifier with multiple control objectives, including DC voltage stability and unity power factor operation.

3.2.3 Analysis of Current Four-Quadrant Characteristics for Interconnection Converter

To simplify the analysis of current four-quadrant operation characteristics during the operation of voltage source inverters for interconnection, the internal resistance of the inductor (R_s) in the three-phase PWM rectifier illustrated in Fig. 3.3 is disregarded, and only phase *j* of the three phases is analyzed, as depicted in Fig. 3.16.

The diagram of four eigenvectors for voltage and current in converters is presented in Fig. 3.17, with the variables represented in vector form: the voltage of the AC power supply ($\dot{U}_{Sj}$), the inductor voltage drop ($\dot{U}_L$), the output port voltage of the VSI converter ($\dot{U}_j$), as well as the inductor current and the converter output current ($\dot{I}_{Sj}$), which are equivalent since the inductor is connected in series between the bridge arm of the converter and the power supply. In this diagram, the positive

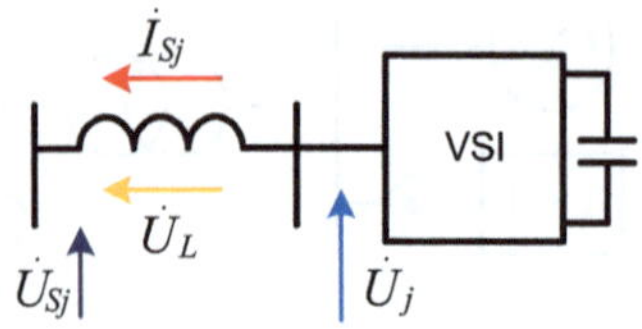

Fig. 3.16 Simplified circuit for quadrant analysis of interconnection converter

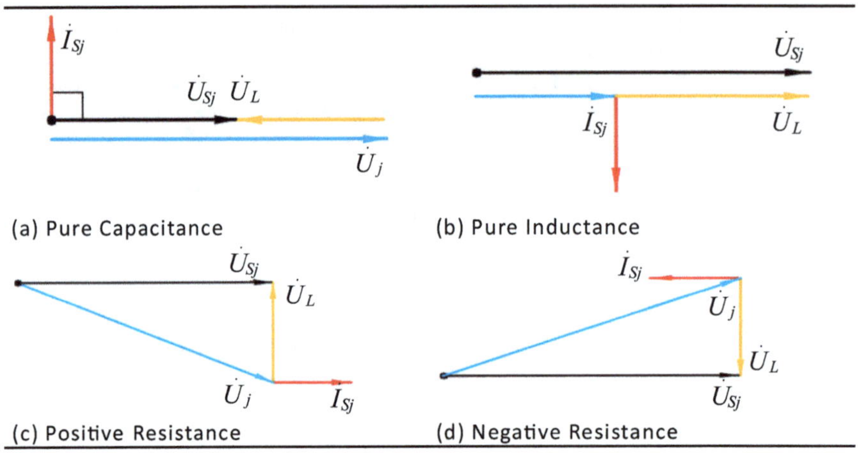

Fig. 3.17 Steady-state vector diagram of single-phase MMC. (**a**) Pure capacitance. (**b**) Pure inductance. (**c**) Positive resistance. (**d**) Negative resistance

direction of the vector is defined as counterclockwise, and only the sinusoidal fundamental component of the AC output is considered, while the harmonic components are neglected. The port voltage can be distributed at four key positions—upper, lower, left, and right—corresponding to the converter's operation under four typical characteristics: pure capacitance, pure inductance, positive resistance, and negative resistance.

Pure capacitance mode: When the port voltage ($\dot{U}_j$) is in phase with the power supply voltage ($\dot{U}_{Sj}$) but in a greater amplitude, the inductor voltage ($\dot{U}_L$) is equal to the vector difference between the two. In this case, $\dot{U}_L$ and $\dot{U}_{Sj}$ are opposite in direction. According to the vector relationship between inductor voltage and current, expressed as $\dot{U}_L = j\omega \dot{I}_{sj}$, the inductor current ($\dot{I}_{Sj}$) lags behind $\dot{U}_L$ by 90° in phase, which means that the converter output current ($\dot{I}_{Sj}$) leads the power supply voltage ($\dot{U}_{Sj}$) by 90° in phase. The converter operates in the pure capacitance mode, where the amplitude of the converter port voltage ($\dot{U}_j$) exceeds that of the power supply voltage ($\dot{U}_{Sj}$), thereby necessitating a higher DC voltage or modulation degree for the converter.

With the phase unchanged, changing the port voltage ($\dot{U}_j$) in amplitude and exceeding $\dot{U}_{Sj}$ result in a corresponding change in the amplitude of the inductor voltage ($\dot{u}_L$) while keeping its direction constant. Altering the operating point of the converter port voltage leads to a variation in the amplitude of the converter current

($\dot{I}_{Sj}$), causing the converter output to produce capacitive reactive current with different amplitudes.

Pure inductance mode: When $\dot{U}_j$ is in phase with $\dot{U}_{Sj}$ but in a lower amplitude, voltage ($\dot{U}_L$) is also equal to the vector difference between them, indicating $\dot{U}_L$ and $\dot{U}_{Sj}$ in the same direction. Since the inductor current ($\dot{I}_{Sj}$) lags behind $\dot{U}_L$ by 90°, the output current ($\dot{I}_{Sj}$) of the converter lags behind $\dot{U}_{Sj}$ by 90°. In this case, the converter operates in the pure inductance mode.

Pure resistance mode: When $\dot{U}_j$ is located under $\dot{U}_{Sj}$, a similar analysis indicates that $\dot{U}_L$ is perpendicular to $\dot{U}_{Sj}$ and 90° ahead of it. Since the inductor current ($\dot{I}_{Sj}$) lags behind $\dot{U}_L$ by 90°, the output current ($\dot{I}_{Sj}$) of the converter is in the same direction as $\dot{U}_{Sj}$. In this case, the converter operates in the pure resistance mode, and the power factor is 1. Generally, PWM rectifiers dominantly operate in this mode, i.e., the unity power factor mode. The rectifier receives active power from the AC power supply and delivers the required active power to DC loads, thus maintaining the stability of DC voltage.

Negative resistance mode: When $\dot{U}_j$ is located over $\dot{U}_{Sj}$, $\dot{U}_L$ is perpendicular to $\dot{U}_{Sj}$ and lags behind it by 90°. Since the inductor current ($\dot{I}_{Sj}$) lags behind $\dot{U}_L$ by 90°, the output current ($\dot{I}_{Sj}$) of the converter lags behind $\dot{U}_{Sj}$ by 180°. In this case, the converter operates in the negative resistance mode. Generally, interconnection converters that output active power to the power grid, such as photovoltaic interconnection converters, mostly operate in this mode, and the converter transmits active power from the DC side to the AC power supply.

Assuming that the AC voltage ($\dot{U}_{Sj}$) is constant, as known from the above analysis, by controlling the amplitude and phase of the port voltage ($\dot{U}_j$) of the interconnection converter (VSI), the amplitude and phase of the converter current ($\dot{I}_{Sj}$) can be controlled. Therefore, the converter can be controlled to perform different functions such as reactive power compensation and active power filtering or to operate as a PWM rectifier or interconnection inverter when being controlled to operate under the four modes, pure capacitance, pure inductance, positive resistance, and negative resistance, or under any combination of active and reactive modes.

When the amplitude of the output current ($\dot{I}_{Sj}$), i.e., the power supply current, is constant, the trajectory of the vector ($\dot{U}_j$) forms a circle with the amplitude of the AC inductor voltage drop ($|\dot{U}_L|$) as its radius. The positive and negative directions of the power supply current ($\dot{I}_{Sj}$) and voltage ($\dot{U}_{Sj}$) divide the circle into four quadrants, as shown in Fig. 3.18. The trajectory endpoints of the vector ($\dot{U}_j$) fall on the boundaries of these quadrants, where point A represents the operating point in the pure inductance mode, point B represents the operating point in the positive resistance mode, point C represents the operating point in the pure capacitance mode, and point D represents the operating point in the negative resistance mode. The operating characteristics of these points correspond to the four eigenvectors shown in Fig. 3.17.

Generally, the converter can operate at any position along a circular trajectory with the amplitude of $|\dot{U}_L|$ as its radius. In Section AB, the converter operates in a rectifier state and outputs inductive reactive power; in Section BC, the converter remains in a rectifier state but outputs capacitive reactive power; in Section CD, the

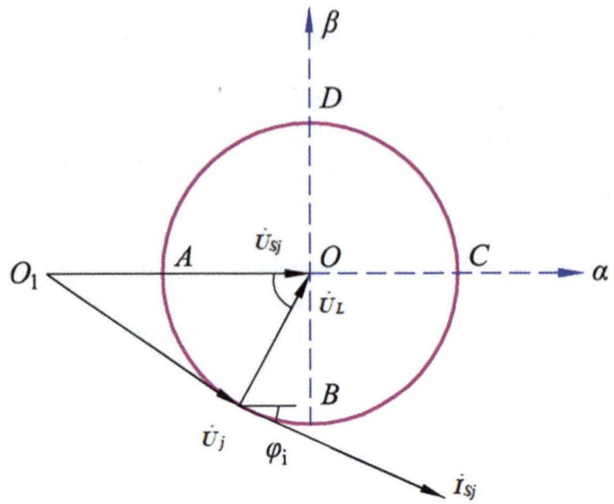

Fig. 3.18 Vector diagram of four-quadrant operation of converter

converter operates in an active inverter state and outputs inductive reactive power; in Section AD, the converter remains in an active inverter state but outputs capacitive reactive power. As shown, the size of the vector ($\dot{U}_j$) in the capacitive operating region is greater than that in the inductive region. Therefore, the converter requires a higher port voltage ($\dot{U}_j$) when outputting capacitive reactive power compared to inductive reactive power. This explains why the capability to output capacitive reactive power is lower than that to output inductive reactive power when the DC voltage and modulation degree remain constant.

3.3 Analysis of Three-Phase Phase-Locked Loop

The active and reactive *dq* decoupling control discussed in the previous section requires a *dq* transformation of the three-phase current. All components in this key transformation matrix, as expressed in Eq. (3.19), represent the real-time phases of the three-phase voltage. Active and reactive decoupling control can be understood as a control method based on the orientation of the power supply voltage. Consequently, the accuracy of the phase information of the power voltage directly influences the correctness of the output from the *dq* transformation and the active/reactive control. Identifying the real-time phase of the three-phase voltage provides a foundation for the active and reactive decoupling control. This section focuses on the real-time phase-locking method for three-phase voltage, which is based on a phase-locked loop (PLL) (Gong et al. 2009; Beig et al. 2007).

3.3.1 Basic Principle and Structure of Three-Phase Phase-Locked Loop

A straightforward method for acquiring the phase information of three-phase voltage is to detect the zero-crossing points of voltage. However, the voltage information between these zero-crossing points is not utilized, which diminishes the alignment effectiveness and results in significant detection deviations when harmonics and other voltage distortions occur.

The three-phase phase-locked loop (PLL) system provides an improved approach for alignment with the real-time phase of voltage, exhibiting strong resistance to interference from abnormal waveforms such as harmonics, unbalances, and DC offsets in the input signals. In the basic structure shown in Fig. 3.19, the system utilizes the three-phase supply voltages V_a, V_b, and V_c as input signals to align with the angular frequency of these signals in the *dq* plane. The result of the voltage phase alignment is output, delivering either the current phase value (θ) or its corresponding sine and cosine values in real time.

Assuming that V_a, V_b, and V_c are ideal three-phase power supply voltage signals, a normalization process yields:

$$V_{abc} = \begin{bmatrix} v_a \\ v_b \\ v_c \end{bmatrix} = \begin{bmatrix} \sin\theta' \\ \sin(\theta' - 2\pi/3) \\ \sin(\theta' + 2\pi/3) \end{bmatrix} \tag{3.28}$$

Three-phase voltages are converted into the two-phase static coordinate system through a three-phase-two-phase static coordinate transformation process:

$$V_{\alpha\beta} = T_{\alpha\beta} \cdot V_{abc} \tag{3.29}$$

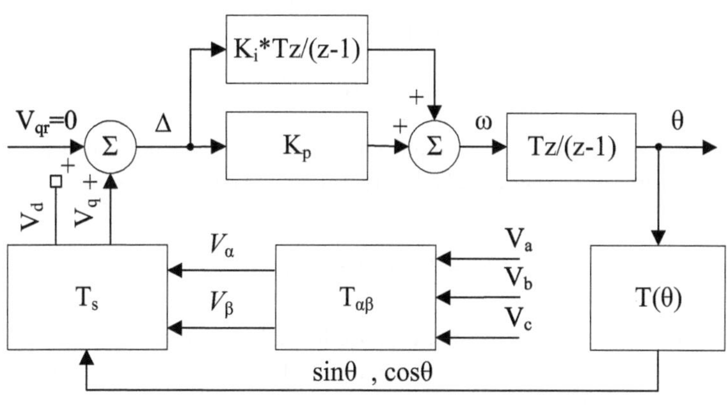

Fig. 3.19 Basic structure of three-phase phase-locked loop

where $V_{\alpha\beta} = [V_\alpha\ V_\beta]^T$, and $T_{\alpha\beta}$ represents the three-phase-two-phase static coordinate transformation matrix:

$$T_{\alpha\beta} = \frac{2}{3} \cdot \begin{bmatrix} 1 & -\frac{1}{2} & -\frac{1}{2} \\ 0 & \frac{\sqrt{3}}{2} & -\frac{\sqrt{3}}{2} \end{bmatrix} \tag{3.30}$$

V_α and V_β are projected into the synchronous rotating frame through a synchronous coordinate transformation process:

$$V_{dq} = T_s \cdot V_{\alpha\beta} \tag{3.31}$$

where $V_{dq} = \begin{bmatrix} V_d & V_q \end{bmatrix}^T$, and T_s represents the synchronous coordinate transformation matrix:

$$T_s = \begin{bmatrix} \cos\theta & -\sin\theta \\ \sin\theta & \cos\theta \end{bmatrix} \tag{3.32}$$

In fact, directly substituting the *dq* transformation results from Eq. (3.18) into Eq. (3.31) yields

$$V_{dq} = \begin{bmatrix} V_d & V_q \end{bmatrix}^T = P \cdot V_{abc} \tag{3.33}$$

where the q-axis component can be expressed as follows:

$$V_q = \frac{2}{3}\left[V_a \cos\theta + V_b \cos\left(\theta - \frac{2}{3}\pi\right) + V_c \cos\left(\theta + \frac{2}{3}\pi\right) \right] \tag{3.34}$$

With the phase-locked objective of maintaining a q-axis deviation of zero, the closed-loop calculation process yields

$$V_q = \sin(\theta' - \theta) \tag{3.35}$$

As the difference between the output phase angle (θ) and input phase angle (θ') approaches zero in the phase-locked state, Eq. (3.35) can be rewritten as follows:

$$V_q \overset{V_{qr}=0}{\rightleftarrows} V_{qr} = \lim_{\theta' \to \theta} \sin(\theta' - \theta) = \theta' - \theta = \Delta \tag{3.36}$$

As indicated above, the three-phase phase-locked loop uses feedback control to drive the q-axis component (V_q) to zero, and V_q follows the reference $V_{qr} = 0$, thus achieving phase-frequency locking.

The calculation of V_q allows for the calculation of alignment deviation, represented by $\Delta = \theta' - \theta$, thus functioning as a phase detector in common PLL structures. Moreover, the output represents the difference between the input voltage phase angle (θ', phase reference) and output phase angle (θ, phase alignment output), which enables deviation calculation within the feedback system. Therefore, no negative value needs to be taken after comparing V_q and V_{qr} in Fig. 3.19. The phase deviation signal (Δ) is locked to the angular frequency (ω) of the input signal through a PI regulator, and the subsequent integration of ω produces the phase-locked output phase angle (θ).

3.3.2 *Influence of Voltage Distortion on Phase-Locked Loop*

In interconnected applications, the supply voltage at the connection point is often nonideal due to the impacts of nonlinear loads and compensation devices. Interferences such as asymmetry, harmonics, and DC offsets in the voltage can lead to harmonic disturbances in the phase deviation detected by the three-phase PLL system. Due to such input voltage distortions, the output of the three-phase phase-locked loop typically affects the compensation performance of the inverter system. This section discusses the output deviation of the three-phase phase-locked loop system under these distorted input conditions.

1. Influence of harmonics

   Assuming that the system voltage mainly contains harmonics at characteristic orders such as 5, 7, 11, and 13, the input to the three-phase phase-locked loop can be expressed as

$$\begin{cases} v_a = v_1 \sin\theta' + v_5 \sin 5\theta' + v_7 \sin 7\theta' + v_{11} \sin 11\theta' \cdots \\ v_b = v_1 \sin\left(\theta' - \dfrac{2\pi}{3}\right) + v_5 \sin 5\left(\theta' - \dfrac{2\pi}{3}\right) + v_7 \sin 7\left(\theta' - \dfrac{2\pi}{3}\right) \\ \quad + v_{11} \sin 11\left(\theta' - \dfrac{2\pi}{3}\right) \cdots \\ v_c = v_1 \sin\left(\theta' + \dfrac{2\pi}{3}\right) + v_5 \sin 5\left(\theta' + \dfrac{2\pi}{3}\right) + v_7 \sin 7\left(\theta' + \dfrac{2\pi}{3}\right) \\ \quad + v_{11} \sin 11\left(\theta' + \dfrac{2\pi}{3}\right) \cdots \end{cases} \tag{3.37}$$

   where v_1, v_5, v_7, v_{11}... represent the amplitudes of the fundamental harmonics and harmonics at characteristic orders such as 5, 7, 11, and 13, respectively. The implementation of the synchronous rotating coordinate transformation based on Eq. (3.37) yields

$$\begin{aligned} V_d &= -v_1 \sin(\theta' - \theta) + v_5 \sin(\theta + 5\theta') + v_7 \sin(\theta - 7\theta' + v_{11} \sin(\theta + 11\theta') \cdots \\ &\approx -v_1 \Delta + (v_5 - v_7) \sin 6\theta' + (v_{11} - v_1 3) \sin 12\theta' \cdots \end{aligned} \tag{3.38}$$

Comparing with Eq. (3.37) yields the phase-locked loop deviation caused by harmonic voltage, which can be expressed as follows:

$$\Delta_h = \frac{v_5 - v_7}{v_1}\sin 6\theta' + \frac{v_{11} - v_{13}}{v_1}\sin 12\theta' + \cdots \tag{3.39}$$

As indicated above, the phase-locked loop deviation caused by harmonic voltage contains components at orders that are six times the fundamental frequency, such as the 6th and 12th orders. Its amplitude is defined by differences in the amplitudes of harmonics at these characteristic orders that are approximately six times the fundamental frequency.

2. Influence of voltage unbalance

Assume that the amplitude unbalance of voltage phases B and C relative to phase A is expressed as follows:

$$\begin{cases} v_a = v_1 \sin\theta' \\ v_b = v_1(1+m)\sin\left(\theta' - \frac{2\pi}{3}\right) \\ v_c = v_1(1+n)\sin\left(\theta' + \frac{2\pi}{3}\right) \end{cases} \tag{3.40}$$

where v_1 denotes the fundamental amplitude of three-phase symmetrical components, and m, n is the phase distortion amplitude of phases A, B and is constant. After synchronous rotating coordinate transformation, the d-axis component is calculated as follows:

$$\begin{aligned} V_d = {} & v_1 \sin(\theta' - \theta) - v_1\Big[\frac{m-n}{2\sqrt{3}}(\cos\theta'\cos\theta - \sin\theta'\sin\theta) \\ & + \frac{m+n}{6}(\sin\theta'\cos\theta + \cos\theta'\sin\theta)\Big] \end{aligned} \tag{3.41}$$

Assuming that the phase-locked deviation ($\Delta = \theta' - \theta$) is small, and letting $\theta' + \theta \approx 2\theta'$, the above equation is equivalent to

$$V_d \approx v_1\Delta' - v_1 v_u \cos(2\theta' + \phi_u) \tag{3.42}$$

where $v_u = \sqrt{\left(\frac{m-n}{2\sqrt{3}}\right)^2 + \left(\frac{m+n}{6}\right)^2}$ and $\phi_u = -\tan^{-1}\left(\frac{m+n}{\sqrt{3}(m-n)}\right)$.

As indicated by comparison with Eq. (3.38), voltage unbalance can cause fluctuating phase-locked deviations of up to two times the fundamental frequency.

3. Influence of DC offset

In real-world systems, more frequent level conversion and processing are typically required to input actual voltage signals into a controller via a sensor. This process usually involves analog signal processing. Due to the influence of design and device parameters, the output signals in Section AD may consequently exhibit a DC offset in the voltage.

Assume that the offset in three-phase voltage can be expressed as

$$\begin{cases} v_a = v_1 \sin\theta' + v_{ao} \\ v_b = v_1 (1+m)\sin\left(\theta' - \dfrac{2\pi}{3}\right) + v_{bo} \\ v_c = v_1 (1+n)\sin\left(\theta' + \dfrac{2\pi}{3}\right) + v_{co} \end{cases} \tag{3.43}$$

where v_{ao}, v_{bo}, v_{co} are the DC offsets in A, B, C phases, respectively. The implementation of the synchronous rotating coordinate transformation based on Eq. (3.43) yields

$$\begin{aligned} V_d &= -v_1 \sin(\theta' - \theta) + v_{\alpha o} \sin\theta + v_{\beta o} \cos\theta \\ &\approx -v_1\Delta + v_o \cos(\theta' + \phi_o) \end{aligned} \tag{3.44}$$

where $v_{\alpha o} = 2(v_{ao} + v_{bo} + v_{co})/3$, $v_{\beta o} = (v_{co} - v_{bo})/\sqrt{3})$, $v_o = \sqrt{v_{\alpha o}^2 + v_{\beta o}^2}$, and $\phi_o = -\tan^{-1}(v_{\alpha o}/v_{\beta o})$. Therefore, the DC offset in the voltage results in fluctuating phase-locked deviations of up to fundamental frequency.

To sum up, harmonics, unbalance, or DC offset in the voltage can affect phase locking in the three-phase PLL system, resulting in fluctuating phase-locking deviations. Due to the closed-loop structure, the angular frequency (ω) is calculated following PI adjustment. The integral and proportional effects of the PI regulator help limit the influence of phase-locking deviations on both the angular frequency (ω) and the real-time output phase angle (θ). As a result, the three-phase phase-locked loop system can achieve effective phase-locking performance even in the presence of harmonics, unbalance, and DC offset in the voltage.

3.3.3 Simulation and Design of Three-Phase Phase-Locked Loop

Based on the control structure in Fig. 3.19, a three-phase phase-locked loop simulation model was established within the Simulink environment. The input voltage is 50 Hz, including third, fifth, and seventh harmonics with an amplitude of 5% and random interferences with a variance of 0.001. The simulation parameters include $T = 1/138\,\mu s$, $K_p = 0.5$, and $K_i = 128$. Figure 3.20 shows the waveform results for a period of 0.05 s following the initiation of the three-phase phase-locked loop simulation. Figure b shows the waveform comparison between the sinusoidal output ($\sin(\theta)$) and the ideal values ($\sin(\theta')$) for phase angle alignment of phase A. Figure

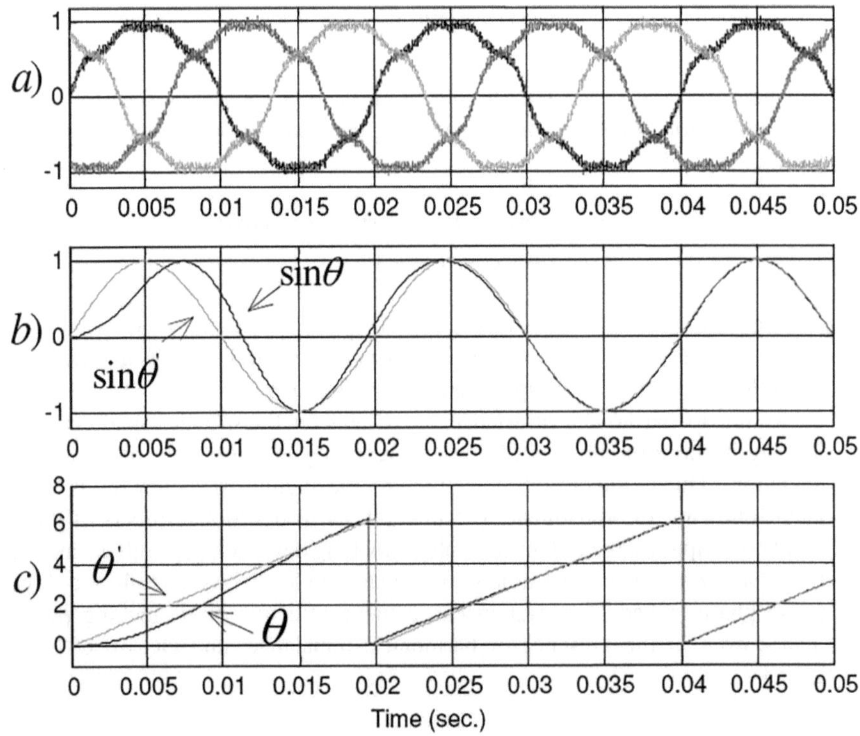

Fig. 3.20 Waveforms from simulation of three-phase phase-locked loop: (**a**) three-phase voltage waveform; (**b**) sinusoidal output ($\sin(\theta)$) and ideal values ($\sin(\theta')$) of phase A; (**c**) comparison between phase angle output (θ) and ideal values (θ') of phase A

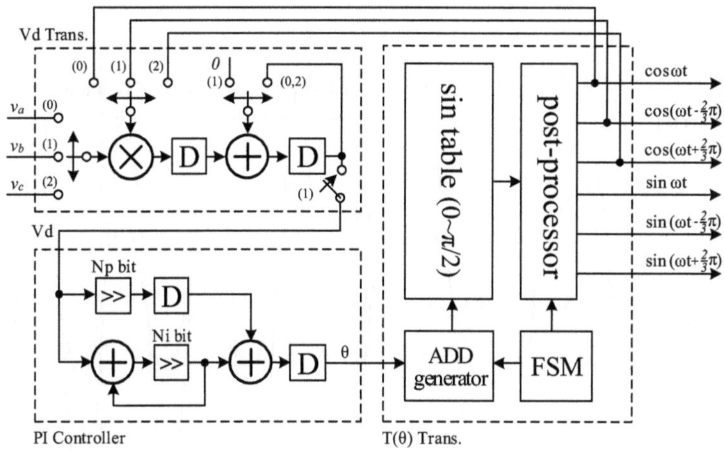

Fig. 3.21 Controller structure of fast algorithm for three-phase phase-locked loop based on FPGA

c shows the comparison between the phase angle output (θ) and the ideal values (θ') of phase A. As illustrated, despite the presence of relatively serious harmonics in the input three-phase voltage, the three-phase phase-locked loop system can accurately align with the phase of the input voltage after less than two fundamental periods.

From the analysis in the previous section, it is understood that the phase deviation (Δ) is solely related to V_q components and is independent of V_d and V_z. Therefore, the detection of phase deviation can rely exclusively on the calculation of V_q, and the calculation coefficient can be integrated into the PI controller parameters. The simplified deviation calculation can thus be rewritten as follows:

$$\Delta = V_q = V_a \cdot \cos\theta + V_b \cdot \cos\left(\theta - \frac{2}{3}\pi\right) + V_c \cdot \cos\left(\theta + \frac{2}{3}\pi\right) \tag{3.45}$$

Based on the simplified calculation model mentioned above, a fast phase-locked loop controller model can be established using field-programmable gate arrays (FPGAs), as illustrated in Fig. 3.21. In the figure, the calculation of the PI controller is simplified by employing Np-bit shifts and Ni-bit accumulation, while trigonometric function operations ($T(\theta)$) rely on a lookup table. Considering the symmetry of sinusoidal functions, the lookup table can be further condensed into a quarter-period table (0~π/2). Furthermore, by implementing time division multiplexing for the lookup table across the six phase outputs required by the synchronous rotating

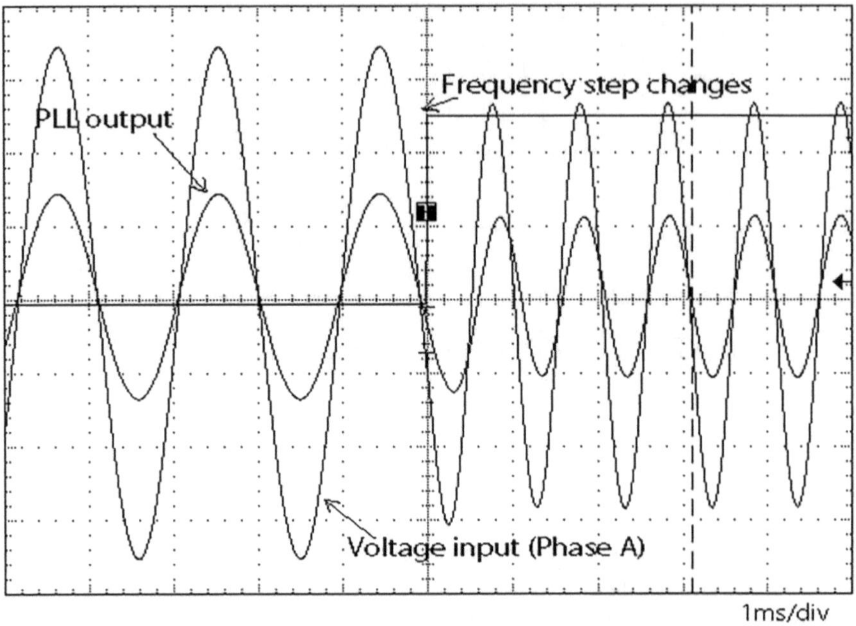

Fig. 3.22 Dynamic system performance in response to sudden changes in input signals from 26.2 to 50.0 Hz

coordinate transformation method, the resource usage of the lookup table can be significantly minimized.

Figure 3.22 illustrates the output waveform of phase A voltage alignment by the phase-locked loop in response to changes in the frequency of the three-phase voltage during the experiment designed to assess the performance of the three-phase phase-locked loop. Specifically, as the voltage frequency changed from 26.2 to 50.0 Hz, the phase-locked loop quickly achieved phase and frequency locking within two cycles, demonstrating strong dynamic performance.

Figure 3.23 presents the output waveform of phase A voltage alignment by the phase-locked loop in response to the presence of harmonic components in the three-phase voltage. Specifically, as the frequency changed abruptly from 26.2 to 50.0 Hz due to significant harmonic components in the voltage, the phase-locked loop quickly achieved phase and frequency locking within three cycles, demonstrating strong performance under harmonic conditions.

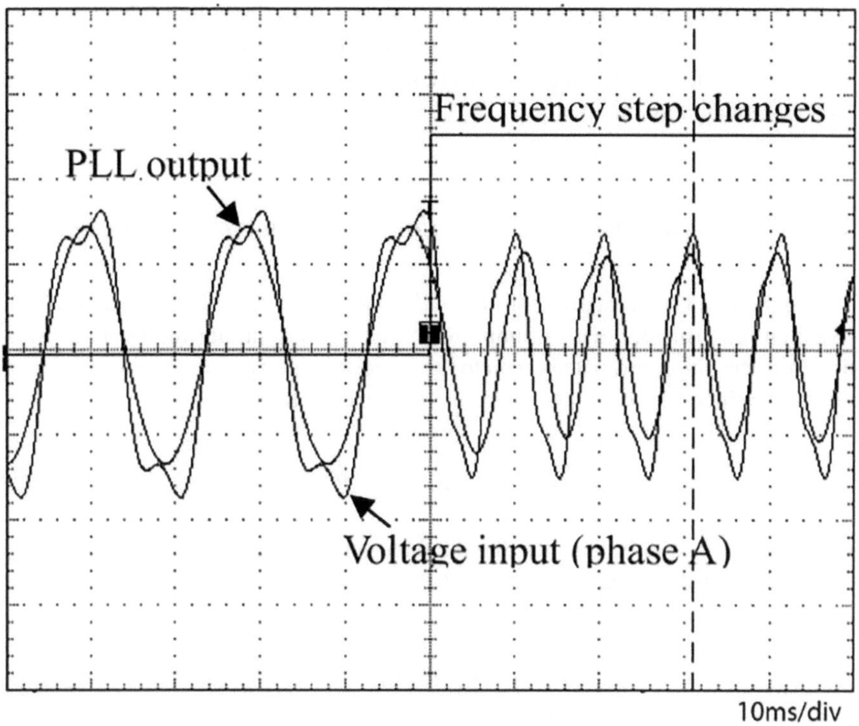

Fig. 3.23 Dynamic system performance in response to sudden changes caused by third harmonic distortion from 26.2 to 50.0 Hz

3.4 Three-Phase SVPWM

The voltage reference output by the converter control algorithm is processed by a PW modulation module before signals are sent to drive the switching tube. This voltage reference can be either real-time sine and cosine waveforms in the three-phase *abc* coordinate system or a reference in the *dq* or $\alpha\beta$ coordinate system, as shown in Fig. 3.15, derived through calculations using the modulation module. This section focuses on the space vector pulse width modulation (SVPWM) method, which offers additional functionalities such as voltage balancing, while common modulation methods like sinusoidal pulse width modulation (SPWM) are omitted (Ge and Feng 2009).

Basic vectors are constructed using a limited number of voltage space vectors output by the converter. By combining these basic vectors and designating various dwell durations, SVPWM equivalently outputs approximately circular rotating voltage vectors, resulting in better driving performance compared to SPWM at lower switching frequencies. Additionally, the output voltage of SVPWM exhibits lower harmonic content than that of SPWM, indicating a higher utilization rate of the DC voltage. Consequently, SVPWM is widely employed in three-phase interconnection converters (Wu and He 2002).

The SVPWM method typically involves several steps, including reference vector calculation, sector determination, vector decomposition, dwell duration calculation, and on-off state output. Below is a brief introduction to a simplified two-level SVPWM calculation method.

3.4.1 Reference Vector Calculation and Sector Determination

Taking the input reference vector in the *abc* coordinate system as an example, and based on the principle that the amplitude remains constant during the transformation from the *abc* coordinate system to the $\alpha\beta$ coordinate system, the $\alpha\beta$ transformation of the three-phase voltage yields the reference voltage as follows:

$$V_{\text{ref}} = \frac{2}{3}\cdot\left(u_{sa} + mu_{sb} + m^2 u_{sc}\right) \tag{3.46}$$

where $m = e^{j\frac{2}{3}\pi}$, which indicates that the three-phase reference voltage (u_{sa}, u_{sb}, u_{sc}) can be expressed by space vector (V_α, V_β) as follows:

$$V_{\text{ref}} = \begin{bmatrix} V_\alpha \\ V_\beta \end{bmatrix} = \frac{2}{3}\cdot\begin{bmatrix} 1 & -\frac{1}{2} & -\frac{1}{2} \\ 0 & \frac{\sqrt{3}}{2} & -\frac{\sqrt{3}}{2} \end{bmatrix}\cdot\begin{bmatrix} u_{sa} \\ u_{sb} \\ u_{sc} \end{bmatrix} \tag{3.47}$$

The voltage space vector output by the converter can be expressed as follows:

$$V_k = \begin{cases} \frac{2}{3}\cdot V_{\text{dc}}\cdot e^{j\frac{k-1}{3}\pi} & k = 0,1,\cdots,7 \\ 0 & k = 7,8 \end{cases} \tag{3.48}$$

As indicated above, eight basic vectors are constructed from the output voltage of the three-phase converter in the $\alpha\beta$ coordinate system, i.e., V_1, V_2, V_3, V_4, V_5, V_6, V_7, and V_8 (V_k, $k = 0, 1, \cdots, 7$), as shown in Fig. 3.24. Among them, two zero space vectors (V_0, V_7) are at the origin of coordinates, and the other six nonzero basic vectors (V_k, $k = 1, 2, \cdots, 6$) form a regular hexagon. The inscribed circle of this regular hexagon defines the SVPWM linear modulation region, as the boundary for the maximum sinusoidal output vector. Any given voltage vector (V_{ref}) can be constructed using these eight space vectors. Symmetrical three-phase voltages can be achieved by rotating V_{ref} at a certain speed within the vector plane. The diagram also shows the waveform shape of pulses output by the two-level SVPWM in six sectors (indicated by I ~ VI) using a single-cycle seven-segment combination.

Sector determination for the reference vector (V_{ref}) follows the criteria listed in Table 3.6, which is based on the amplitude relationship of V_α, V_β.

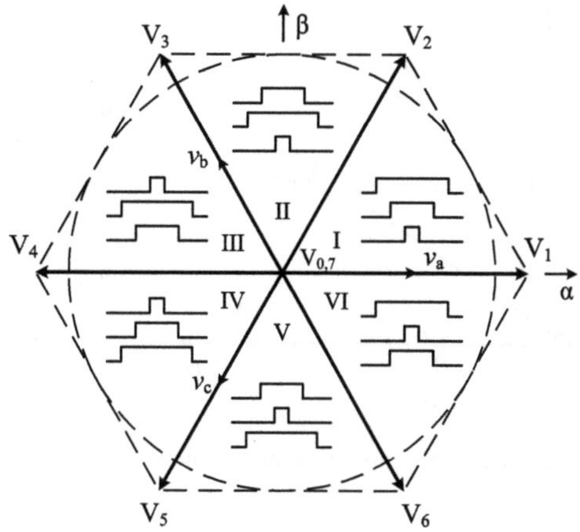

Fig. 3.24 Voltage vector space and SVPWM output pulse waveforms

Table 3.6 Criteria of sector determination in $\alpha\beta$ coordinate system

Sector	Vector angle	V_α, V_β Determination criteria
I	$[0^\circ, 60^\circ)$	$0 \le V_\beta < \sqrt{3}V_\alpha$
II	$[60^\circ, 120^\circ)$	$0 \le \sqrt{3}V_\alpha \le V_\beta$ *or*, $0 < -\sqrt{3}V_\alpha < V_\beta$
III	$[120^\circ, 180^\circ)$	$0 < V_\beta \le -\sqrt{3}V_\alpha$
IV	$[180^\circ, 240^\circ)$	$\sqrt{3}V_\alpha < V_\beta \le 0$
V	$[240^\circ, 300^\circ)$	$V_\beta \le \sqrt{3}V_\alpha \le 0$ *or*, $V_\beta < -\sqrt{3}V_\alpha < 0$
VI	$[300^\circ, 360^\circ)$	$-\sqrt{3}V_\alpha \le V_\beta < 0$

The result of sector determination identifies the two adjacent basic vectors used in the composition of the reference vector, which serves as the foundation for vector output.

3.4.2 Vector Decomposition of Reference Voltage and Dwell Duration Calculation

According to the principle of volt-second balance, any voltage vector (V_{ref}) inside the hexagon can be decomposed using the parallelogram law, and V_{ref} is equivalent to the result of two adjacent nonzero basic vectors and zero vectors acting together over different durations in each switching cycle.

Figure 3.25 illustrates the vector decomposition principle by taking the decomposition of the reference vector in Sector I as an example. Specifically, seven-segment waveforms can be formed for V_{ref}, which are composed of V_1, V_2, V_0, and V_7. Each cycle begins and ends with V_0, resulting in a V_0 vector [0, 0, 0] between different switching cycles, ensuring that the transition between switching cycles is continuous without any level jumps. Additionally, V_7 serves as an intermediate

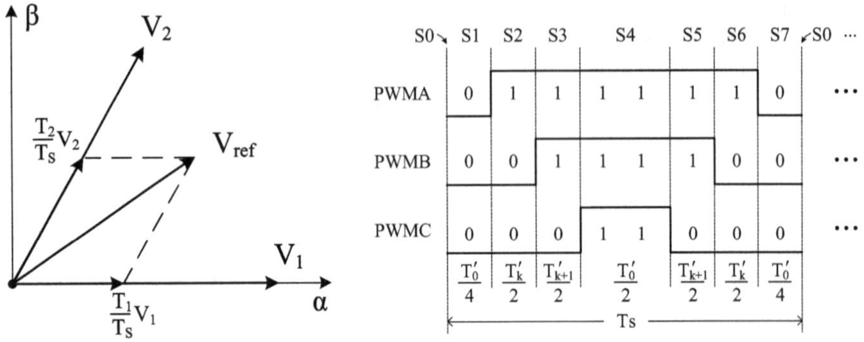

Fig. 3.25 Decomposition of reference voltage (V_{ref}) in sector I and seven-segment switching output solution

vector between the switching cycles, allowing for state change in only one switching during the switching process of the other two nonzero vectors. This configuration prevents simultaneous state changes among the switches.

In short, the seven-segment switching output provides a smooth switching action solution. A similar seven-segment combination is applied to the other five sectors. The waveform diagram is shown in Fig. 3.24.

To enable seven-segment output, it is necessary to calculate the dwell durations for four switching vectors, i.e., V_k, V_{k+1}, V_0, and V_7. Among them, both V_0 and V_7 are zero vectors, and their dwell durations can be obtained by equally dividing the dwell duration (T_0) of the zero vectors. Therefore, the dwell duration calculation is simplified into T_k, T_{k+1}, and T_0. The implementation of the volt-second balance principle yields

$$\begin{cases} V_1T_1 + V_2T_2 + V_{0,7}T_3 = V_{ref}T_s \\ T_3 = T_s - V_1 - V_2 \end{cases} \tag{3.49}$$

where T_1, T_2, and T_3 indicate the dwell durations of vectors V_1, V_2, and V_0 or V_7, respectively; T_s indicates the carrier period of the converter.

The volt-second integral of the reference input (V_{ref}) and each switching sampling period (T_s) can be expressed using the dwell durations of its adjacent vectors V_k and V_{k+1} as follows:

$$V_{ref} = \frac{T_k}{T_s}V_k + \frac{T_{k+1}}{T_s}V_{k+1} \tag{3.50}$$

where T_k and T_{k+1} are the dwell durations of vectors V_k and V_{k+1} in each switching cycle, respectively, and k is the sector number of the reference vector. For example, if the reference vector (V_{ref}) is located in sector I, i.e., k = 1, its action during a switching period is equivalent to the sum of actions of V_1 (T_1) and V_2 (T_2). Considering that the amplitude of nonzero space vectors based on the SVPWM method also meets $V_k = 2V_{dc}/3$, the dwell duration can be expressed as follows:

$$\begin{cases} T_k = \dfrac{\sqrt{3}T_s}{V_{dc}}\left[\sin\left(\dfrac{\pi}{3}k\right)V_\alpha - \cos\left(\dfrac{\pi}{3}k\right)V_\beta\right] \\ T_{k+1} = \dfrac{\sqrt{3}T_s}{V_{dc}}\left\{-\sin\left[\dfrac{\pi}{3}(k-1)\right]V_\alpha + \cos\left[\dfrac{\pi}{3}(k-1)\right]V_\beta\right\} \\ T_0 = T_s - T_k - T_{k+1} \end{cases} \tag{3.51}$$

where V_{dc} is the DC voltage of the converter and T_0 is the dwell duration of the zero vector during each switching cycle. As the sector number must be determined for calculations using Eq. (3.51), it can be known that the sector number (k) is related to the angle of the reference vector. The calculations of T_k, T_{k+1}, and T_0 support the correct selection of the PWM waveform output.

If the reference vector remains within the linear modulation region, which is defined by the boundary of the maximum sinusoidal output vector, Eq. (3.51) meets the requirement of $T_0 > 0$, and the locus of the converter output voltage vector is circular. Converters operating in an overmodulation region can cause serious distortions in the converter current, which is beyond the scope of this book.

3.4.3 A Fast SVPWM Algorithm and Expansion in Multilevel Scenarios

As analyzed in the previous section, the traditional algorithm for SVPWM implementation includes several functions, such as vector transformation, sector determination, vector decomposition, dwell duration calculation, and PWM waveform generation. These functions necessitate various operations, such as the multiplication and division of irrational numbers. To facilitate the development of a fast SVPWM algorithm, intermediate variables of $X_\alpha X_\beta$ are introduced, which are defined as follows:

$$\begin{cases} X_\alpha = v_a - v_b/2 - v_c/2 \\ X_\beta = v_b - v_c \end{cases} \tag{3.52}$$

Then,

$$V_\alpha = 2X_\alpha/3 \quad \text{and} \quad V_\beta = X_\beta/\sqrt{3} \tag{3.53}$$

In this equation, the reference vectors of V_α, V_β are replaced by X_α, X_β. To use Eq. (3.53) for calculating the sector number, the relationship for sector determination of

Table 3.7 Calculation criteria corresponding to sector number in $\alpha\beta$ coordinate system

Sector	$X_\alpha > 0$	$X_\beta > 0$	$\|X_\alpha\| > \|X_\beta/2\|$
I	1	1	1
II	×	1	0
III	0	1	1
IV	0	0	1
V	×	0	0
VI	1	0	1

Table 3.8 Relationships between values of dwell duration matrix (T) and sector numbers (k)

k	1	2	3	4	5	6
T_{00}	1	1	0	−1	−1	0
T_{01}	−1/2	1/2	1	1/2	−1/2	−1
T_{10}	0	−1	−1	0	1	1
T_{11}	1	1/2	−1/2	−1	−1/2	1/2

$\sqrt{3}V_\alpha$ and V_β given in Eq. (3.53) is replaced by $|X_\alpha|$ and $|X_\beta/2|$ indicated in Table 3.7. This approach allows for sector determination to be achieved solely by comparing $|X_\alpha|$ and $|X_\beta/2|$ in both magnitudes and signs.

Meanwhile, by substituting Eqs. (3.52) and (3.53) into Eq. (3.51), the dwell duration expression can be simplified as follows:

$$\left[T_k \ T_{k+1}\right]^{\mathrm{T}} = \frac{T_s}{V_{\mathrm{dc}}} \cdot T \cdot \left[X_\alpha \ X_\beta\right]^{\mathrm{T}} \tag{3.54}$$

where matrix T is the dwell duration matrix:

$$T = \begin{bmatrix} T_{00} & T_{01} \\ T_{10} & T_{11} \end{bmatrix} = \begin{bmatrix} \frac{2}{\sqrt{3}}\sin\left(\frac{\pi}{3}k\right) & -\cos\left(\frac{\pi}{3}k\right) \\ -\frac{2}{\sqrt{3}}\sin\left[\frac{\pi}{3}(k-1)\right] & \cos\left[\frac{\pi}{3}(k-1)\right] \end{bmatrix} \tag{3.55}$$

where the four trigonometric function coefficients are all related to the sector number (k). For example, the location of sector I, i.e., $k = 1$, indicates $T_{00} = 1$, $T_{01} = -1/2$, $T_{10} = 0$, and $T_{11} = 1$. The values of T for other sectors are shown in Table 3.8.

It is apparent that Eq. (3.54) only involves simple algebraic operations, including $-X_\alpha$, $-X_\beta$, $X_\beta/2$, and $-X_\beta/2$.

In the linear modulation region, the maximum modulation coefficient and output phase voltage can reach 1.15 and $V_{\mathrm{dc}}/\sqrt{3}$. By letting the maximum three-phase voltage be the rated reference, the DC voltage (V_{dc}) is represented by $\sqrt{3}$ p.u. Letting $T_s = \sqrt{3}$ p.u., $T_s/V_{\mathrm{dc}} = 1$ is established in Eq. (3.54). Equation (3.51) is then simplified as follows:

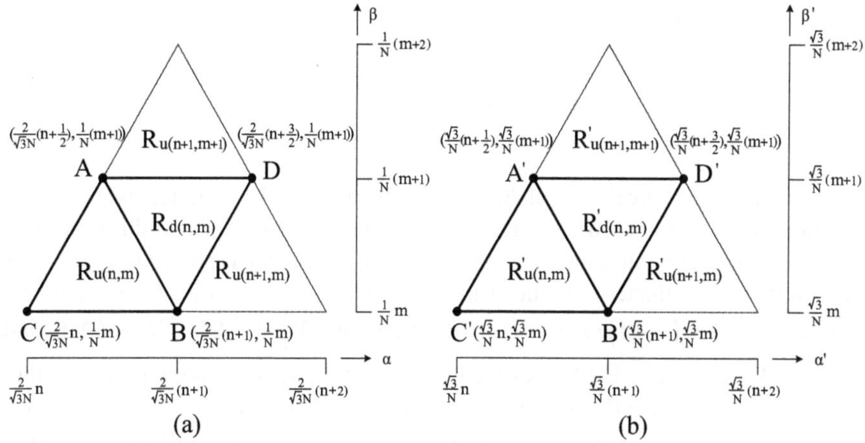

Fig. 3.26 Region division diagram: (**a**) $\alpha\beta$ coordinate system; (**b**) $\alpha'\beta'$ coordinate system

$$\begin{cases} \left[T_k' T_{k+1}'\right]^{\mathrm{T}} = T \cdot \left[X_\alpha \;\; X_\beta\right]^{\mathrm{T}} \\ T_0' = \sqrt{3} - T_k' - T_{k+1}' \end{cases} \tag{3.56}$$

where T_k', T_{k+1}', and T_0' are the per-unit values of T_k, T_{k+1}, and T_0 normalized by T_s, respectively. The relationships between the matrix values (T) and the sector numbers (k) are shown in Table 3.8. Only five values are taken for all elements of the matrix (T), specifically 1, 1/2, 0, −1/2, and −1, which can be determined using fast algorithms such as negation and shifting.

The N-level converter outputs $3N^2 - 3N + 1$ voltage vectors, indicating the association with the increasing number of voltage space vectors. The $\alpha\beta$ plane is divided into $6N^2$ unit triangles as shown in Fig. 3.26, with the vertices of these triangles corresponding to different on-off states. These triangles are further divided into upper triangles ($R_{u(n,m)}$) and lower triangles ($R_{d(n,m)}$) based on their direction, with n, $m = -N+2, -N+3, \cdots, N-2$. The increase in the number of levels complicates the determination of the reference vector position and redundant on-off state. As a result, the computational intensity of algorithms for multilevel SVPWM is significantly higher than that for two-level SVPWM.

The simplified calculation method for the dwell duration of N-level SVPWM in the $\alpha\beta$ coordinate system is discussed below. Letting the amplitude of the output voltage vector of N-level SVPWM within the linear modulation region be reference (Liu 2010) p.u., the side length of the maximum space hexagon is expressed as $\frac{2}{\sqrt{3}}$ p.u. and the DC voltage as $\sqrt{3}$ p.u.

By defining the side length as $\frac{2}{\sqrt{3}N}$ p.u. and assuming that the reference vector lies within the triangle ABC, i.e., the $R_{u(n,m)}$ area shown in Fig. 3.26, selecting the neighboring vectors V_{OA}, V_{OB}, and V_{OC} to compose the reference voltage vector aligns with the volt-second balance principle:

$$\begin{cases} V_{A\alpha}T_A + V_{B\alpha}T_B + V_{C\alpha}T_C = V_\alpha T_s \\ V_{A\beta}T_A + V_{B\beta}T_B + V_{C\beta}T_C = V_\beta T_s \\ T_A + T_B + T_C = T_s \end{cases} \tag{3.57}$$

where T_A, T_B, and T_C are the dwell durations of vectors V_{OA}, V_{OB}, and V_{OC}, respectively.

Equation (3.53) is used to convert the parameters of the $\alpha\beta$ coordinate system shown in Fig. 3.26a into those of the $\alpha'\beta'$ coordinate system as shown in Fig. 3.26b. The vertex coordinates of the triangles in the $\alpha\beta$ coordinate system are $A\left((2n+1)/\sqrt{3}N,(m+1)/N\right)$, $B\left((2n+2)/\sqrt{3}N,m/N\right)$, $C\left(2n/\sqrt{3}N,m/N\right)$, and $D\left((2n+3)/\sqrt{3}N,(m+1)/N\right)$, respectively. Substituting the coordinate values into Eq. (3.57) can indicate the location within the upper triangle $A'B'C'$, i.e., the reference vector is located in the $R'_{u(n,m)}$ area. Based on Eq. (3.53), the dwell duration of resultant vectors $V_{OA'}$, $V_{OB'}$, and $V_{OC'}$ can be expressed as follows:

$$\begin{cases} T_{A'} = N \cdot X_\beta \cdot T_s / \sqrt{3} - m \cdot T_s \\ T_{B'} = N \cdot X_\alpha \cdot T_s / \sqrt{3} - n \cdot T_s - T_A' / 2 \\ T_{C'} = T_s - T_{A'} - T_{B'} \end{cases} \tag{3.58}$$

Similarly, in the lower triangle $A'B'D'$, i.e., the reference vector is located in the $R'_{d(n+1,m)}$ area, the dwell durations of the resultant vectors $V_{OA'}$, $V_{OB'}$, and $V_{OD'}$ are expressed as follows:

$$\begin{cases} T_{B'} = (m+1) \cdot T_s - N \cdot X_\beta \cdot T_s / \sqrt{3} \\ T_{D'} = N \cdot X_\alpha \cdot T_s / \sqrt{3} - (n+1/2) \cdot T_s - T_B' / 2 \\ T_{A'} = T_s - T_{B'} - T_{D'} \end{cases} \tag{3.59}$$

Vertex coordinates are transformed into the $\alpha'\beta'$ coordinate system as follows: $A'\left(\sqrt{3}(2n+1/2)/N,\sqrt{3}(m+1)/N\right)$, $B'\left(\sqrt{3}(2n+1)/N,\sqrt{3}m/N\right)$, $C'\left(\sqrt{3}n/N,\sqrt{3}m/N\right)$, and $D'(\sqrt{3}(2n+3/2)/N,\left(\sqrt{3}(m+1)/N\right)$.

In addition, letting $T_s = \sqrt{3}/N$ p.u., $T_1 = X_\beta - \sqrt{3}m/N$, and $T_2 = X_\alpha - \sqrt{3}n/N$, and based on Eq. (3.58), the dwell durations of the vectors in the upper triangle $A'B'C'$ can be expressed as follows:

$$\begin{cases} T_{A'} = T_1 \\ T_{B'} = T_2 - T_1 / 2 \\ T_{C'} = \sqrt{3}/N - (T_2 + T_1/2) \end{cases} \tag{3.60}$$

Based on Eq. (3.59), the dwell durations of the vectors in the lower triangle $A'B'D'$ can be expressed as follows:

$$\begin{cases} T_{A'} = \sqrt{3}/N - (T_2 - T_1/2) \\ T_{B'} = \sqrt{3}/N - T_1 \\ T_{D'} = (T_2 + T_1/2) - \sqrt{3}/N \end{cases} \tag{3.61}$$

where the coefficients of $\sqrt{3}/N$, $\sqrt{3}m/N$, and $\sqrt{3}n/N$ are constants. Therefore, in the $\alpha'\beta'$ coordinate system, the multilevel SVPWM dwell duration is only related to the reference vectors (X_α, X_β) and the triangle number (n, m). Both Eqs. (3.60) and (3.61) solely require simple operations such as addition, subtraction, and shifting.

The triangle number (n, m) can be calculated from the ratio of X_α and X_β to the side length of the unit triangle. For V_{ref} located in sector I to VI as shown in Fig. 3.24 and with $N = 1$, a two-level SVPWM fast algorithm can be derived from the above multilevel algorithm, which can be expressed as follows:

Sector I, $m = 0$, $n = 0$, and in $A'B'C'$: $T_1 = T_{B'}$, $T_2 = T_{A'}$, and $T_0 = T_{C'}$
Sector II, $m = 0$, $n = -1$, and in $A'B'D'$: $T_2 = T_{D'}$, $T_3 = T_{A'}$, and $T_0 = T_{B'}$
Sector III, $m = 0$, $n = -1$, and in $A'B'C'$: $T_3 = T_{A'}$, $T_4 = T_{C'}$, and $T_0 = T_{B'}$
Sector IV, $m = -1$, $n = -3/2$, and in $A'B'D'$: $T_4 = T_{A'}$, $T_5 = T_{B'}$, and $T_0 = T_{D'}$
Sector V, $m = -1$, $n = -1/2$, and in $A'B'C'$: $T_5 = T_{C'}$, $T_6 = T_{B'}$, and $T_0 = T_{A'}$
Sector VI, $m = -1$, $n = -1/2$, and in $A'B'D'$: $T_6 = T_{B'}$, $T_1 = T_{D'}$, and $T_0 = T_{A'}$

The dwell durations of the three voltage vectors in the above sectors, calculated using Eqs. (3.60) and (3.61), were consistent with those listed in Table 3.8. This consistency indicates that the two-level fast algorithm discussed in the previous section is a special case of the multilevel fast SVPWM algorithm discussed in this section.

3.4.4 FPGA Design and Implementation for Fast SVPWM Algorithm

The aforementioned fast SVPWM algorithm can be fully implemented using simple operations such as addition, subtraction, and shifting, thereby eliminating the need for complex operations like multiplication, division, and trigonometric functions. This section discusses the fixed-point design and implementation of the fast SVPWM algorithm based on FPGA, taking into account both computing speed and complexity, in an effort to optimize performance in terms of speed and area (Shu et al. 2007).

The hardware architecture of the two-level fast SVPWM algorithm can be established according to Eqs. (3.52) and (3.56), as well as Tables 3.7 and 3.8, as shown in Fig. 3.27. The entire architecture can be divided into the following signal processing modules:

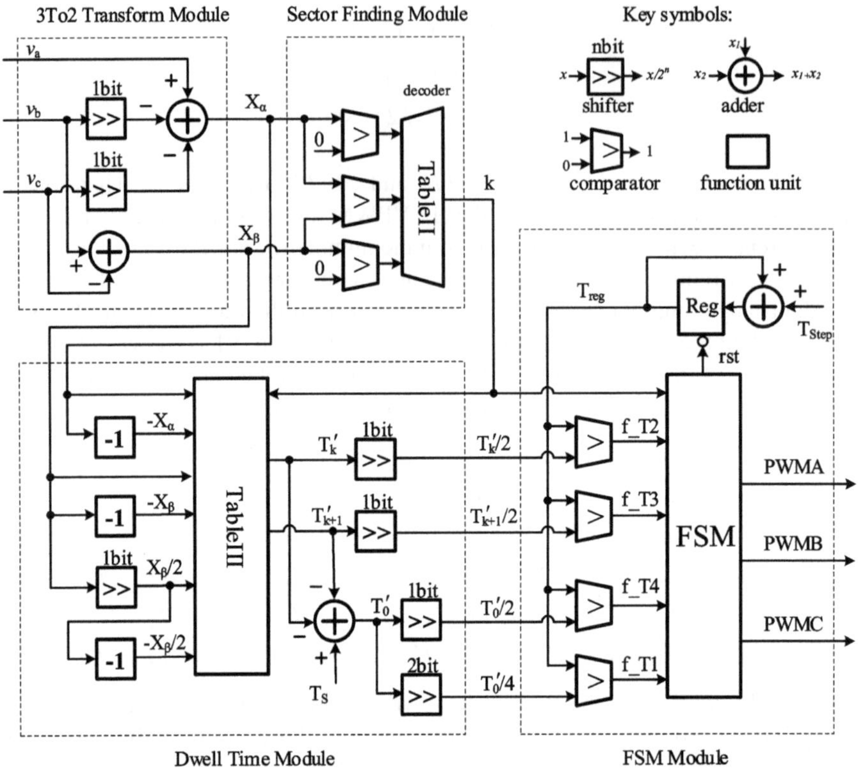

Fig. 3.27 Architecture of two-level fast SVPWM algorithm

1. 3-to-2 transformation module: transform the three-phase reference input to the $\alpha'\beta'$ plane.
2. Sector finding module: determine the sector number that corresponds to the position of the reference vector.
3. Dwell duration module: calculate the dwell duration of the resultant vector during a switching cycle.
4. Finite-state machine module (FSM): output symmetric SVPWM pulses under the finite-state machine control.

3-to-2 conversion module:

As shown in Fig. 3.27, the 3-to-2 transformation module converts the three-phase reference inputs (v_a, v_b, and v_c) into intermediate variables (X_α and X_β). Only addition and shifting operations are required in this module.

Sector determination module:

The following three relationships are used to identify the sector number that corresponds to the position of the reference vector, including $X_\alpha > 0$, $X_\beta > 0$, and $|X_\alpha| > |X_\beta/2|$. According to the straightforward logical relationships presented in Table 3.7, a 3-input decoder can output the correct sector number.

Dwell duration calculation module:

Equation (3.56) is used to calculate the dwell durations ($T_k^{'}$ and $T_{k+1}^{'}$) of two vectors (V_k and V_{k+1}) adjacent to the reference vector and the dwell duration ($T_0^{'}$) of the zero vector (V_0 or V_7). The input signals are multiplexed and output, depending on the sector number (k) selected from Table 3.8. All elements of the matrix (T) merely include five values: 1, 1/2, 0, −1/2, and −1, which can be determined using fast algorithms such as negation and shifting.

Finite-state machine module:

On-off state output is generated based on the sector number. This module employs an eight-state FSM as shown in Fig. 3.28 to output the correct switching waveforms. For example, in sector I, the PWM waveform output during one switching cycle is decomposed into seven parts. The dwell duration is counted by an accumulator to enable the timing function. T_{Step} represents the step length of the accumulator, and changing its value leads to alterations to the switching frequency of the SVPWM output. According to Fig. 3.27, the setting of T_{Step} can be expressed as follows:

$$T_{\text{Step}} = \frac{f_{\text{cnt}}}{2 \cdot f_{\text{car}} \cdot m}, \quad T_{\text{Step}} \ll T_{\text{reg}} \tag{3.62}$$

where f_{cnt} represents the clock of the accumulator, f_{car} represents the carrier frequency, and m represents the frequency modulation ratio. Once the accumulator output (T_{reg}) is greater than $T_0/4$, $T_k/2$, $T_{k+1}/2$, or $T_0/2$, FSM state flags (f_T1, f_T2, f_T3, f_T4) are set, respectively. These flags are reset following state transitions in FSM. Throughout S0 → S1 → S2 → S3 → S4 → S5 → S6 → S7 → S0, the FSM experiences one full switching cycle of the PWM output.

Based on the above two-level SVPWM algorithm architecture, an 8/12/16-bit intellectual property (IP) core is designed using Verilog HDL. Table 3.9 presents the resource utilization for the modules of this IP core, which are 372, 515, and 643 logic elements (LEs) corresponding to the bit widths of 8, 12, and 16, respectively.

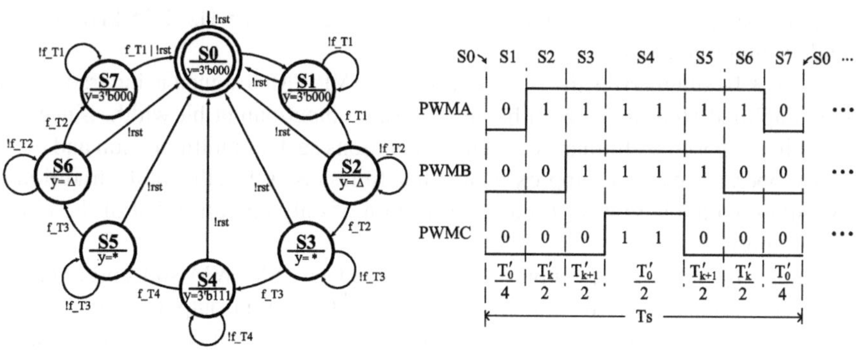

Fig. 3.28 FSM diagram of symmetric PWM

Table 3.9 Resource utilization of SVPWM in 8/12/16-bit configurations

Word length	Entity name	Logic elements	LC registers	LUT/Reg LCs	Carry chain LCs
8-bit	3/2 Trans.	26	26	26	26
	Sector Find.	26	3	2	7
	Dwell Time	199	75	75	89
	FSM	121	53	35	28
	Total	372	157	138	150
12-bit	3/2 Trans.	38	38	38	38
	Sector Find.	38	3	2	11
	Dwell Time	279	107	107	125
	FSM	160	69	47	40
	Total	515	217	194	214
16-bit	3/2 Trans.	50	50	50	50
	Sector Find.	47	3	2	15
	Dwell Time	341	139	139	161
	FSM	202	85	59	52
	Total	643	277	250	278

Table 3.10 Results of 12-bit two-level SVPWM based on traditional and simplified algorithms

Architecture	Multiplier	Divider	Adder	Delay (clock)	Speed (MHz)	LEs	Utilization (%)
Traditional algorithm	8	1	24	26	33	1796	15
Simplified algorithm	0	0	12	11	151	515	5

This utilization represents 3%, 4%, and 5% of the LE resources of the EP1C12Q240C8 chip. The core supports operation at system frequencies of 162, 151, and 148 MHz, respectively. The delay from input to output for the entire architecture requires only 11 system clocks, and the computational delay is less than 100 ns.

Based on Eq. (3.62), with $m = 400$ and assuming that f_{cnt} fluctuates within the range from 10 kHz to 1 MHz and T_{Step} fluctuates within the range of 1–500, the IP core can be applied across a wide frequency range of 0.025 Hz–1.25 kHz for the fundamental frequency.

Table 3.10 summarizes the IP core results of SVPWM designed using traditional and simplified algorithms, both having the same input/output bit width. Since the simplified algorithm eliminates complex operations such as multiplication and division, it operates at a speed approximately five times faster than that of the traditional algorithm, with a hardware resource utilization of only one-third that of the traditional algorithm.

Figure 3.29 shows the output of the designed fast SVPWM soft core. Due to space limitations, only the waveform corresponding to an output switching frequency of 1 kHz is displayed, and the three-phase output switching waveforms in the six sectors align with the illustration of Fig. 3.28.

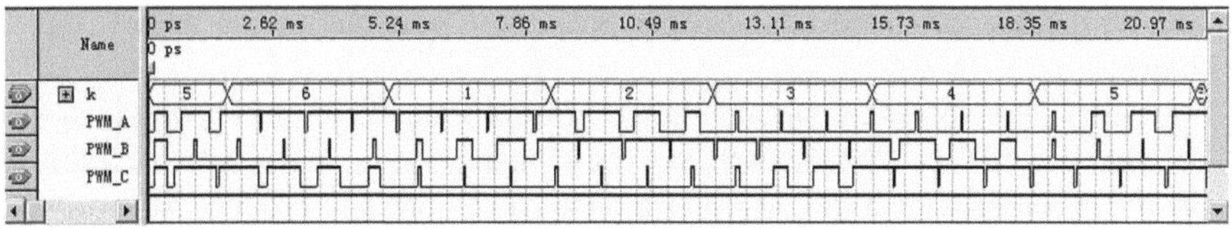

Fig. 3.29 Waveform of three-phase output pulse width based on fast SVPWM

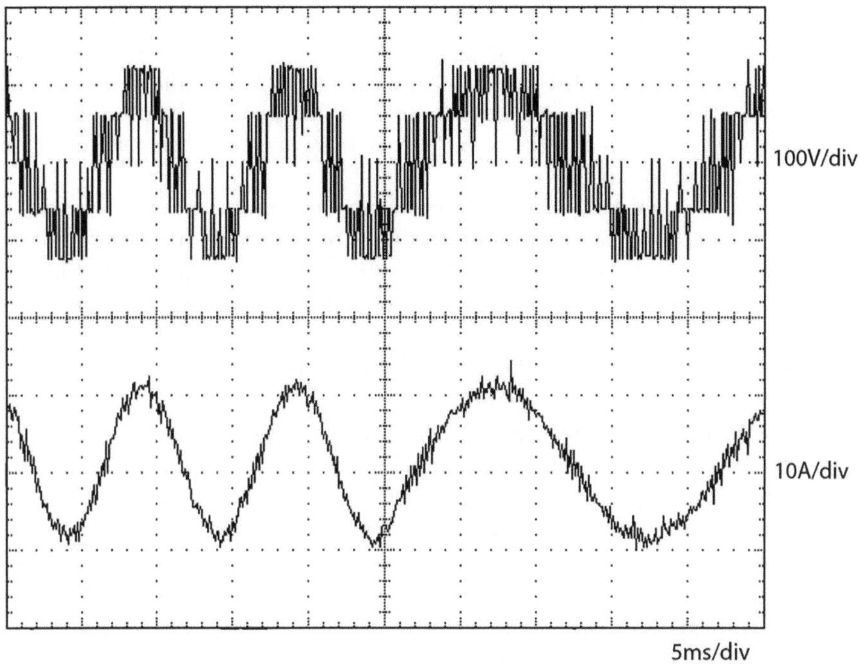

Fig. 3.30 Voltage and current waveforms at output port with sudden change of frequency from 50 Hz to 25 Hz

The dynamic waveforms in response to the reference frequency changing from 50 Hz to 25 Hz during the inverter experiment using the aforementioned SVPWM with the 12-bit IP core are presented in Fig. 3.30.

3.5 Simulation and Experiment of Three-Phase Converter

This section presents the verification of the correctness of the circuit, control method, and modulation, using diode-clamped cascaded hybrid three-phase converters as an example. A diode-clamped three-level H-bridge circuit is utilized with multiple modules, allowing for a maximum output of five levels at each module port. With three cascaded modules per phase, the port output can achieve up to 13 levels.

Figure 3.31 displays the simulation waveforms of the three-phase port voltage of the cascade rectifiers. The three-phase port voltages (V_a, V_b, and V_c) each exhibit 13 levels. Therefore, only a low AC filter inductor is required to produce AC grid-side currents with satisfactory sinusoidal performance.

Figure 3.32 shows the experimental waveforms. Channel 1 displays the port level (uab) between phases A and B, represented as seven-level step waves with sinusoidal variations. Channel 2 indicates the port level (u_{ano}) of phase A relative to the midpoint (n_0) in the converter's star connection, presented as five-level step waves with sinusoidal variations. Channel 3 shows the port level (u_{ao}) of the left bridge arm of phase A's H-bridge module relative to the midpoint (n_0), depicted as three-level step waves with sinusoidal variations. Channel 4 displays the filtered waveform of the phase A current output.

Figure 3.33 shows the level waveforms of the output ports between phases A, B, and C. Channel 1 indicates the output port level (u_{ab}) of phase A relative to phase B. Channel 2 shows the output port level (u_{bc}) of phase B relative to phase C. Channel 3 displays the output port level (u_{ca}) of phase C relative to phase A. These port levels of u_{ab}, u_{bc}, and u_{ca} are represented as seven-level step waves with sinusoidal variations, demonstrating a mutual difference of 120°. Channel 4 presents the filtered waveform of phase A current output.

Figure 3.34 illustrates the voltage and current simulation waveforms of the cascaded rectifiers, with Fig. 3.34a presenting the supply voltage (u_{abc}) and Fig. 3.34b displaying the supply current (i_{abc}). During the simulation, no load was applied from $t = 0$ to $t = 1.5$ s, resulting in a grid-side current of zero. From $t = 1.5$ s to $t = 4$ s, each module was subjected to a 100 Ω load, causing the grid-side current to rise to 82 A (RMS). A minimal current overshoot occurred upon system startup and when the loads were applied.

Figure 3.35 illustrates the curves of active power (P) and reactive power (Q) received from the grid during the operation of the cascaded rectifiers. As shown, the

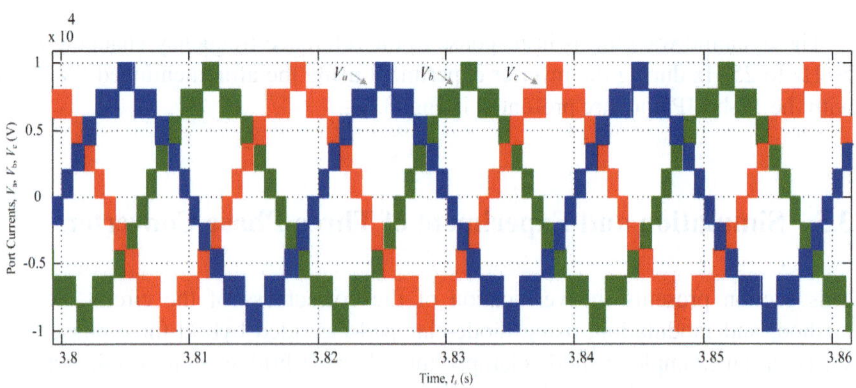

Fig. 3.31 Three-phase port voltages, V_a, V_b, and V_c

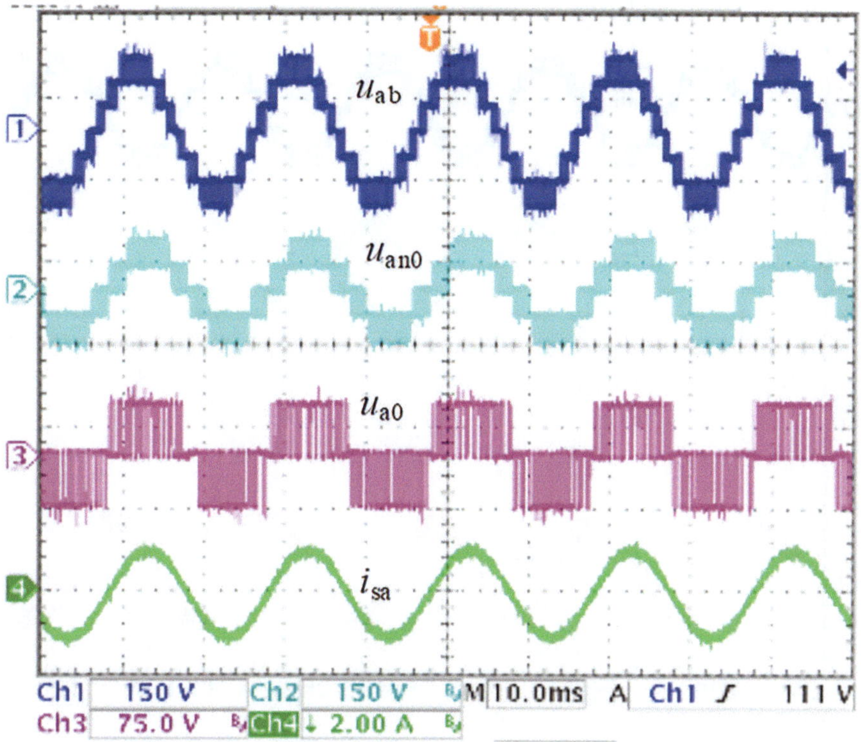

Fig. 3.32 Comparison of output port level across intra-phase inverters

reactive power (Q) stabilizes and maintains a value near zero after a brief adjustment period, around the time when the load is connected at $t = 1.5$ s. Meanwhile, the active power (P) fluctuates in response to changes in load, becoming zero in the no-load condition and responding quickly to load changes once the load is applied. These results indicate that the cascaded rectifiers operate in a unit power mode.

Figure 3.36 illustrates the output simulation waveform of the rectifier DC voltage controlled by a PI regulator. As shown, when the system is started and the load changes, the controller quickly stabilizes after a brief adjustment period.

Figure 3.37a shows the experimental waveforms of the cascaded rectifiers.

Stage 1: The DC voltage stabilizes at the reference voltage set by the control system when the system is under the no-load closed-loop condition, as well as the balanced load condition.

Stage 2: The DC voltage drops by about 20 V when the system switches from the no-load state to the on-load state; it subsequently returns to the reference voltage value and remains stable after 1.6 s.

Stage 3: The DC voltage rises by about 5 V, with an overshoot of about 7%, during the dynamic switching process from a balanced load state to an unbalanced state; it subsequently returns to the reference voltage value after about 2.5 s. Throughout

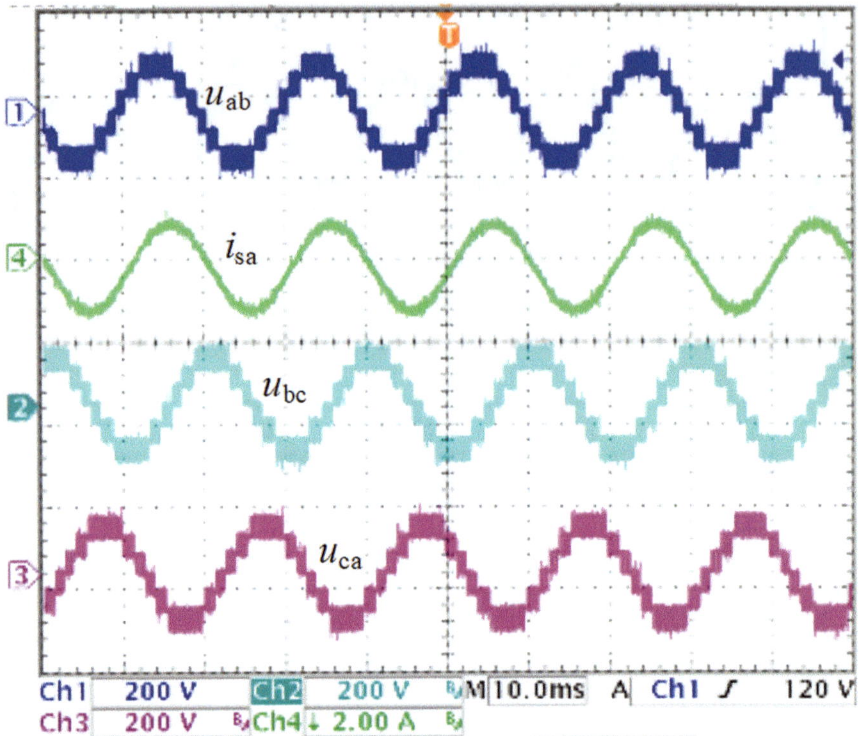

Fig. 3.33 Port level of inverter output line

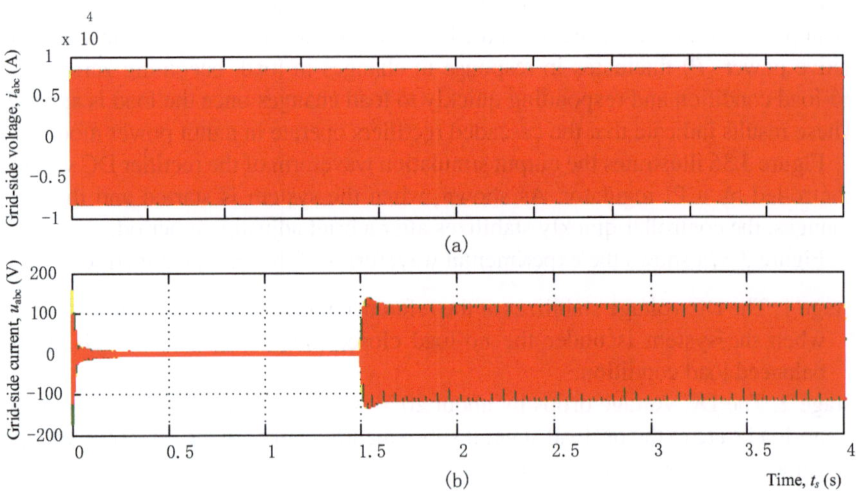

Fig. 3.34 Grid-side voltage (u_{abc}) and grid-side current (i_{abc})

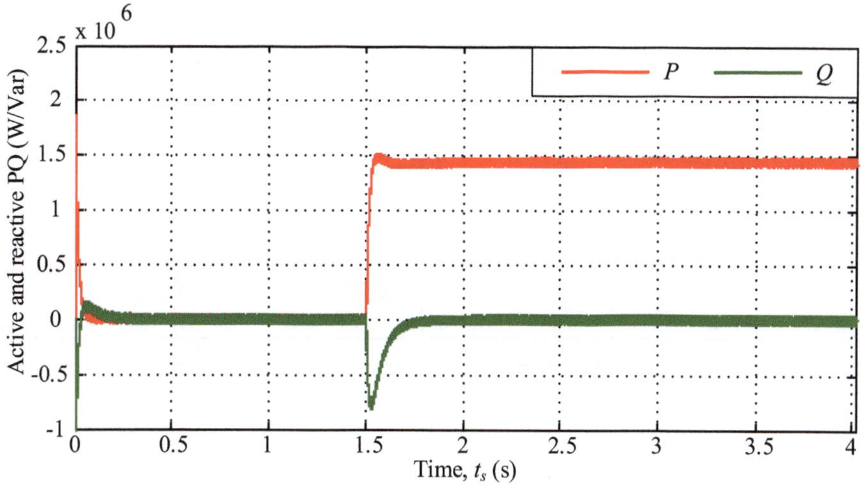

Fig. 3.35 Active and reactive *PQ*

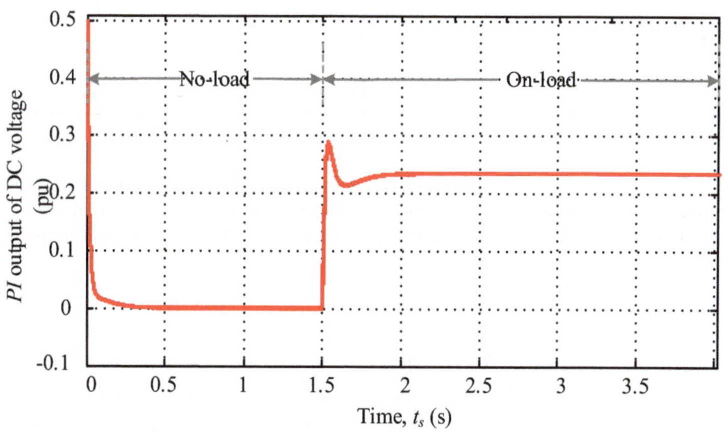

Fig. 3.36 PI output of DC voltage

this dynamic switching process, alternating current on the grid side transitions smoothly to the next state without noticeable oscillations.

As shown in Fig. 3.37b, when the system is switched from an unbalanced state to a no-load unbalanced state, intra-phase cascaded module 1 is switched to a no-load condition. During the dynamic switching process, each DC voltage in the system rises by approximately 5 V, with an overshoot of around 7%. After about 2.6 s, the DC voltage returns to the reference voltage value, without oscillations.

In Fig. 3.37c, d, channel 1 displays the phase A voltage (u_{sa}), channels 2–4 indicate the DC-side output voltages (U_{dca1}–U_{dca3}) of phase A modules 1–3,

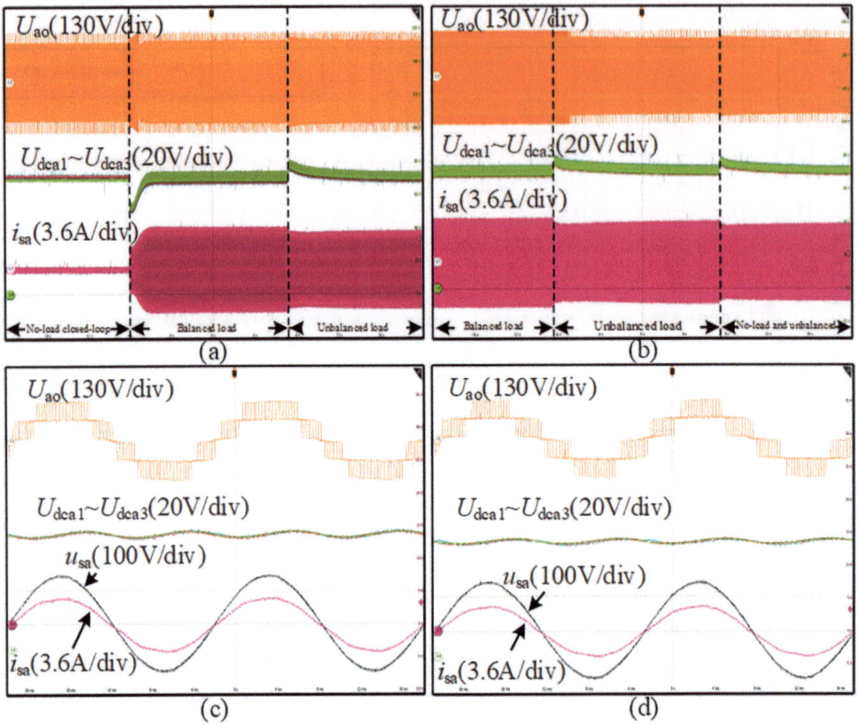

Fig. 3.37 Experimental waveforms of voltage balancing and load switching within intra-phase wide on-load range with mixed modulation: (**a**)–(**b**) experimental waveforms of load switching; (**c**) steady-state experimental waveforms in unbalanced state; (**d**) steady-state experimental waveforms in no-load unbalanced state

channel 5 represents the total port level (U_{ao}) of the phase A cascade modules, and channel 7 shows the phase A grid-side alternating current (i_{sa}). Figure 3.37c presents the waveform when the load operates steadily in the unbalanced state, while Fig. 3.37d exhibits the waveform when the load operates steadily in the no-load unbalanced state. It is evident that the DC voltage of each module in the three-phase cascaded rectifier remains stable, and the grid-side alternating current (i_{sa}) shows satisfactory sinusoidal performance, maintaining phase alignment with the grid-side voltage (u_{sa}).

3.6 Summary

This chapter introduces the operating principles, control, and modulation of three-phase converters. The subsequent analysis of three-phase converters is divided into four topologies, focusing on their operating modes and establishing mathematical models in the *abc* coordinate system using a deductive approach. Additionally, the active and reactive decoupling control of three-phase converters is analyzed and discussed. In particular, the principle and implementation of the three-phase phase-locked loop module, which is essential for this control function, along with the modulation method, are explored. Furthermore, the simulation analysis and experiments of the three-phase converter system are detailed, demonstrating the dynamic and static performance of three-phase converters in rectifier applications.

References

Liu Neng. Study on Compensation Optimization Scheme for Co-phase Traction Substation [D]. Master's Degree Dissertation of Southwest Jiaotong University, 2010.

Jie Feng. Three-Phase PWM Rectifier and Its Control [D]. Master's Degree Dissertation of Zhejiang University, 2006.

Zheng Zhongjiu. Research on Control Strategy and Application of Three-Phase Voltage-Source PWM Rectifier [D]. Doctor's Degree Dissertation of Dalian University of Technology, 2011.

Wang Xiaofeng. Research on Voltage Balance, Clamping Strategy and Modulation Technology of Three-Phase Multilevel Converter [D]. Doctor's Degree Dissertation of Zhejiang University, 2007.

Shu Zeliang, Guo Yuhua, Tang Jian. Implementation of Three-Phase Phase-Locked Loop Based on FPGA [J]. 2005, 39(6): 126–128.

Gong Jinxia, Xie Da, Chang Yanchi. Principle and Performance of Three-Phase Digital Phase-Locked Loop [J]. Transactions of China Electrotechnical Society, 2009, 24(10): 94–99.

Beig A R, Narayanan G, Ranganathan V T. Modified SVPWM algorithm for three level VSI With Synchronized and Symmetrical Waveform [J]. IEEE Transactions on Industrial Electronics, 2007, 54(1):486–494.

Ge Xinglai, Feng Xiaoyun. SVPWM Control of Three-Level Inverter Based on FPGA and DSPACE [J]. Power Electronics, 2009, 43(12): 10–11.

Wu Hongyang, He Xiangning. Essential Relationship Between Multilevel Carrier PWM Method and SVPWM Method and Their Applications [J]. Proceedings of the CSEE, 2002, 22(5): 10–15.

Shu Z, Tang J, Guo Y, et al. An Efficient SVPWM Algorithm With Low Computational Overhead for Three-Phase Inverters [J]. IEEE Transactions on Power Electronics, 2007, 22(5): 1797–1805.

Chapter 4
Control and Modulation of Single-Phase Converter in Traction Power Supply

4.1 Brief Introduction to Single-Phase Converter on Traction Network Side

The traction power supply network operates in a single-phase mode, and the converter on the traction side features a single-phase topology. Single-phase conversion circuits are widely utilized in various applications, such as power factor correction (PFC) and single-phase PWM converters (Zheng 2009; Xu 2013). To align with the traction network side, which typically operates at high voltages of up to 25 kV and supports traction load power ranging from several megavolt-amperes to tens of megavolt-amperes, most power converters for traction applications are designed with high-voltage, high-power single-phase conversion circuits. Applicable conversion circuit topologies include the single-phase two-level H-bridge, single-phase three-level H-bridge, and single-phase cascaded multilevel configurations. This chapter examines the aforementioned single-phase AC-DC conversion topologies and their operating principles. It also discusses the modulation and control strategies for typical single-phase converters, analyzes the unique characteristics of DC voltage ripple inherent to single-phase AC-DC topologies, and explores control methods to mitigate the effects of DC voltage ripple.

4.1.1 Two-Level Single-Phase Circuit Topology

The single-module two-level topology is primarily realized through half-bridge and H-bridge circuits. The half-bridge topology employs fewer switching devices, operates at a higher DC voltage, and has a lower DC voltage utilization rate, making it more suitable for low-voltage and low-power applications. In contrast, the H-bridge topology offers more voltage levels, operates at a lower DC voltage, and

Z. Shu, *Power Electronic Converters in Advanced Co-Phase Traction Power Supply System*, https://doi.org/10.1007/978-981-95-3276-6_4

achieves a higher DC voltage utilization rate, making it better suited for high-voltage and high-power applications. This section provides a brief overview of the operating principles of these two topologies (Li and Zhang 2002; Song 2008; Gao and Liu 2006).

4.1.1.1 Single-Phase Half-Bridge Topology

Within the single-phase half-bridge circuit, as shown in Fig. 4.1, two capacitors connected in series on the DC side act as a power bridge arm. The midpoint of the capacitors and the midpoint of the half-bridge together form the AC terminal. In a simple topology, this circuit serves as the foundation for constructing and analyzing cascaded H-bridges and other single-phase AC-DC circuits. The following analysis examines two switch states:

1. Mode 1: S1 is turned on, and S2 is turned off

   In the flow direction from the DC upper capacitor to the AC power grid, as shown in Fig. 4.1a, current flows from the positive terminal of the DC upper capacitor into the single-phase grid through S1 and LC and returns to the negative terminal of the DC upper capacitor. During this process, the converter ab port is connected to the DC capacitor, generating a positive level at the output.

   In the flow direction from the AC power grid to the DC upper capacitor, as shown in Fig. 4.1b, current flows from the power supply to the positive terminal of the DC upper capacitor through LC and the body diode of S1. During this process, the converter ab port remains connected to the DC capacitor, generating a positive level at the output. Since S1 is turned on in this mode, the switching tube S1 actually operates in a synchronous rectification state. Even if S1 is turned off, the body diode of S1 provides a flow path for current from the AC side to the DC upper capacitor. Therefore, when both S1 and S2 are turned off (e.g., during the dead time), the inductor current flows from the AC side (a) through the capacitor to terminal b in this mode.

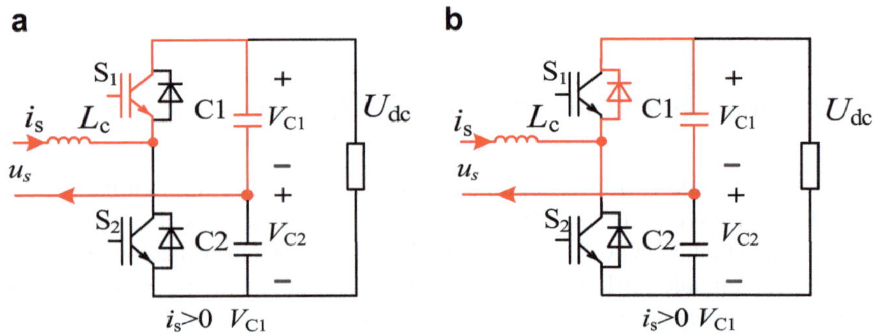

Fig. 4.1 Operating path of single-phase half-bridge circuit generating positive output level. (**a**) Flow direction from the DC upper capacitor to the AC power grid, (**b**) Flow direction from the AC power grid to the DC upper capacitor

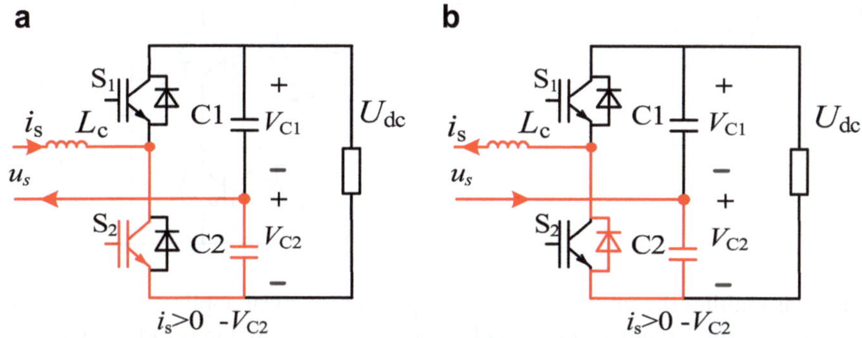

Fig. 4.2 Operating path of single-phase half-bridge circuit generating negative output level. (**a**) Flow direction from the AC end to the lower capacitor, (**b**) Flow direction from the lower capacitor to the AC end

2. Mode 2: S1 is turned off, and S2 is turned on

   In the flow direction from the AC end to the lower capacitor, as shown in Fig. 4.2a, current flows from the AC power supply to the negative terminal of the lower capacitor through LC and S2 and returns to the AC end from the positive terminal of the DC lower capacitor. During this process, the converter ab port is connected to the lower DC capacitor, generating a negative level at the output, which is equal to that of the negative voltage of the lower capacitor.

   In the flow direction from the lower capacitor to the AC end, as shown in Fig. 4.2b, current flows from the negative terminal of the lower capacitor into the AC power supply through the body diode of S2 and LC and returns to the positive terminal of the lower capacitor from the AC end. During this process, the converter ab port on the AC side generates a negative level at the output.

Similar to mode 1, since S2 is turned on in this mode, the switching tube S2 actually operates in a synchronous rectification state. Even if both S1 and S2 are turned off (e.g., during the dead time), the body diode of S2 provides a reverse flow path for current from the negative terminal of the DC lower capacitor to the AC power supply through S2. If the drive signal of S2 is conducting, S2 operates in a synchronous rectification state.

Operating with only half of the switching tubes compared to the full-bridge topology, the half-bridge topology offers advantages such as fewer switching devices and simpler switching states. However, the following disadvantages need to be considered:

1. This topology provides only two types of output voltage levels: positive and negative. The limited number of levels necessitates cascading more modules in high-voltage applications.
2. The DC voltage of the module is twice that of the full-bridge topology, resulting in higher withstand voltage requirements for the capacitors and increased DC voltage ripple on the sub-capacitors.
3. Fewer current flow paths and modes through the DC capacitors make voltage equalization across the capacitors a challenging task.

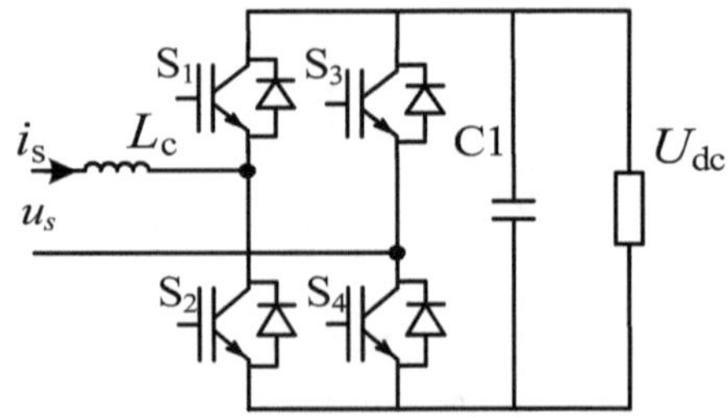

Fig. 4.3 Topology of single-module H-bridge circuit

Modular multilevel converters (MMCs) represent a typical application of the cascaded half-bridge multi-module topology, which is discussed in detail in Chap. 6.

4.1.1.2 Single-Phase Full-Bridge Topology

Figure 4.3 shows a two-level full-bridge circuit. Compared to the half-bridge topology, the full-bridge topology contains more port levels, and the available vectors and modulation modes are increased accordingly. Additionally, the DC voltage of the module is reduced, with a significantly improved utilization rate of DC voltage.

The H-bridge port voltage can be controlled using either unipolar modulation or bipolar modulation. In unipolar modulation, the AC-side voltage (u_s) switches between U_{dc} and 0 or between $-U_{dc}$ and 0. In bipolar modulation, the AC-side voltage (u_s) switches between U_{dc} and $-U_{dc}$. However, significant ripple is observed at the zero-crossing point of the current. Since unipolar modulation provides an additional output level of zero compared to bipolar modulation, it results in lower harmonic content in the grid current (Zheng 2009). Therefore, this section focuses on the unipolar modulation method.

Ignoring the fully open and fully closed conditions, a half-bridge circuit has two possible switching combinations: 10 and 01. Since a full-bridge circuit consists of two half-bridges, there are four possible switching states: 1001, 1010, 0101, and 0110. Taking into account the current flow in both forward and reverse directions, the single H-bridge PWM converter operates in eight distinct modes, corresponding to eight current paths between the AC side and the DC capacitor. These modes are illustrated in Fig. 4.4.

Without considering the influence of DC-side loads on the DC-link capacitor, the changing trend of DC voltage (U_{dc}) depends on different circuit modes. Specifically, when current (i_s) flows into the positive terminal of the DC capacitor, the capacitor voltage (U_{dc}) gradually increases as AC power is gradually stored in the DC capacitor. On the contrary, when current (i_s) flows out of the positive terminal of the DC capacitor, the capacitor voltage (U_{dc}) gradually decreases as energy stored in the capacitor is released to the AC power supply.

Mode 1: The capacitor is charged by i_s (>0) through D1 and D4, and the voltage (U_{dc}) rises.

Mode 2: i_s (>0) flows back into the grid through inductors L, S3, and D4, and C1 supports energy to the load.

Mode 3: L is charged by i_s (>0) through capacitors C1, S2, and S3, and U_{dc} decreases.

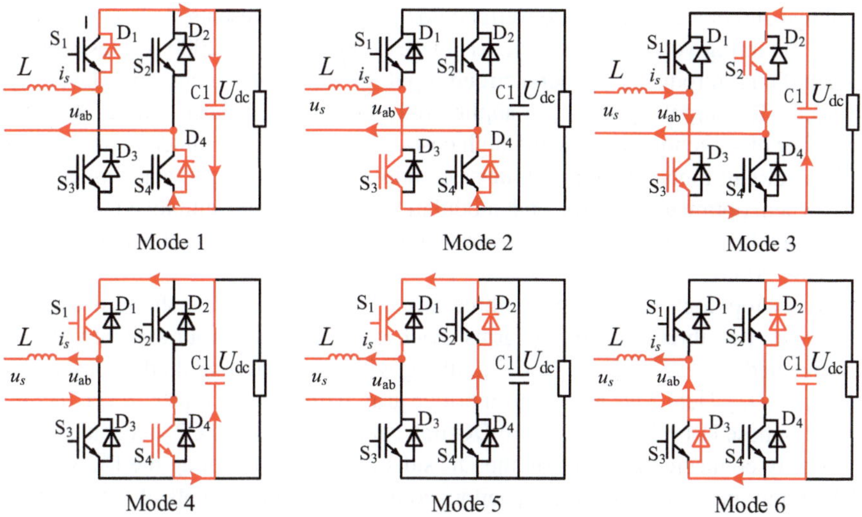

Fig. 4.4 Operating mode of single-phase full-bridge single-module circuit

Table 4.1 Relationships between switching states and output voltage levels

Operating state	S1	S2	S3	S4	us
1	1	0	0	1	U_{dc}
0	1	0	1	0	0
−1	0	1	1	0	$-U_{dc}$
0	0	1	0	1	0

Mode 4: L is charged by i_s (<0) through capacitors C1, S1, and S4, and U_{dc} decreases.
Mode 5: i_s (<0) flows back into the AC grid through D2 and S1, and capacitor C1 discharges to loads.
Mode 6: Capacitor C is charged by i_s (<0) through D2 and D3, and U_{dc} rises.

With the switching states of closing and opening defined as 1 and 0, respectively, the relationships of the operating states and the switching states with the output voltage levels are summarized in Table 4.1.

Taking operating state 1 as an example, when switching tubes S1 and S4 are turned on while S2 and S3 are turned off, the port voltage level (U_{ab}) equals the DC voltage (U_{dc}). In other operating states, similar relationships can be observed among the three voltage levels at the port: U_{dc}, $-U_{dc}$, and 0.

4.1.2 *Single-Phase Three-Level Topology*

Diode-clamped three-level H-bridge converters are widely used in multilevel conversion circuits. This topology was first proposed by Japanese scholars Akira Nabae, Isao Takahashi, and Hirofumi Akagi in 1981. It significantly enhances the voltage and power levels of converters; however, it also presents challenges related to maintaining voltage balance across the DC capacitors. In 2000, Professor Fang ZhengPeng introduced a general circuit topology for self-balancing multilevel converters. This topology has been applied only in limited instances due to the large number of switching devices and high costs. Nevertheless, the diode-clamped multilevel circuit topology demonstrates a certain degree of voltage-balancing capability when optimized control and modulation strategies are employed. Thus, the analysis in this section focuses on single-phase three-level H-bridge circuits (Zhang 2003).

In the topology of single-phase three-level H-bridge circuits, as shown in Fig. 4.5, there are two three-level bridge arms. Each arm consists of four switching tubes, designated as *Sm1*, S_{m2}, S_{m3}, and S_{m4}, along with two clamping diodes, designated D_{m1} and D_{m2}. Here, "*m*" represents a bridge arm, with *m* being either "*a*" or "*b*" in a single-phase H-bridge circuit. Two DC-link capacitors, C1 and C2, of equal size are typically connected in series on the DC side, effectively doubling the DC voltage. The midpoint O of these two capacitors connects to the midpoints of the clamping diodes D_{m1} and D_{m2} in the two bridge arms. These diodes facilitate additional paths for current flow and ensure that the voltage across the bridge arm switches does not exceed the DC capacitor voltage. If the voltage across a bridge arm switch surpasses the DC capacitor voltage—such as when the device voltage exceeds V_{dc1} after S_{a1} is turned off—diode D1 provides an electric path to mitigate the voltage difference between S_{a1} and C1 until the voltage of the switching tube aligns with the DC capacitor voltage. This functionality is what gives this topology its name: diode-clamped multilevel circuits.

Given that the upper and lower capacitors are equal in size, that is, C1 = C2 = C, and assuming that the voltages of V_{C1} and V_{C2} are equal, that is, $V_{C1} = V_{C2} = E$, and if a single bridge arm can be connected to the positive terminal of the upper capacitor, the midpoint, and the negative terminal of the lower capacitor, the AC-side port voltage (u_{ab}) of the converter will exhibit five different levels: −2*E*, −*E*, 0, *E*, and

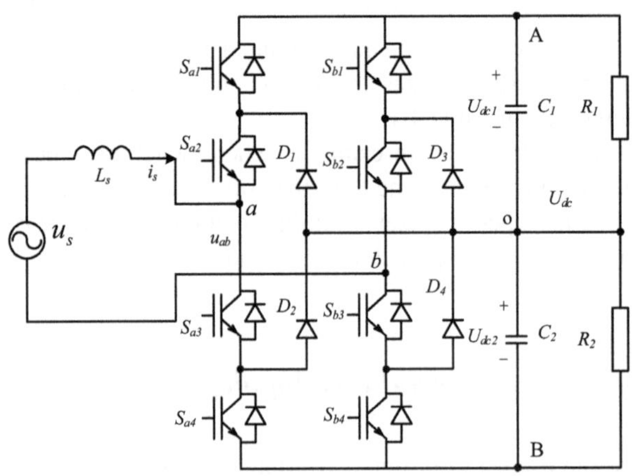

Fig. 4.5 Single-phase three-level H-bridge converter

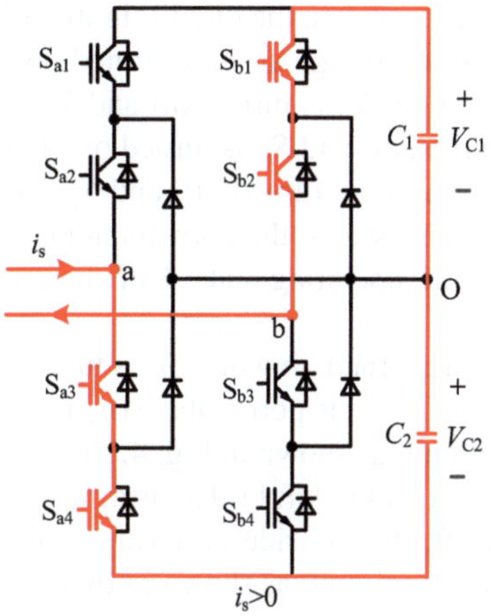

Fig. 4.6 Circuit operating state 1 (S_{a1} and S_{a2} off, S_{b1} and S_{b2} on)

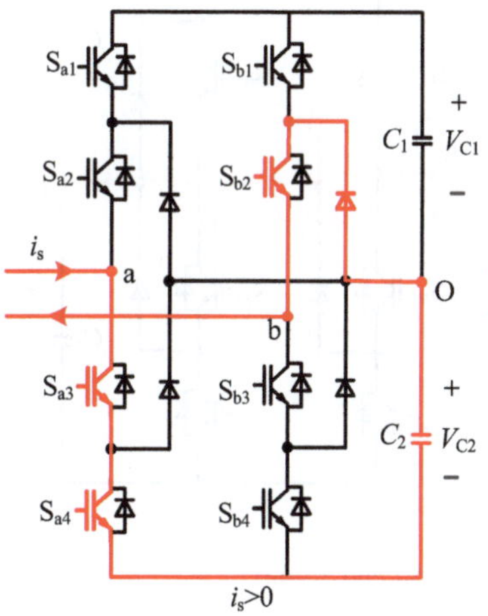

Fig. 4.7 Circuit operating state 2 (S_{a1} and S_{a2} off, S_{b1} off, and S_{b2} on)

$2E$. The relationships between the on-off state of its switching tube and the corresponding u_{ab} are described as follows:

1. When S_{a1} and S_{a2} on the left bridge arm are turned off and S_{b1} and S_{b2} on the right bridge arm are turned on, the AC-side port voltage (u_{ab}) is $-2E$, which represents the operating state 1 of the circuit, as shown in Fig. 4.6.
2. When S_{a1} and S_{a2} on the left bridge arm are turned off while S_{b1} is turned off and S_{b2} is turned on on the right bridge arm, the AC-side port voltage (u_{ab}) is $-E$, which represents the operating state 2 of the circuit, as shown in Fig. 4.7.
3. When S_{a1} is turned off and S_{a2} is turned on on the left bridge arm, while S_{b1} and S_{b2} on the right bridge arm are turned on, the AC-side port voltage (u_{ab}) is $-E$, which represents the operating state 3 of the circuit, as shown in Fig. 4.8.

4. There is no circuit formed with the DC capacitor in the circuit, in the following three switching states: S_{a1} and S_{a2} on the left bridge arm are turned on, and S_{b1} and S_{b2} on the right bridge arm are turned on; S_{a1} is turned off and S_{a2} is turned on on the left bridge arm, while S_{b1} is turned off and S_{b2} is turned on on the right bridge arm; S_{a1} and S_{a2} on the left bridge arm are turned off, and S_{b1} and S_{b2} on the right bridge arm are turned off. In these states, the corresponding AC-side port voltage (u_{ab}) is zero, representing the operating state 4 of the circuit, as shown in Fig. 4.9.
5. When S_{a1} and S_{a2} on the left bridge arm are turned on while S_{b1} is turned off and S_{b2} is turned on on the right bridge arm, the AC-side port voltage (u_{ab}) is E, which represents the operating state 5 of the circuit, as shown in Fig. 4.10.
6. When S_{a1} is turned off and S_{a2} is turned on on the left bridge arm, while S_{b1} and S_{b2} on the right bridge arm are turned off, the AC-side port voltage (u_{ab}) is E, which represents the operating state 6 of the circuit, as shown in Fig. 4.11.

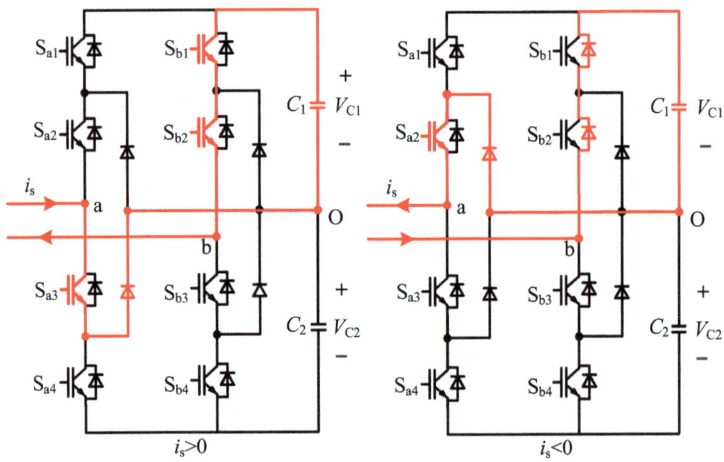

Fig. 4.8 Circuit operating state 3 (S_{a1} off, S_{a2} on, and S_{b1} and S_{b2} on)

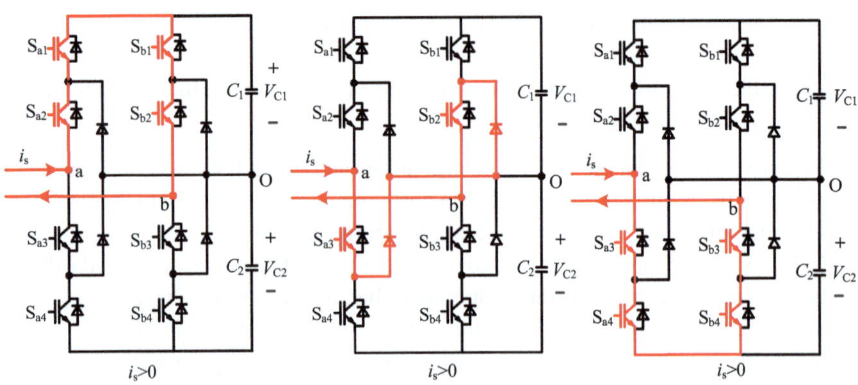

Fig. 4.9 Circuit operating state 4 (DC capacitor without charging and discharging circuits)

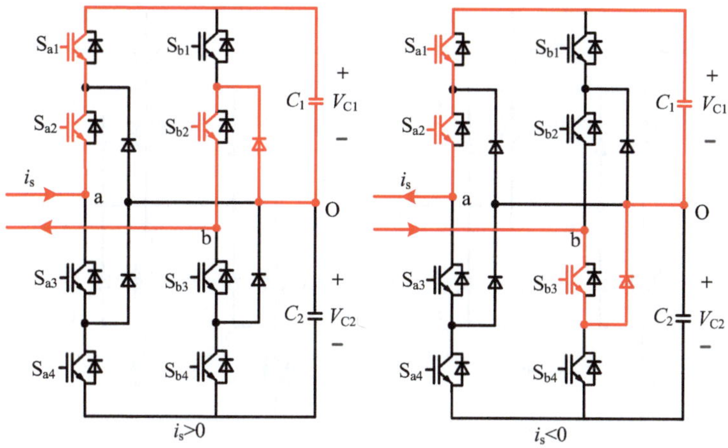

Fig. 4.10 Circuit operating state 5 (S_{a1} and S_{a2} on, S_{b1} off, and S_{b2} on)

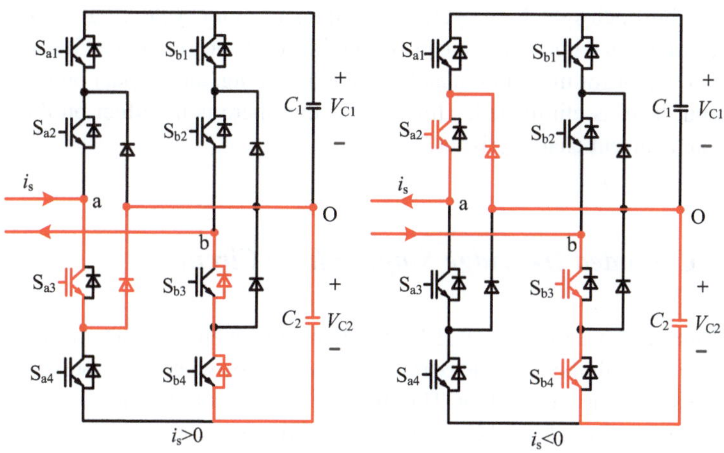

Fig. 4.11 Circuit operating state 6 (S_{a1} off, S_{a2} on, and S_{b1} and S_{b2} off)

7. When S_{a1} and S_{a2} on the left bridge arm are turned on, and S_{b1} and S_{b2} on the right bridge arm are turned off, the AC-side port voltage (u_{ab}) is $2E$, which represents the operating state 7 of the circuit, as shown in Fig. 4.12.

Controllers can be designed to select among these switching states, i.e., redundant switching vectors, to enable modulation voltage equalization while meeting the AC port voltage requirements. The modulation voltage equalization principle for this selection process is as follows. When the capacitor voltage exceeds the voltage reference, the redundant switching vector for DC capacitor discharging is selected; on the contrary, when the capacitor voltage falls below the voltage reference, the redundant switching vector for DC charging is selected. Since the DC capacitor voltage cannot change suddenly, the amplitude of the capacitor voltage rise or fall is

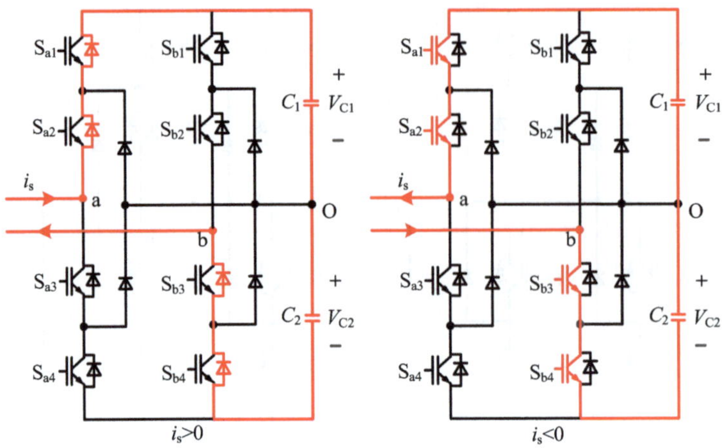

Fig. 4.12 Circuit operating state 7 (S_{a1} and S_{a2} on, S_{b1} and S_{b2} off)

limited within each switching cycle. When the switching cycle is sufficiently short, the measured capacitor voltage can be stabilized at the reference. By selecting a redundant vector for modulation and voltage balancing, any impact on the modulation output level is eliminated. This ensures that there is no effect on the AC port voltage and current characteristics.

4.1.3 Cascaded H-Bridge Single-Phase Circuit

Several single-phase three-level H-bridge converters mentioned above are cascaded (number = n), with the AC-side ports labeled a_i and b_i, where i denotes the number of modules, ranging from 1 to n. The AC-side port (a_1) of module 1 is connected to the positive terminal of the AC power supply through a connecting reactor, and the AC-side port (b_1) of module 1 is connected to the AC-side port (a_2) of module 2, and this sequence continues until the AC-side port (b_{n-1}) of module $n-1$ is connected to the AC-side port (a_n) of module n. Finally, the AC port point (b_n) of module n is connected to the negative terminal of the AC power supply. This configuration results in a single-phase three-level H-bridge cascaded converter (n-module), as shown in Fig. 4.13.

The symbols are defined as follows:

R_s, L_s: the internal resistance and inductance of the connecting reactor on the AC side; $Ci1$, $Ci2$: the upper and lower DC-link capacitors in the ith module; L, C: the inductor and capacitor in the secondary filter circuit; R_i: the DC-side load of the ith module; u_s: the voltage of the AC-side power supply; i_s: the AC-side current; u_{aibi}: the AC-side port voltage of the ith module; u_{a1bn}: the total AC-side port voltage of n modules cascaded; V_{Ci1}, V_{Ci2}: the voltages of the upper and lower DC-link capacitors in the ith module.

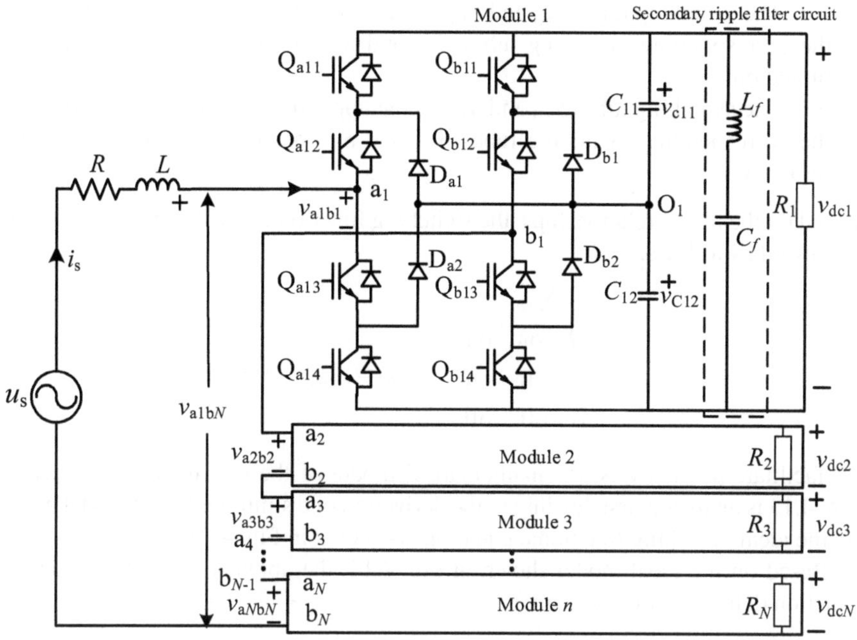

Fig. 4.13 Single-phase three-level H-bridge cascaded converter (n-module)

V_{dci}: the DC-side voltage of the ith module. For an n-module cascaded converter, its total port voltage (u_{a1bn}) is the sum of the port voltages of the n modules, which is expressed as follows:

$$u_{a1bn} = u_{a1b1} + u_{a2b2} + \cdots + u_{aibi} + u_{anbn} \tag{4.1}$$

This topology allows for the superposition of voltages in series, resulting in a high-voltage output when low-withstand-voltage switching devices are employed. Consequently, this configuration is commonly utilized in medium- to high-voltage and high-power applications. In an n-module cascaded converter, if the DC-side capacitor voltage of each module is equal, expressed as $V_{Ci1} = V_{Ci2} = E$, the number of levels in the total port voltage (u_{a1bn}) varies from $(-2n)E$ to $(2n)E$, that is, a total of $4n + 1$ levels.

To establish a mathematical model of single-phase cascaded converters, the first step is to simplify the converter model, by assuming that all switching devices of the converter are ideal and defining the switching function for the bridge arm in each module as shown in Fig. 2.9 as S_{mi}. For example, the switching function for the left bridge arm in module 1 is defined as S_{a1}, while that for the right bridge arm as S_{b1}. Based on the on-off state of the four switching tubes on each bridge arm, S_{mi} is defined in three states, i.e., 0, 1, and 2, with specific conditions detailed as follows:

$S_{mi} = 2$: The switching tubes S_{mi1} and S_{mi2} on the upper bridge arm are turned on, and the corresponding switching tubes S_{mi3} and S_{mi4} on the lower bridge arm are turned off.

$S_{mi} = 1$: The switching tubes S_{mi1} and S_{mi2} on the upper bridge arm are turned off, and the corresponding switching tubes S_{mi3} and S_{mi4} on the lower bridge arm are turned on.

$S_{mi} = 0$: The switching tubes S_{mi1} and S_{mi2} on the upper bridge arm are turned off, and the corresponding switching tubes S_{mi3} and S_{mi4} on the lower bridge arm are turned on.

Based on the above relationships, the switching function for each bridge arm can be expressed as follows:

$$S_{mi} = \begin{cases} & S_{mi1} & S_{mi2} \\ 2 & \text{on} & \text{on} \\ 1 & \text{off} & \text{on}, \\ 0 & \text{off} & \text{off} \end{cases} \qquad m = a, b; i = 1, 2 \ldots n \tag{4.2}$$

To facilitate the analysis, the mathematical model of the unit module depicted in Fig. 4.13 is analyzed first, leading to the derivation of a simplified ideal model based on the ideal switching functional equation, as shown in Fig. 4.14.

Based on the ideal model shown in Fig. 4.14, the application of the switching function defined in Eq. (4.2) gives

$$u_{ao} = \frac{(S_a - 1)S_a}{2} v_{C1} - \frac{(S_a - 1)(S_a - 2)}{2} v_{C2} \tag{4.3}$$

$$u_{bo} = \frac{(S_b - 1)S_b}{2} v_{C1} - \frac{(S_b - 1)(S_b - 2)}{2} v_{C2} \tag{4.4}$$

Based on Eqs. (4.3) and (4.4), the derivation gives

$$u_{ab} = u_{ao} - u_{bo} = \frac{S_a - S_b}{2}(v_{C1} + v_{C2}) + \frac{(S_a - 1)^2 - (S_b - 1)^2}{2}(v_{C1} - v_{C2}) \tag{4.5}$$

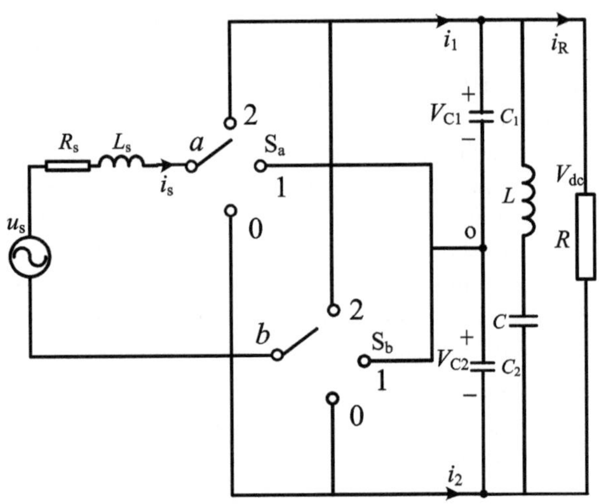

Fig. 4.14 Simplified model of single-phase three-level H-bridge converter

It is assumed that there is no energy loss in the converter in the analysis leading to the ideal model; therefore, the instantaneous power can be expressed as follows:

$$u_{ab}i_{s} = v_{C1}i_{1} - v_{C2}i_{2} \tag{4.6}$$

Substituting Eq. (4.5) into Eq. (4.6) results in

$$i_{1} = \left[\frac{(S_{a}-1)S_{a}}{2} - \frac{(S_{b}-1)S_{b}}{2}\right]\cdot i_{s} \tag{4.7}$$

$$i_{2} = \left[\frac{(S_{a}-1)(S_{a}-2)}{2} - \frac{(S_{b}-1)(S_{b}-2)}{2}\right]\cdot i_{s} \tag{4.8}$$

The load-side current shown in Fig. 4.14 can be expressed as follows:

$$i_{R} = \frac{v_{C1}+v_{C2}}{R} \tag{4.9}$$

Applying Kirchhoff's voltage laws (KVL) and Kirchhoff's current laws (KCL) to the ideal model shown in Fig. 4.14 gives

$$\begin{cases} u_{s} = L_{s}\dfrac{\mathrm{d}i_{s}}{dt} + R_{s}i_{s} + u_{ab} \\ C_{1}\dfrac{\mathrm{d}v_{\mathrm{dc1}}}{\mathrm{d}t} = i_{1} - i_{R} \\ C_{2}\dfrac{\mathrm{d}v_{\mathrm{dc2}}}{\mathrm{d}t} = -i_{2} - i_{R} \end{cases} \tag{4.10}$$

Substituting Eqs. (4.5), (4.7), (4.8), and (4.9) into (4.10) results in

$$\begin{cases} \dfrac{\mathrm{d}i_{s}}{\mathrm{d}t} = \dfrac{u_{s}}{L_{s}} - \dfrac{R_{s}}{L_{s}}i_{s} - \dfrac{(S_{a}-1)S_{a}-(S_{b}-1)S_{b}}{2L_{s}}v_{C1} \\ \qquad - \dfrac{(S_{a}-1)(S_{a}-2)-(S_{b}-1)(S_{b}-2)}{2L_{s}}v_{C2} \\ \dfrac{\mathrm{d}v_{C1}}{\mathrm{d}t} = \dfrac{(S_{a}-1)S_{a}-(S_{b}-1)S_{b}}{2C_{1}}\cdot i_{s} - \dfrac{v_{C1}+v_{C2}}{RC_{1}} \\ \dfrac{\mathrm{d}v_{C2}}{\mathrm{d}t} = \dfrac{(S_{a}-1)(S_{a}-2)-(S_{b}-1)(S_{b}-2)}{2C_{2}}\cdot i_{s} - \dfrac{v_{C1}+v_{C2}}{RC_{2}} \end{cases} \tag{4.11}$$

Based on Eq. (4.11), derivation gives the state equation of the simplified model as follows:

$$\dot{X} = AX + Bu \tag{4.12}$$

where

$$X=\begin{bmatrix} i_s \\ v_{C1} \\ v_{C2} \end{bmatrix}, B=\begin{bmatrix} 1/L_s \\ 0 \\ 0 \end{bmatrix}, u=u_s,$$

$$A=\begin{bmatrix} -\frac{R_s}{L_s} & & \frac{(S_a-1)S_a-(S_b-1)S_b}{2L_s} \\ & & \frac{(S_a-1)(S_a-2)-(S_b-1)(S_b-2)}{2L_s} \\ \frac{(S_a-1)S_a-(S_b-1)S_b}{2C_1} & -\frac{1}{RC_1} & -\frac{1}{RC_1} \\ \frac{(S_a-1)(S_a-2)-(S_b-1)(S_b-2)}{2C_2} & -\frac{1}{RC_2} & -\frac{1}{RC_2} \end{bmatrix}$$

The mathematical equation for single-phase three-level H-bridge cascaded converters can be expressed using KVL, following a derivation process similar to that for single-phase H-bridge converters:

$$\begin{cases} u_s = L_s \frac{di_s}{dt} + R_s i_s + u_{a1bn} \\ u_{a1bn} = u_1 + u_2 + \ldots + u_n = \sum_{i=1}^{n} u_i \\ u_i = (S_{ai} - S_{bi}) \cdot \frac{V_{dci}}{2} \end{cases} \tag{4.13}$$

The equivalent circuit of these cascaded converters can be derived from Eq. (4.13), as shown in Fig. 4.15. The port level (u_{a1bn}) is generated by the input voltage source (u_s) through the grid-side inductor (L_s) and parasitic resistor (R_s). Consequently, the difference between the total output port voltage (u_{a1bn}) of the cascaded converters and the power supply voltage (u_s) influences the line reactance, and this voltage difference determines the magnitude of the converter current (i_s).

Therefore, assuming that us is stable, by controlling the amplitude and phase of the total port voltage (u_{a1bn}) of the cascaded converters, both the amplitude and

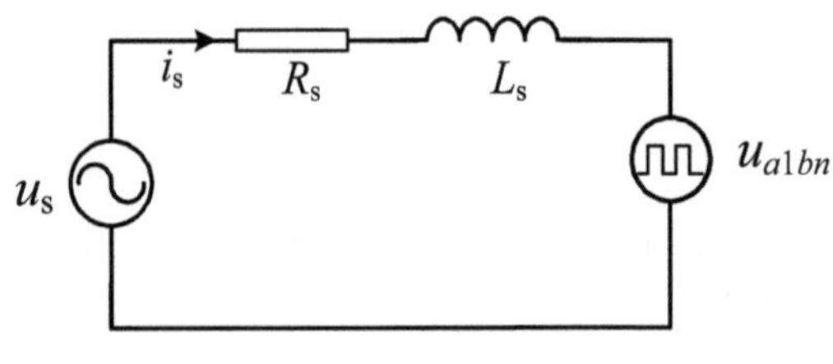

Fig. 4.15 Equivalent circuit of three-phase three-level H-bridge cascaded converter

phase of the converter current (i_s) can be controlled. This control enables controlling the active current, reactive current, and harmonic current of the cascaded converters, thus allowing for controlling the active power, reactive power, and harmonic compensation functions of the converters, which correspond to PWM converters, reactive compensators, or active filters.

4.2 Single-Phase PWM Rectifier Converter and Control Method

4.2.1 Single-Phase DQ *Decoupling Control*

Single-phase *dq* decoupling control offers a method for detecting the *d*-axis and *q*-axis components of the current (i_s) in a single-phase *dq* coordinate system, allowing for closed-loop control of each component separately. In the *dq* coordinate system, the *d*-axis component corresponds to the active component of the AC current (i_s), while the *q*-axis component corresponds to the reactive component of i_s. Therefore, this method is also referred to as active/reactive decoupling control, enabling the independent and stable control of the active power and reactive power of the converters (Peng et al. 2007; Song and Feng 2012; Chen et al. 2007).

First, the AC current (i_s) is tested to distinguish between active and reactive power. The AC current $i_s(t)$ can be expressed as the sum of its active, reactive, and harmonic components:

$$\begin{cases} i_s(t) = i_d(t) + i_q(t) + i_h(t) \\ i_d(t) = I_d \sin \omega t \\ i_q(t) = I_q \cos \omega t \\ i_h(t) = \sum_{i=2}^{\infty} I_i \sin(i\omega t + \phi_i) \end{cases} \tag{4.14}$$

where ω, $\sin\omega t$, and $\cos\omega t$ represent the angular velocity of the AC voltage (u_s) and the sine and cosine values of the current voltage detected by the single-phase phase-locked loop (PLL); I_d and I_q are the *d*-axis and *q*-axis components of the AC current, respectively, representing the amplitudes of the active and reactive currents; I_i denotes the amplitude of the *i*th harmonic. The detection method is divided into three steps as follows:

1. Active current detection:

   First, multiplying the AC current $i_s(t)$ component by the sine value of the voltage ($\sin\omega t$) detected by the single-phase PLL results in

$$P(t) = i_s(t) \sin \omega t \tag{4.15}$$

Substituting Eq. (4.14) into Eq. (4.15) yields

$$P(t)=\frac{I_d}{2}-\frac{I_d}{2}\cos 2\omega t+\frac{I_q}{2}\sin 2\omega t+ \\ \sum_{i=2}^{\infty}\frac{I_i}{2}\left\{\left[\cos\left(i-1\right)\omega t+\phi_i\right]-\left[\cos\left(i+1\right)\omega t+\phi_i\right]\right\} \tag{4.16}$$

As indicated in Eq. (4.16), except $I_d/2$ being a DC component, all the other components are AC components with frequencies equal to or greater than twice the fundamental frequency. Therefore, when $P(t)$ passes through a low-pass filter (LPF), the AC components are filtered out, leaving the DC component $I_d/2$. Since the frequency of the AC component before filtering is at least twice the fundamental frequency, the cutoff frequency of the LPF should be below 100 Hz. Considering the dynamic response characteristics of the LPF and the boundary effect of the cutoff frequency, it is advisable to select a cutoff frequency ranging from 40 to 60 Hz for the LPF. Then, the filtering result is multiplied by 2 to yield the active component (I_d) of is. Finally, the real-time value ($i_d(t) = I_d \sin \omega t$) of the active current is yielded by using the PLL output.

2. Reactive current detection:

   Similar to the active current detection, after deducting the active real-time value component ($i_d(t)$) in the AC current $i_s(t)$, multiplying by the cosine value of the voltage ($\cos\omega t$) detected by single-phase PLL results in

$$Q(t)=\left[i_s(t)-i_d(t)\right]\cos\omega t \tag{4.17}$$

Substituting Eq. (4.15) into Eq. (4.17) yields

$$Q(t)=\frac{I_q}{2}+\frac{I_q}{2}\cos 2\omega t+\frac{I_d}{2}\sin 2\omega t+ \\ \sum_{i=2}^{\infty}\frac{I_i}{2}\left\{\left[\cos\left(i-1\right)\omega t+\phi_i\right]-\left[\cos\left(i+1\right)\omega t+\phi_i\right]\right\} \tag{4.18}$$

In Eq. (4.18), except $I_q/2$ being a DC component, all the other components are AC components with frequencies equal to or greater than twice the fundamental frequency. Therefore, after the low-pass filtering of $Q(t)$, all AC components are filtered out, leaving the DC component $I_q/2$. Then, the filtering result is multiplied by 2 to yield the active component (I_q) of is. Finally, according to the definition of reactive current and using the PLL cosine output, the real-time value of reactive current is yielded.

3. Harmonic current detection:

   Since harmonic currents typically contain components at multiple frequencies, only their sum is detected. Based on the real-time values of active and reactive currents ($i_d(t)$, $i_q(t)$) yielded from the abovementioned two steps, using Eq. (4.14) that defines the AC currents $i_s(t)$, the real-time value of harmonic currents can be obtained, i.e., $i_h(t) = i_s(t) - i_d(t) - i_q(t)$.

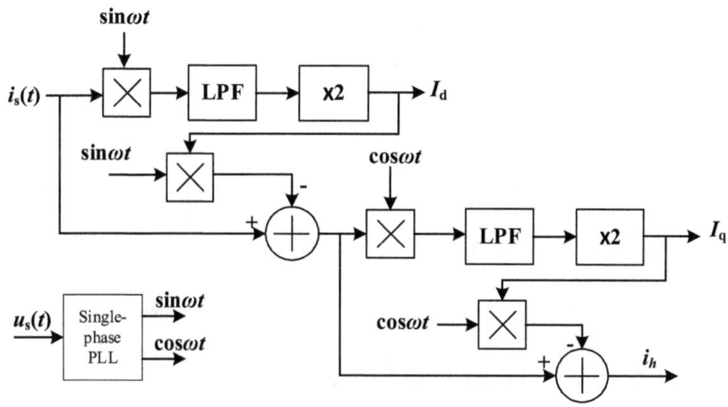

Fig. 4.16 Block diagram for implementation of active and reactive current detection

To sum up, the process of single-phase active, reactive, and harmonic detection is shown in Fig. 4.16.

The AC-side voltage (u_s) and current (i_s), as well as the AC port voltage (u_{ab}), can also be expressed by *d*-axis and *q*-axis components, respectively, as follows:

$$u_s(t) = U_s \sin \omega t \tag{4.19}$$

$$i_s(t) = I_d \sin \omega t + I_q \cos \omega t \tag{4.20}$$

$$u_{ab}(t) = U_d \sin \omega t + U_q \cos \omega t \tag{4.21}$$

where U_d and U_q represent the *d*-axis and *q*-axis components of the AC port voltage (u_s), respectively. Substituting Eqs. (4.19)–(4.21) into Eq. (4.10) results in

$$\begin{cases} U_d = U_s - (R_s + pL_s) I_d + \omega L_s I_q \\ U_q = -(R_s + pL_s) I_q - \omega L_s I_d \end{cases} \tag{4.22}$$

where p is the differential operator, taken as $p = \mathrm{d}/\mathrm{d}t$.

According to Eq. (4.22), the single-phase d_q decoupling control can be implemented as shown in Fig. 4.17, where V_{dc}* denotes the DC voltage reference, and I_d* and I_q* indicate the active and reactive power references.

In AC-DC converter applications, the consumption of active power by DC loads causes drops in DC voltage. To maintain DC voltage stability, a PI controller is designed for alignment with the DC voltage reference (V_{dc}*) as shown in Fig. 4.17, with its output component serving as an active reference (I_d*). Converters generally operate at a unity power factor; therefore, the controller needs to ensure that the reactive power remains at zero, which is achieved by setting I_q* = 0. In applications involving reactive power compensation, it is necessary to detect the reactive component of load currents and use it as a reactive reference (I_q*) to achieve dynamic reactive power compensation.

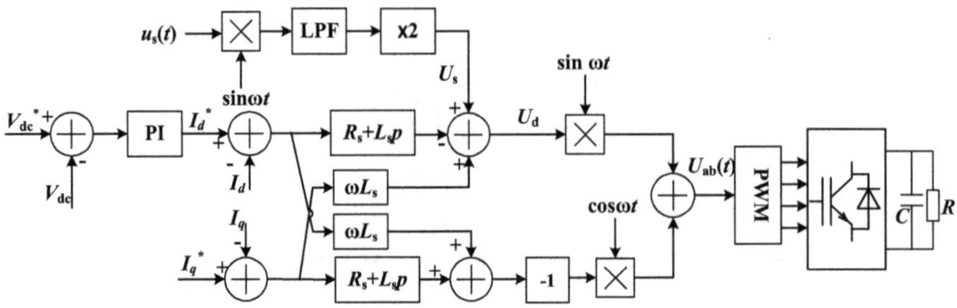

Fig. 4.17 Block diagram for implementation of single-phase *dq* decoupling control

4.2.2 *Predictive Current Control*

Due to current sampling delays, analog-to-digital conversion delays, and controller calculation delays within the control loop of PWM converters, the controllers exhibit response delays. They require at least one PWM switching cycle to react to the current operating state of the converters. Furthermore, a specific duration is necessary from the initiation of a switching action to the converter's response in generating the expected output. As a result, converter response delays are typically longer than one switching cycle, which significantly impacts applications where the rate of change in current is high, such as harmonic compensation, or low switching frequency applications (He et al. 2012; Borghetti et al. 2008).

Predictive current control helps mitigate the impact of these delays, by predicting at the current moment (t_k) that the current $i_s(t_k + T_s)$ at the next switching cycle moment ($t_k + T_s$) is equal to the current $i_s*(t_k)$ at the current moment. This approach helps ensure that the actual current $i_s(t_k + T_s)$ aligns with the current reference $i_s*(t_k)$ within a switching cycle, thereby achieving the control purpose of fast response. Therefore, within one switching cycle (t_k, $t_k + T_s$) of the predictive current control, the following equation is satisfied:

$$i_s(t_k + T_s) = i_s^*(t_k) \tag{4.23}$$

In a switching cycle (T_s), the following equation is also satisfied:

$$L_s \frac{di_s}{dt} = L_s \frac{i_s(t_k + T_s) - i_s(t_k)}{T_s} = L_s \frac{i_s^*(t_k) - i_s(t_k)}{T_s} \tag{4.24}$$

Substituting Eq. (4.24) into Eq. (4.10) yields

$$u_{ab} = u_s - R_s i_s - L_s \frac{i_s^*(t_k) - i_s(t_k)}{T_s} \tag{4.25}$$

Since the internal resistance of the connecting reactance on the grid side of rectifier converters is generally low, a simplified expression for predictive current control can be derived by neglecting R_s as follows:

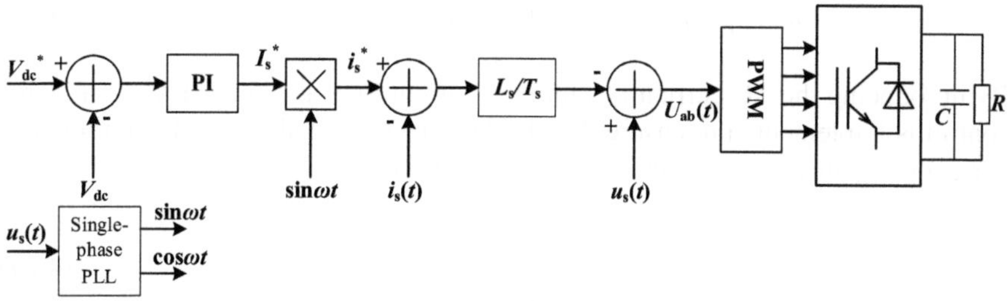

Fig. 4.18 Block diagram for implementation of predictive current control

$$\begin{cases} I_s^* = K_P\left(V_{dc}^* - V_{dc}\right) + \dfrac{1}{T_i}\int\left(V_{dc}^* - V_{dc}\right)dt \\ u_{ab}(t) = u_s(t) - \dfrac{L_s}{T_s}\left[I_s^*\sin\omega t - i_s(t)\right] \end{cases} \tag{4.26}$$

The predictive current control, as expressed in Eq. (4.26), is shown in Fig. 4.18.

In this control system, K_p and $1/T_i$ represent the gains of the proportional controller and integral controller in the PI controller, respectively; V_{dc}* is the DC-side voltage reference of the converter, while V_{dc} is the actual voltage on the DC side; I_s*sinωt is the grid-side current reference, with ω being the angular frequency of the grid-side voltage and $i_s(t)$ being the grid-side current. It is evident that the calculation of the voltage reference based on the current loop only involves a proportional relationship, indicating that a proportional controller is utilized in this component. Although this approach may result in static errors in control results, it offers fast response times and good dynamic performance, making it well suited for applications characterized by rapid changes in current.

4.2.3 Transient Current Control

Transient current controllers, also known as voltage and current double closed-loop control, consist of two loops: one for voltage and the other for current. In single-phase H-bridge converters, this control strategy ensures that the AC-side current $i_s(t)$ and voltage remain aligned in phase to operate at a unity power factor. This process can be expressed as follows:

$$\begin{cases} u_s(t) = U_s\sin\omega t \\ i_s(t) = I_s\sin\omega t \end{cases} \tag{4.27}$$

where U_s and I_s are AC voltage and current amplitudes, and ω is the angular frequency of the AC voltage (u_s), which can be obtained by single-phase PLL. Substituting Eq. (4.27) into Eq. (4.10) yields

$$u_{ab}(t) = u_s(t) - (R_s + L_s p) I_s \sin \omega t - \omega L_s I_s \cos \omega t \tag{4.28}$$

By introducing the AC current $i_s(t)$ into the current loop as a feedback signal, error control is implemented to ensure that the AC current $i_s(t)$ aligns with the reference current ($I^*s \sin\omega t$), which can be expressed as follows:

$$\begin{cases} I_s = I_s^* \\ pI_s = \left(I_s^* - I_s\right) / \Delta t \end{cases} \tag{4.29}$$

where Δt indicates the control period. By neglecting the internal resistance (R_s) of the connecting reactance, substituting Eq. (4.29) into Eq. (4.28) yields

$$u_{ab}(t) = u_s(t) - \frac{L_s}{\Delta t}\left[I_s^* \sin \omega t - i_s(t)\right] - \omega L_s I_s^* \cos \omega t \tag{4.30}$$

The reference current magnitude, denoted as I_s^*, is obtained from the voltage control loop and can be expressed as follows:

$$I_s^* = K_P\left(V_{dc}^* - V_{dc}\right) + \frac{1}{T_i}\int\left(V_{dc}^* - V_{dc}\right)dt \tag{4.31}$$

In Eq. (4.31), V_{dc} is the actual DC-side voltage that is used as a feedback signal to align with the reference voltage (V_{dc}^*). Combining Eqs. (4.30) and (4.31) results in the mathematical expression of transient current control as follows:

$$\begin{cases} I_s^* = K_P\left(V_{dc}^* - V_{dc}\right) + \frac{1}{T_i}\int\left(V_{dc}^* - V_{dc}\right)dt \\ u_{ab}(t) = u_s(t) - K\left[I_s^* \sin \omega t - i_s(t)\right] - \omega L_s I_s^* \cos \omega t \end{cases} \tag{4.32}$$

where $K = L_s/\Delta t$. Transient current control is shown in Fig. 4.19:

As indicated, the controller consists of two control loops for voltage and current, respectively. In the voltage loop control, the actual DC voltage (V_{dc}) is used as a feedback signal, and its error (ΔV_{dc}) with the DC voltage reference (V^*_{dc}) is used as the amplitude reference (I_s^*) of the AC current after passing through the PI controller. I_s^* is multiplied by the sine value ($\sin\omega t$) synchronized with the AC voltage output by the PLL to yield the real-time reference ($I_s^*\sin\omega t$) of the AC current. In the current loop control, the actual AC current (i_s) is used as a feedback signal and error control is implemented with the abovementioned real-time reference ($I_s^*\sin\omega t$) of the AC current to align with the reference value. Then, $K[I_s^*\sin\omega t - i_s(t)]$ is obtained through the proportional controller. Finally, the modulation wave signal (u_{ab}) is obtained by subtracting $K[I_s^*\sin\omega t - i_s(t)]$ and $I_s^*\cos\omega t$ from the AC voltage (u_s). As indicated in Fig. 4.19, there is only one PI controller in this control strategy, offering a straightforward solution. However, in the absence of a reactive component control loop, this controller can only operate at a unity power factor, resulting in an unadjustable power factor.

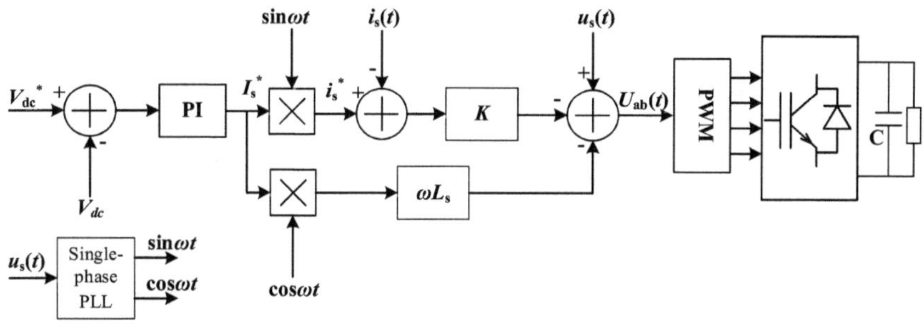

Fig. 4.19 Block diagram for implementation of transient current control

4.2.4 Simulation of Active and Reactive Decoupling Control System for Single-Phase Converter

Based on the single-phase three-level topology shown in Fig. 4.5 and the single-phase *dq* decoupling control strategy shown in Fig. 4.17, a closed-loop simulation model of the main circuit and control system is established in the MATLAB/Simulink simulation environment, with the input AC voltage of 380 V and the output DC target voltage of 700 V.

Figure 4.20 illustrates the simulation waveform of the DC-side voltage of a single-phase converter operating under the single-phase *dq* decoupling control. During the 5-s simulation, multiple sudden load changes occur. The no-load startup stage lasts from $t = 0$ to $t = 2$ s, followed by a resistive load stage from $t = 2$ to $t = 3$ s, then a resistive load stage with capacitive reactive power output from $t = 3$ to $t = 4$ s, and finally a resistive load stage with inductive reactive power output from $t = 4$ to $t = 5$ s.

As indicated, during the no-load startup stage, an overshoot in voltage occurs, which is linked to the capacitance of the main circuit and the parameters of the controller. In this simulation, after approximately 1.5 s, the DC-side voltage of the system stabilizes at 700 V.

Following a sudden change in the resistive load, due to the converter's response delay, the load energy is directly supplied by the DC capacitor, resulting in a small amplitude drop in the DC voltage. The system stabilizes at 700 V again after approximately 0.5 s, without displaying a significant overshoot, which indicates the rapid response of the system.

During the capacitive output and inductive reactive stages, the DC-side voltage of the system shows no significant change. The system operating under the single-phase *dq* decoupling control exhibits a fast response speed and minimal overshoot in all the aforementioned conditions. This performance ensures that the DC-side voltage remains stable at the reference level, providing a consistent and reliable DC voltage for DC loads.

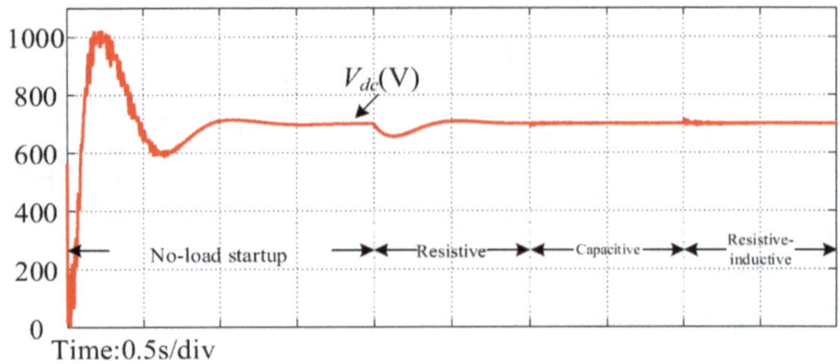

Fig. 4.20 DC-side voltage waveform of single-phase *dq* decoupling control

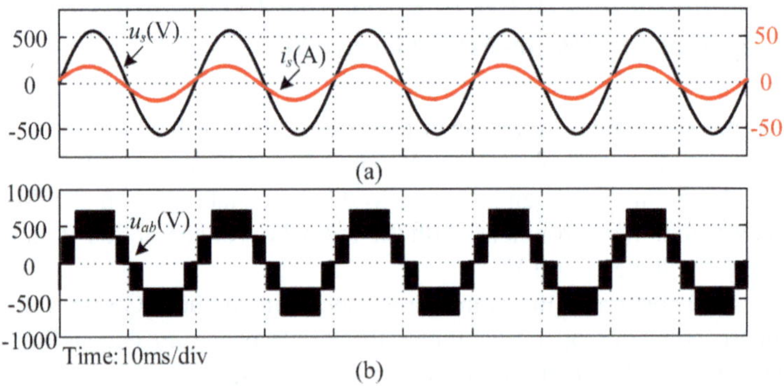

Fig. 4.21 Voltage and current waveforms of single-phase converter with resistive load

Figure 4.21 presents the detailed voltage and current waveforms of a single-phase converter under a resistive load. The input-side voltage and current waveforms are shown in Fig. 4.21a, while the converter port voltage is depicted in Fig. 4.21b. As indicated, the voltage (u_s) and current (i_s) are aligned in phase, meaning the system operates at a unity power factor. The port voltage exhibits a five-level stepwise waveform. Consequently, in the case of resistive loads, the converter behaves as if it is a resistive load from the input side.

Figure 4.22 shows the detailed voltage and current waveforms from the simulation of a single-phase converter operating under a resistive load with capacitive reactive power output. The input-side voltage waveform is shown in Fig. 4.22a, while the converter port voltage is shown in Fig. 4.22b. As indicated, the current (i_s) leads the voltage (u_s) in phase by less than 90°. From the input side of the converter, it behaves as if it is a resistor-capacitor load, supplying a stable DC voltage for the DC load while consuming active power. Furthermore, the converter delivers capacitive reactive power to the single-phase AC grid.

Figure 4.23 shows the detailed voltage and current waveforms from the simulation of a single-phase converter operating under a resistive load with inductive reactive power output. The input-side voltage waveform is shown in Fig. 4.23a, while

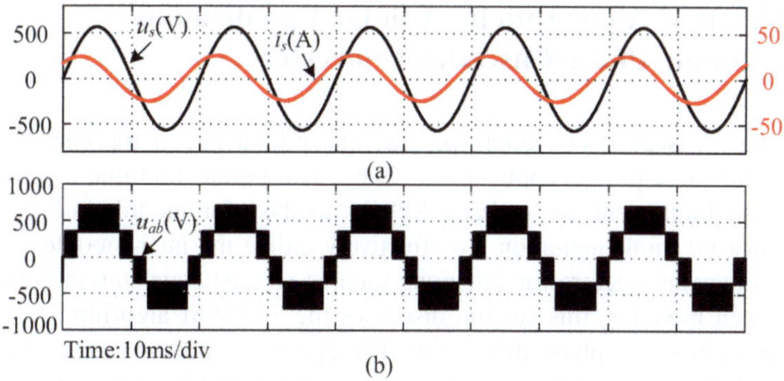

Fig. 4.22 Voltage and current waveforms of single-phase converter with resistive-capacitive load

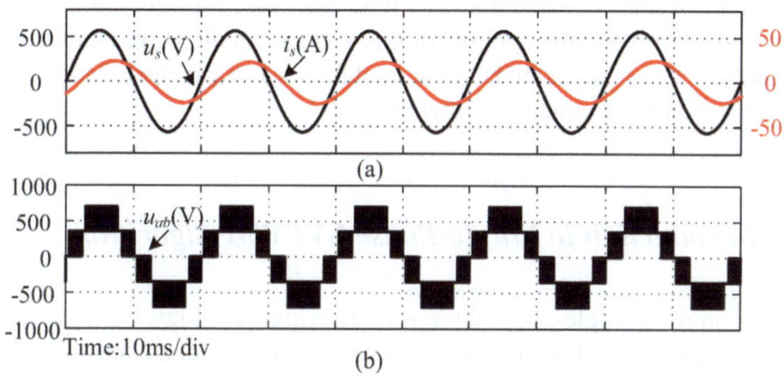

Fig. 4.23 Voltage and current waveforms of single-phase converter with resistive-inductive load

the converter port voltage is shown in Fig. 4.23b. As indicated, the current (i_s) lags behind the voltage (u_s) in phase. From the input side of the converter, the converter behaves as if it is a resistive-inductive load, supplying a stable DC voltage for the DC load while consuming active power. Furthermore, the converter provides inductive reactive power to the single-phase AC grid.

Under the single-phase *dq* decoupling control, single-phase converters can exhibit resistive, capacitive, inductive, resistive-capacitive, and resistive-inductive characteristics, depending on the control requirements. These converters can operate at any operating point within these four quadrants. As a result, this control strategy facilitates the bidirectional transmission control of active energy for future single-phase interconnection and enables reactive power compensation into the grid.

In summary, transient current control offers a simpler topology and easier implementation, making it suitable for applications that require operation solely under unity power factor conditions. In contrast, single-phase *dq* decoupling control provides greater controllability for single-phase converters, allowing them not only to support rectification under unity power factor conditions but also to facilitate applications that require reactive power output and adjustable power factors.

4.3 SVPWM Algorithm for Voltage Equalization of Single-Phase Cascaded Converter

Single-phase cascaded converters provide an optimal circuit topology for high-voltage and large-power applications. They incorporate multiple modules and switches to implement the cascade SPWM method for modulation, offering a straightforward implementation. To effectively utilize the inter-module and intra-module redundant vectors for functions such as voltage equalization both within and between modules, this section discusses the SVPWM algorithm for voltage equalization in single-phase three-level H-bridge multi-module cascaded converters. This algorithm facilitates module voltage equalization through software modulation, eliminating the need for any additional hardware voltage equalizer circuits.

Following a brief introduction to the single-phase SVPWM algorithm, this study focuses on the voltage equalization SVPWM algorithm for three-module cascaded converters to summarize the patterns of this voltage equalization modulation algorithm. Further discussion extends to the modulation voltage equalization algorithm applicable to n-module cascade configurations.

4.3.1 Introduction to Single-Phase SVPWM Algorithm

Figure 4.24 shows a single-phase three-level H-bridge topology, where v_a represents the voltage at the port of bridge arm a relative to neutral point O, and v_b indicates the voltage at the port of bridge arm b relative to neutral point O.

Table 4.2 presents the relationships between the output switch states and output levels of single-phase three-level H-bridge cascade modules. The output levels of this topology include $2E$, E, 0, $-E$, and $-2E$, which together form the voltage vector spaces for single-phase three-level H-bridge circuits. The modulation method that

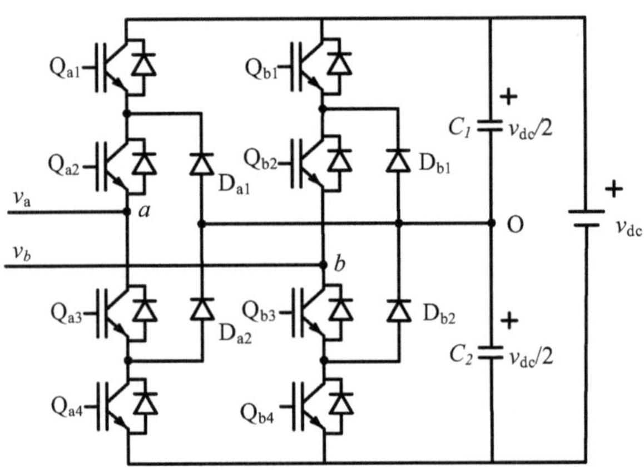

Fig. 4.24 Single-phase three-level H-bridge inverter

Table 4.2 Switching states of three-level H-bridge converter

S_{an}	2	2	1	0	1	2	1	0	0
S_{bn}	0	1	0	0	1	2	2	1	2
V_{AnBn}	$V_{Cn1} + V_{Cn2}$	V_{Cn1}	V_{Cn2}	0			$-V_{Cn1}$	$-V_{Cn2}$	$-(V_{Cn1} + V_{Cn2})$
$V_{AnBn}(V_{Cn1} = V_{Cn2} = E)$	$2E$	E	E	0			$-E$	$-E$	$-2E$

utilizes a combination of different switch states corresponding to these voltage vector spaces is referred to as the space vector pulse width modulation (SVPWM) algorithm.

Among the voltage vector spaces of single-phase three-level H-bridge circuits, the levels E, 0, and $-E$ correspond to multiple switch combinations on average. These vectors associated with these levels are considered redundant vectors within the modules. For instance, the level E corresponds to two switch combinations: {2, 1} and {1, 0}. The {2, 1} switch combination is connected to Cn1, while the {1, 0} switch combination is connected to Cn2. Based on this, different vectors can be selected depending on the direction of the AC current, leading to various outcomes for capacitor voltage equalization. Furthermore, cascade multi-module circuits accommodate a greater number of redundant vectors between modules. These redundant vectors produce the same output level and maintain unchanged output characteristics. However, different redundant vectors create distinct current paths, resulting in varying charging and discharging effects on the capacitors. Therefore, under the condition that the output level remains constant, various redundant vectors can be utilized to achieve capacitor voltage equalization both within and between modules. The SVPWM algorithm for single-phase voltage equalization is discussed below.

As indicated in Table 4.1, there are nine combinations of port levels for the left and right bridge arms, each corresponding to a unique voltage vector. All of these voltage vectors can be represented in the form of a matrix:

$$\begin{bmatrix} v_{ab} \\ v_{ba} \end{bmatrix} = \begin{bmatrix} 1 & -1 \\ -1 & 1 \end{bmatrix} \begin{bmatrix} v_{aO} \\ v_{bO} \end{bmatrix} \tag{4.33}$$

The distribution in the two-dimensional space coordinate system of the voltage vector $u = [v_{ab}, v_{ba}]^{\mathrm{T}}$ derived from Eq. (4.33) is shown in Fig. 4.25. All the vectors are located along the axis α.

The coordinate transformation is defined as follows:

$$u' = \begin{bmatrix} v_{\alpha} \\ v_{\beta} \end{bmatrix} = 2U \begin{bmatrix} \sin \omega t \\ 0 \end{bmatrix} \tag{4.34}$$

Coordinate transformation is performed based on Eq. (4.33), to yield a new voltage vector as follows:

$$u' = \sqrt{2} \begin{bmatrix} 1 & -1 \\ 0 & 0 \end{bmatrix} \begin{bmatrix} v_{aO} \\ v_{bO} \end{bmatrix} \tag{4.35}$$

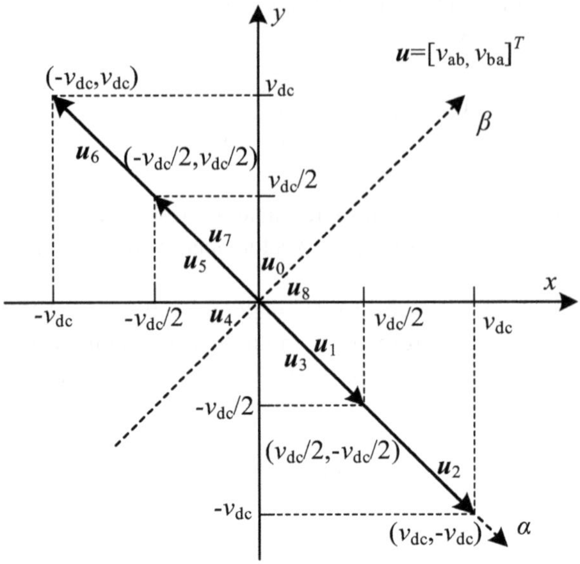

Fig. 4.25 Primary voltage vector diagram of single-phase three-level H-bridge inverter

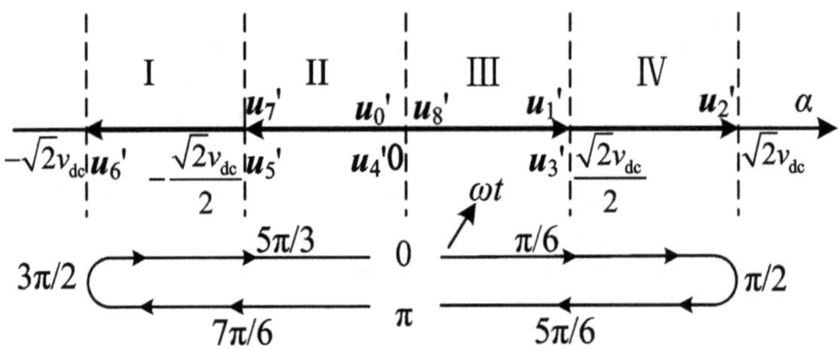

Fig. 4.26 Single-phase one-dimensional space voltage vector diagram after rotational transformation

The voltage vector transformed by Eq. (4.35) is shown in Fig. 4.26. Using a method similar to the two-dimensional space voltage vector division for three-phase systems, the one-dimensional voltage vector space is divided as shown in Fig. 4.26:

According to the vector positions along the axis α, the one-dimensional space is divided by five levels of $2e$, e, 0, $-e$, and $-2e$ into four intervals: i, ii, iii, and iv. The ratio of the expected output voltage vector modulus to the actual output voltage vector modulus of the inverter is defined as the modulation degree (m) for single-phase inverters. This can be expressed as follows:

$$m = \frac{2U}{\sqrt{2}v_{dc}} = \frac{\sqrt{2}U}{v_{dc}} \leq 1 \tag{4.36}$$

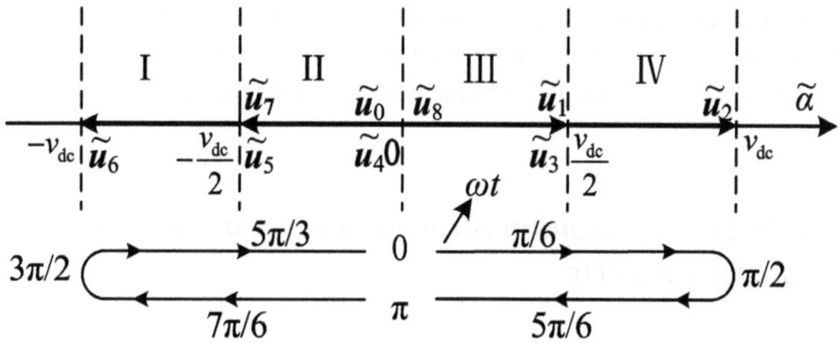

Fig. 4.27 Single-phase one-dimensional space voltage vector diagram after reduction of vector modulus by $\sqrt{2}$

Comparing Eqs. (4.33) and (4.35) indicates that after the coordinate transformation of the primary voltage vector, the vector modulus is multiplied by $\sqrt{2}$. To facilitate the analysis of v_{ab}, the α-axis is transformed into the $\tilde{\alpha}$-axis, the modulus is reduced by $\sqrt{2}$, and the corresponding vector is expressed as $\tilde{u}$. The reduced voltage vector space is shown in Fig. 4.27, with the sector size actually reflecting the amplitude of v_{ab}.

In each interval shown in Fig. 4.27, the desired voltage (v_{ab}) can be synthesized using its two adjacent vectors. For example, when v_{ab} falls within the interval III, vector $\tilde{u}_1$ may be synthesized with $\tilde{u}_0$, $\tilde{u}_4$, or $\tilde{u}_8$.

The dwell duration needs to be determined for each vector within the resultant vectors. Assuming that the vector corresponding to v_{ab} is V_{ref}, the two resultant vectors are defined as V_{down} and V_{up}, respectively; the switching cycle of the converter as T_s; the dwell duration of Vector V_{down} as T_{down}; and the dwell duration of Vector V_{up} as T_{up}. According to the impulse balance principle, the relationship of these parameters can be expressed as follows:

$$\begin{cases} V_{\text{down}} T_{\text{down}} + V_{\text{up}} T_{\text{up}} = V_{\text{ref}} T_s \\ T_{\text{down}} + T_{\text{up}} = T_s \end{cases} \tag{4.37}$$

Solving the above equation yields the dwell duration of each vector, which can be expressed as follows:

$$\begin{cases} T_{\text{down}} = \dfrac{V_{\text{ref}} - V_{\text{up}}}{V_{\text{down}} - V_{\text{up}}} T_s \\ T_{\text{up}} = T_s - T_{\text{down}} \end{cases} \tag{4.38}$$

The SVPWM algorithm for single-module single-phase three-level H-bridge inverters can be derived from the aforementioned principle of space voltage vector synthesis. This process involves several steps, including dwell duration calculation, time allocation, and vector selection. During the vector selection step, a vector with an optimization target—such as capacitor voltage equalization—may be chosen from the redundant vectors. The minimum energy property, discussed in the next section, provides a criterion for selecting the vector with the most effective voltage equalization capability from the redundant vectors for optimal selection.

4.3.2 Voltage Equalization Strategy Based on Minimum Energy Property

In N-module cascaded converters, the total energy (E_{c1}) of capacitors on the DC side of these modules can be expressed as follows:

$$E_{c1} = \frac{1}{2}\sum_{i=1}^{N} C_i v_{dci}^2 \tag{4.39}$$

where C_i denotes the equivalent capacitance on the DC side of module i and v_{dci} denotes the actual value of the DC voltage of module i. To simplify the analysis, it is assumed that the following two conditions hold:

1. The total voltage on the DC side has reached a stable value so that voltage fluctuations can be neglected.
2. The size of all the DC-link capacitors is identical, that is, $C_N = \ldots = C_1 = C$.

A higher E_{c1} value indicates better capacitor voltage equalization on the DC side of the cascaded converter. When voltage balance is achieved among all the capacitors, i.e., $v_{dc1} = v_{dc2} = \ldots = v_{dcN} = E$, E_{c1} reaches its minimum value, that is, $0.5CU_{dc}^2 / N$. When E_{c1} attains this minimum value, voltage is balanced among DC capacitors within the system. This characteristic, which reflects the voltage equalization of DC capacitors, is referred to as the minimum energy property.

In the design of a voltage-balancing strategy for cascaded converters, the value of E_{c1} can be used as a measure of the degree of voltage equalization. An effective strategy should focus on minimizing the E_{c1} value. In voltage-equalizing modulation, the vector that can reduce the E_{c1} value is considered a voltage-equalizing vector, as it promotes the balancing of capacitor voltages. Among the numerous redundant vectors available, the vector that results in the maximum decrease in the E_{c1} value is referred to as the optimal voltage-equalizing vector, which achieves the fastest voltage equalization speed.

The deviation (Δv_{dci}) of the actual DC voltage of module i ($i = 1,\ldots, N$) from the average DC voltage is defined as follows:

$$\Delta v_{dci} = v_{dci} - U_{dc} / N \tag{4.40}$$

The capacitor voltage deviation energy is used to establish an objective function F1 that is always positive, as shown in Eq. (4.41):

$$F_1 = \frac{1}{2}C\sum_{i=1}^{N}\left(v_{dci} - \frac{U_{dc}}{N}\right)^2 \tag{4.41}$$

A smaller value of the objective function F1 indicates a better voltage balance on the DC side of the system and reflects a more stable system. When F1 is zero, the DC voltages of all the modules are balanced. The voltage equalization SVPWM algorithm studied in this section enables voltage equalization by selecting a voltage-equalizing redundant vector based on the objective function expressed in Eq. (4.41).

To achieve voltage equalization, the strategy for selecting a voltage-equalizing redundant vector should ensure a negative increment of Eq. (4.41), i.e., it should approach zero. To this end, the variation trend of F1 over time is analyzed. When its rate of change over time is less than zero, it indicates that the selected redundant vector contributes to voltage balance; otherwise, it does not. The derivation of F1 yields

$$\frac{\mathrm{d}F_1}{\mathrm{d}t} = \sum_{i=1}^{N} \Delta v_{\mathrm{dc}i} C \frac{\mathrm{d}v_{\mathrm{dc}i}}{\mathrm{d}t} \tag{4.42}$$

To achieve voltage equalization, Eq. (4.42) must be less than zero, with a smaller value indicating a better voltage equalization capacity. Therefore, the voltage equalization target can be further transformed into a solving process of Eq. (4.42) for its negative minimum value. In the absence of a negative value, the switching vector corresponding to the smallest positive value is also chosen to minimize the deviation of the DC voltage from the equalizing tendency.

The value of $C(\mathrm{d}v_{\mathrm{dc}i}/\mathrm{d}t)$ in Eq. (4.42) reflects the current of the DC-link capacitor of module i, which is linked to the path and direction of the grid-side current. The following sections provide a detailed analysis in six specific cases.

1. Port voltage v_{aibi} of module i being $2E$

   Ignoring the influence of the DC-side secondary filter circuit, the current path of module i can be obtained as shown in Fig. 4.28, where the direction indicated by arrows on each branch represents the reference positive direction of current in that branch. Assuming that the capacitance of capacitors $Ci1$ and $Ci2$ in module

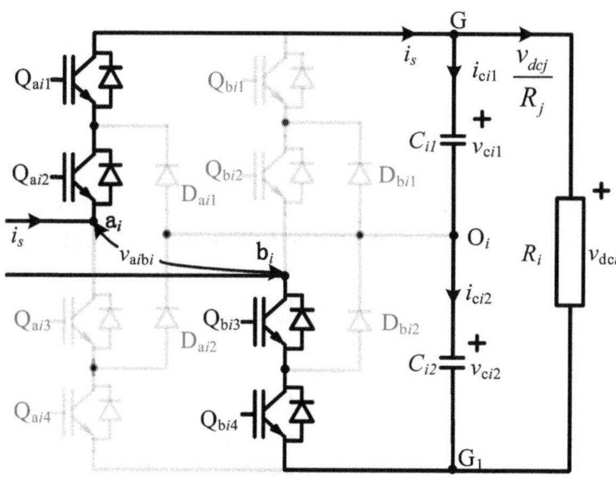

Fig. 4.28 Current flow diagram of module i in the case of v_{aibi} being $2E$

i is identical, both being 2*C*, the KCL equation of node *G* in Fig. 4.28 can be established as follows:

$$2C\frac{\mathrm{d}v_{ci1}}{\mathrm{d}t} = i_s - \frac{v_{dci}}{R_i} \tag{4.43}$$

As shown in Fig. 4.28, the DC-link capacitors $Ci1$ and $Ci2$ are connected in series, with identical currents passing through them. This current can be expressed as follows:

$$2C\frac{\mathrm{d}v_{ci1}}{\mathrm{d}t} = 2C\frac{\mathrm{d}v_{ci2}}{\mathrm{d}t} \tag{4.44}$$

Combining Eqs. (4.43) and (4.44), the $C(\mathrm{d}v_{dci}/\mathrm{d}t)$ value when the port voltage v_{aibi} is 2E can be derived from Eq. (4.45):

$$C\frac{\mathrm{d}v_{dci}}{\mathrm{d}t} = C\frac{\mathrm{d}\left(v_{ci1} + v_{ci2}\right)}{\mathrm{d}t} = 2C\frac{\mathrm{d}v_{ci1}}{\mathrm{d}t} = \frac{2}{2}i_s - \frac{v_{dci}}{R_i} \tag{4.45}$$

2. Port voltages v_{aibi} being $1E$, v_{aioi} $1E$, and v_{bioi} $0E$

   The current path is shown in Fig. 4.29a, and the KCL equation for node *G* can be established as follows:

$$i_s - \frac{v_{dci}}{R_i} = i_{ci1} = 2C\frac{\mathrm{d}v_{ci1}}{\mathrm{d}t} \tag{4.46}$$

   The current i_{ci2} on the capacitor C*i*2 can be expressed as follows:

$$i_{ci2} = 2C\frac{\mathrm{d}v_{ci2}}{\mathrm{d}t} = -\frac{v_{dci}}{R_i} \tag{4.47}$$

   The voltage on capacitors C*i*1 and C*i*2 can be expressed as follows:

$$\frac{\mathrm{d}\left(v_{ci1} + v_{ci2}\right)}{\mathrm{d}t} = \frac{\mathrm{d}v_{dci}}{\mathrm{d}t} \tag{4.48}$$

   Combining Eqs. (4.46), (4.47), and (4.48), the $C(\mathrm{d}v_{dci}/\mathrm{d}t)$ value when the port voltage v_{aibi} is *E* can be derived from the following equation:

$$C\frac{\mathrm{d}v_{dci}}{\mathrm{d}t} = \frac{i_s}{2} - \frac{v_{dci}}{R_i} \tag{4.49}$$

3. Port voltages v_{aibi} being $1E$, v_{aioi} $0E$, and $v_{bioi} - 1E$

   The current path is shown in Fig. 4.29b. Similarly, the KCL equation for node *G* of the circuit can be established as follows:

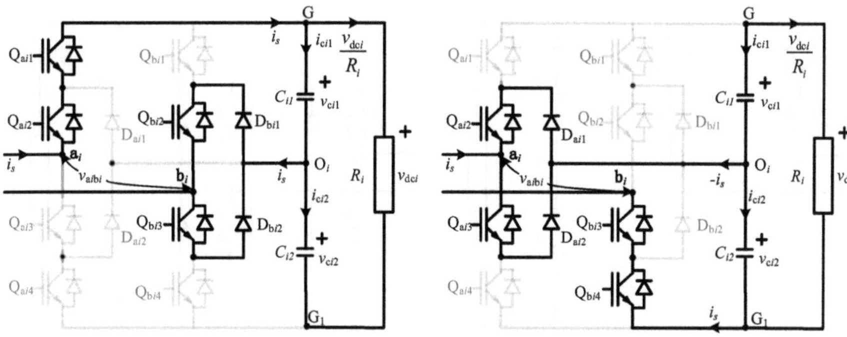

(a) Bridge Arm a Voltage Being 1E and Bridge Arm b Voltage Being 0*E*

(b) Bridge Arm a Voltage Being 0E and Bridge Arm b Voltage Being-1*E*

Fig. 4.29 Current flow diagram of module *i* in the case of v_{aibi} being 1*E*

$$i_s - \frac{v_{dci}}{R_i} = i_{ci2} = 2C\frac{dv_{ci2}}{dt} \tag{4.50}$$

The current i_{ci1} on capacitor C*i*1 can be expressed as follows:

$$i_{ci1} = 2C\frac{dv_{ci1}}{dt} = -\frac{v_{dci}}{R_i} \tag{4.51}$$

Combining Eqs. (4.48), (4.50), and (4.51), the same results can be obtained as those derived from Eq. (4.49).

4. Port voltage v_{aibi} being 0*E*

   The current path is shown in Fig. 4.30. When the port voltage is 0*E*, the grid-side current (i_s) bypasses the DC side. The KCL equation for node *G* can be easily established to yield $C(dv_{dci}/dt)$ as follows:

$$C\frac{dv_{dci}}{dt} = -\frac{v_{dci}}{R_i} = \frac{0}{2}i_s - \frac{v_{dci}}{R_i} \tag{4.52}$$

5. Port voltage v_{aibi} being −1*E*

   The current path is shown in Fig. 4.31, and an analysis process similar to that conducted when the port voltage v_{aibi} is 1*E* yields

$$C\frac{dv_{dci}}{dt} = -\frac{i_s}{2} - \frac{v_{dci}}{R_i} = \frac{-1}{2}i_s - \frac{v_{dci}}{R_i} \tag{4.53}$$

6. Port voltage v_{aibi} being −2*E*

   The current path is shown in Fig. 4.32, and an analysis process similar to that conducted when the port voltage v_{aibi} is 2*E* yields

$$C\frac{dv_{dci}}{dt} = -i_s - \frac{v_{dci}}{R_i} = \frac{-2}{2}i_s - \frac{v_{dci}}{R_i} \tag{4.54}$$

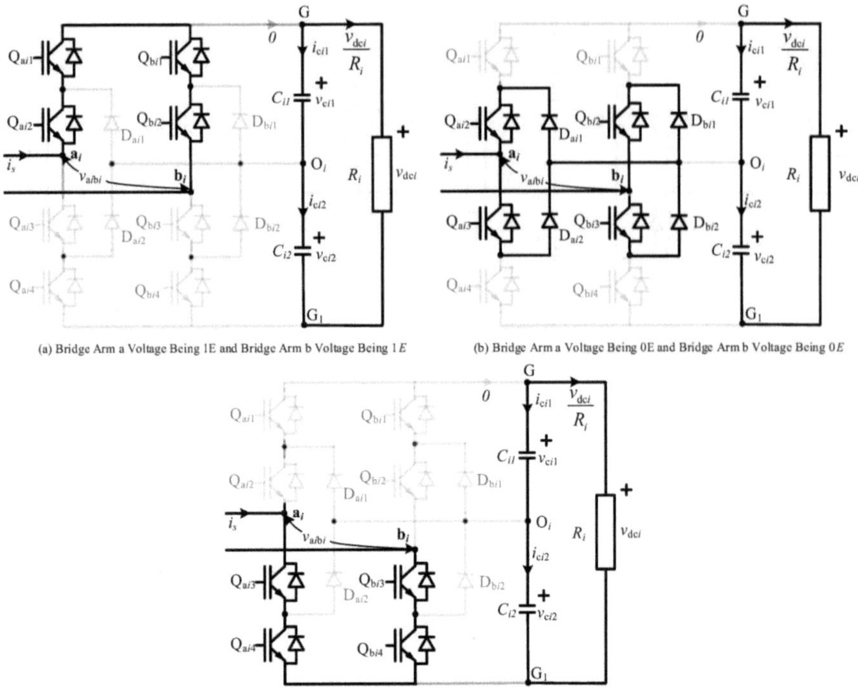

(a) Bridge Arm a Voltage Being 1E and Bridge Arm b Voltage Being 1*E*

(b) Bridge Arm a Voltage Being 0E and Bridge Arm b Voltage Being 0*E*

(c) Bridge Arm a Voltage Being -1E and Bridge Arm b Voltage Being -1*E*

Fig. 4.30 Current flow diagram of module *i* in the case of v_{aibi} being 0*E*

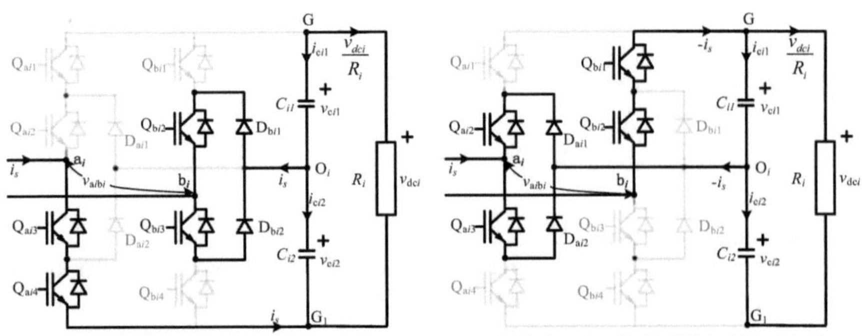

(a) Bridge Arm a Voltage Being -1*E* and Bridge Arm b Voltage Being 0*E*

(b) Bridge Arm a Voltage Being 0*E* and Bridge Arm b Voltage Being 1*E*

Fig. 4.31 Current flow diagram of module *i* in the case of v_{aibi} being −1*E*

In summary, letting x_i represent the port level value corresponding to the port voltage v_{aibi} of module *i*, defined as $x_i = v_{aibi}/E$, based on the above analysis, a general expression for the $C(\mathrm{d}v_{dci}/\mathrm{d}t)$ value of module *i* can be established as follows:

$$C\frac{\mathrm{d}v_{\mathrm{dc}i}}{\mathrm{d}t} = \frac{x_i}{2}i_s - \frac{v_{\mathrm{dc}i}}{R_i} \tag{4.55}$$

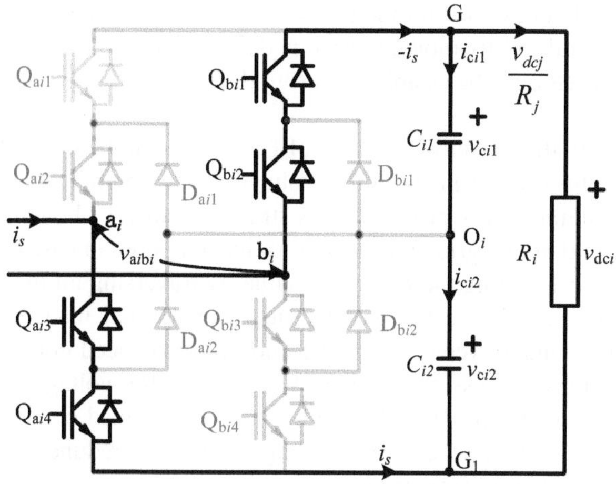

Fig. 4.32 Current flow diagram of module i in the case of v_{aibi} being $-2E$

Substituting Eq. (4.55) into Eq. (4.42) yields

$$\frac{\mathrm{d}F_1}{\mathrm{d}t} = \sum_{i=1}^{N} \Delta v_{\mathrm{dc}i} \left(\frac{x_i}{2} i_{\mathrm{s}} - \frac{v_{\mathrm{dc}i}}{R_i} \right) \tag{4.56}$$

Equation (4.56) can be expressed piecewise as follows:

$$\frac{\mathrm{d}F_1}{\mathrm{d}t} = \sum_{i=1}^{N} \frac{\Delta v_{\mathrm{dc}i} i_{\mathrm{s}} x_i}{2} - \sum_{i=1}^{N} \frac{v_{\mathrm{dc}i} \Delta v_{\mathrm{dc}i}}{R_i} \tag{4.57}$$

Assuming that the switching frequency is high and that $v_{\mathrm{dc}i}$ and R_i remain constant during a switching cycle, $\Delta v_{\mathrm{dc}i}$ also remains constant and the value of expression $\sum_{i=1}^{N} \frac{v_{\mathrm{dc}i} \Delta v_{\mathrm{dc}i}}{R_i}$ remains unchanged. Subsequently, the minimum of $\mathrm{d}F1/\mathrm{d}t$ is calculated, leading to the derivation of the minimum of the objective function S_{cri} in Eq. (4.58) through a further process of equivalence:

$$S_{\mathrm{cri}} = \frac{\Delta v_{\mathrm{dc}1} i_{\mathrm{s}} x_1}{2} + \cdots + \frac{\Delta v_{\mathrm{dc}i} i_{\mathrm{s}} x_i}{2} + \cdots + \frac{\Delta v_{\mathrm{dc}N} i_{\mathrm{s}} x_N}{2} = \sum_{i=1}^{N} \frac{\Delta v_{\mathrm{dc}i} i_{\mathrm{s}} x_i}{2} \tag{4.58}$$

where S_{cri} represents the objective function used to select the optimal voltage-equalizing redundant vector for the SVPWM algorithm in single-phase voltage equalization. Since x_{i} denotes the port level value corresponding to the port voltage v_{aibi} of module i, different values of S_{cri} are calculated using various x_i, based on the given total output port level during a switching cycle, using Eq. (4.58). The switch vector composed of x_i corresponding to the minimum value of S_{cri}—specifically, the current x_i—is selected as the optimal inter-module voltage-equalizing vector for the SVPWM algorithm. If the individual modules employ a multilevel topology, once x_i is determined, the intra-module voltage-equalizing vector can be established using the method discussed in the previous section.

To sum up, the objective function of Eq. (4.58) can be utilized to determine the strategy for selecting the optimal voltage-equalizing redundant vector based on the minimum energy property. This strategy involves obtaining the DC-side voltage deviation and grid-side current for each module in cascaded converters, followed by calculating the minimum value by traversing all redundant vectors according to Eq. (4.58). These steps yield the optimal inter-module voltage-equalizing redundant vector under the current unbalanced conditions. The following sections discuss the SVPWM algorithm for voltage equalization in single-phase three-level H-bridge multi-module cascaded converters, based on the strategy of selecting the optimal voltage-equalizing redundant vector according to the minimum energy property.

4.3.3 Circuit Analysis and Voltage Space Vector Distribution in Cascaded Converter

4.3.3.1 Two-Module Cascaded Converter

The circuit analysis of cascaded converters begins with single-phase three-level H-bridge two-module cascaded converters, whose main circuit topology is shown in Fig. 4.33. This circuit topology consists of two diode-clamped three-level H-bridge converters connected in cascade. The port voltage of module 1 is denoted as ua1b1, while that of module 2 is denoted as u_{a2b2}. Therefore, the total AC-side port voltage of these two cascaded modules can be expressed as $u_{a1b2} = u_{a1b1} + u_{a2b2}$. When the DC-side capacitor voltages of the two modules are balanced, i.e., V $V_{C11} = V_{C12} = V_{C21} = V_{C22} = E$, u_{a1b2} can output nine voltage levels, ranging from $-4E$ to $4E$, corresponding to different switch combinations.

With the number of levels in the port voltage u_{a1b1} of module 1 denoted as x, and that in the port voltage u_{a1b1} of module 2 as y, the relationships of $x = u_{a1b1}/E$ and $y = u_{a2b2}/E$ are established. With the port voltage vector expressed as (x, y) and knowing that $u_{a1b2} = u_{a1b1} + u_{a2b2}$, the total port voltage can be expressed as follows:

$$k = x + y \tag{4.59}$$

Although each module can output levels ranging from $-2E$ to $2E$, i.e., x, $y \in \{-2, -1, 0, 1, 2\}$, the limitations imposed by Eq. (4.59) narrow the range of x and y as follows:

$$\max(k-2,-2) \le x, y \le \min(k+2,2) \tag{4.60}$$

As shown in Fig. 4.34, given $k = -3$ and the constraints -2 $\le x$ and $y \le -1$, the redundant vectors corresponding to the output level of -3 include $(-2, -1)$ and

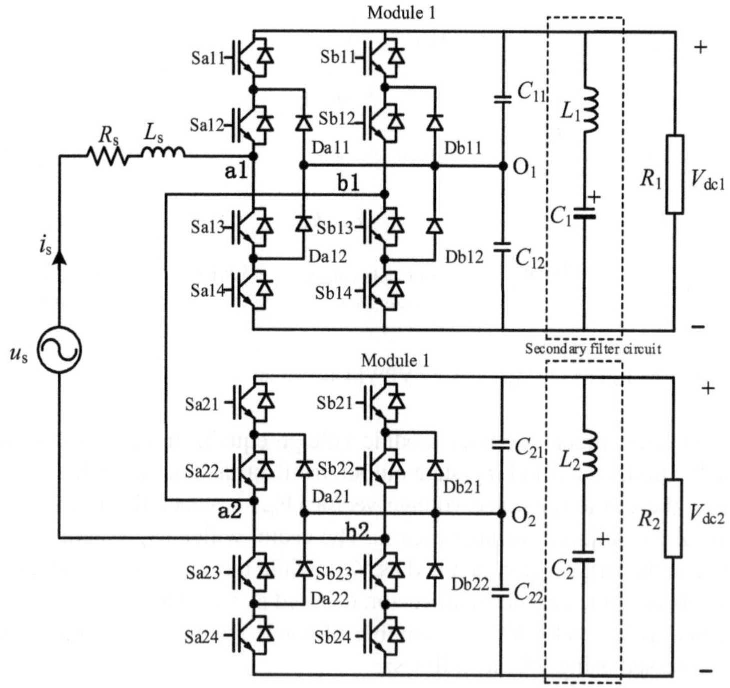

Fig. 4.33 Topology of single-phase three-level H-bridge two-module cascaded converter

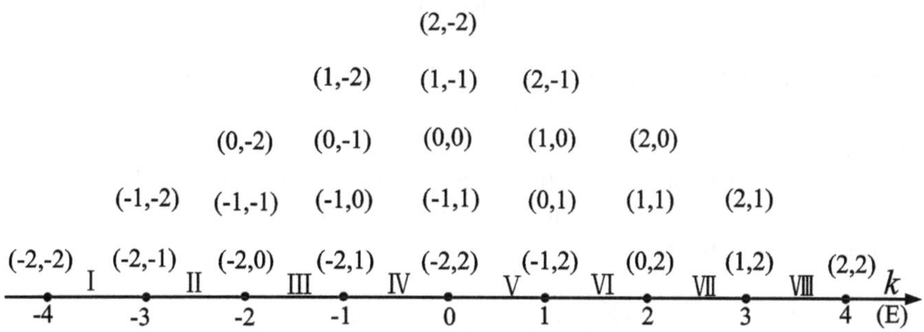

Fig. 4.34 One-dimensional space vector diagram of single-phase two-module cascaded converter

(−1, −2), using Eq. (4.59). Consequently, the one-dimensional spatial distribution of redundant vectors is illustrated in Fig. 4.34, comprising a total of $5^2 = 25$ vectors. These vectors are divided into eight intervals labeled from I to VIII, based on the different values of k:

This circuit topology consists of two three-level H-bridge modules that require inter-module voltage equalization. Each module includes two DC-link capacitors, necessitating the balancing of intra-module capacitor voltages as well. Consequently, the SVPWM algorithm for this topology needs to ensure both inter- and intra-module voltage equalization. The algorithm typically encompasses vector selection, time

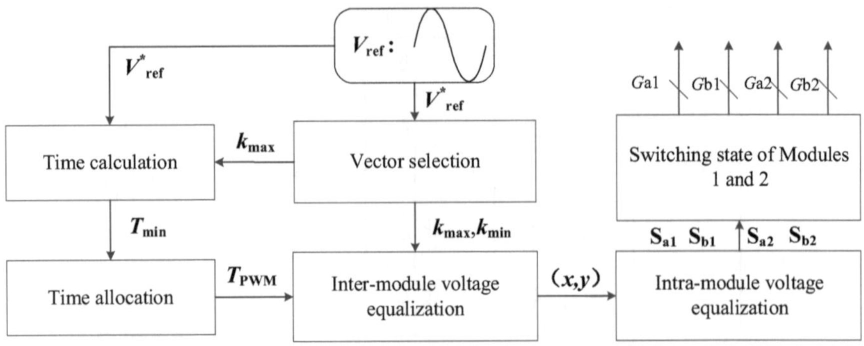

Fig. 4.35 Block diagram of single-phase SVPWM algorithm

calculation, time allocation, inter-module voltage equalization, intra-module voltage equalization, and switching-state output, as illustrated in Fig. 4.35.

Among these, the reference voltage vector (V_{ref}) is generally derived from the synthesis of two adjacent voltage vectors. The vector with a larger number of levels is defined as the large vector, denoted as V_{max}, while the vector with a smaller number of levels is defined as the small vector, denoted as V_{min}. The per-unit value of V_{ref} is expressed as $V^*_{ref} = |V_{ref}|/E$. The number of levels for these two adjacent vectors can be expressed using V^*_{ref} as follows:

$$\begin{cases} k_{max} = \mathrm{floor}\left(V^*_{ref}\right) + 1 \\ k_{min} = \mathrm{floor}\left(V^*_{ref}\right) \end{cases} \tag{4.61}$$

where k_{max} indicates the number of levels at V_{max}, k_{min} is the number of levels at V_{min}, and the function floor(*) is the floor function, which rounds down to the nearest integer.

With the dwell duration of the reference vector (V^*_{ref}) as T_s, the dwell duration of the resultant vectors, according to the principle of impulse conservation, satisfies the following equation:

$$\begin{cases} k_{min} \cdot T_{min} + k_{max} \cdot T_{max} = V^*_{ref} \cdot T_s \\ T_{max} + T_{min} = T_s \end{cases} \tag{4.62}$$

where T_{max} denotes the dwell duration of the large vector and T_{min} denotes the dwell duration of the small vector, yielding

$$\begin{cases} T_{min} = \dfrac{k_{max} - V^*_{ref}}{k_{max} - k_{min}} \cdot T_s \\ T_{max} = T_s - T_{min} \end{cases} \tag{4.63}$$

Since k_{max} and k_{min} are adjacent, with the level difference of 1, i.e., $k_{max} - k_{min} = 1$, Eq. (4.63) can be simplified as follows:

$$\begin{cases} T_{min} = \left(k_{max} - V^*_{ref}\right) \cdot T_s \\ T_{max} = T_s - T_{min} \end{cases} \tag{4.64}$$

This book adopts a three-segment option for the allocation of vector dwell duration. As illustrated, within a triangular carrier period represented as T_s, T_{min} is divided into two parts. The vector V_{min} is active during the first $T_{min}/2$ time period, followed immediately by V_{max} during the T_{max} time period, and then V_{min} is active again during the second $T_{min}/2$ time period. In this manner, the equivalent reference resultant vector (V^*_{ref}) can be obtained using the neighboring vectors V_{max} and V_{min}.

4.3.3.2 Three-Module Cascaded Converter

The main circuit topology of a single-phase three-level H-bridge three-module cascaded converter consists of three diode-clamped three-level H-bridge converters connected in cascade, as shown in Fig. 4.36. This topology is formed by cascading

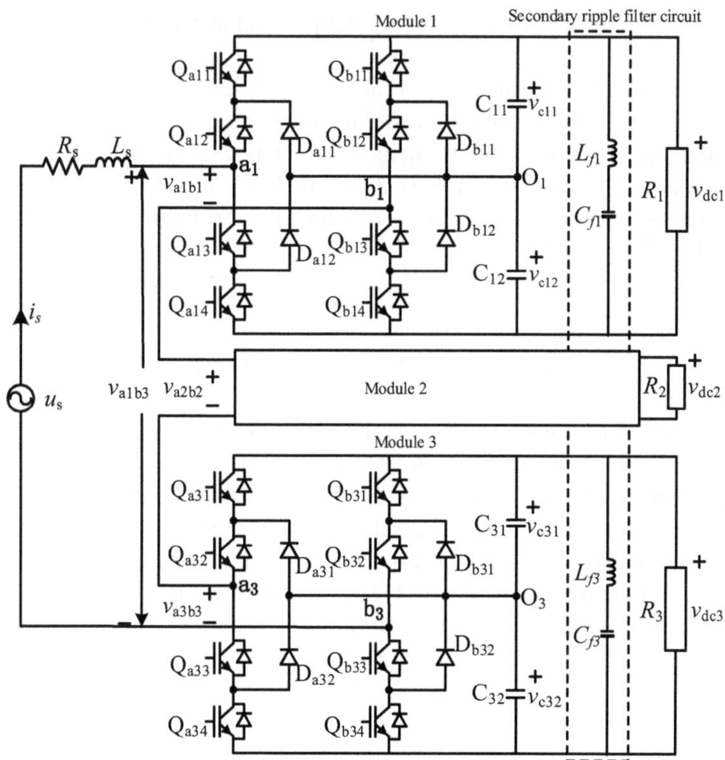

Fig. 4.36 Topology of single-phase three-module three-level H-bridge cascaded converter

the AC terminals of the three H-bridge converters. The port voltage of module 1 is denoted as u_{a1b1}, that of module 2 as u_{a2b2}, and that of module 3 as u_{a3b3}. Therefore, the total AC-side port voltage of the cascaded three modules can be expressed as $u_{a1b3} = u_{a1b1} + u_{a2b2} + u_{a3b3}$.

When the DC-side capacitor voltages of the three modules are balanced, that is, $V_{C11} = V_{C12} = V_{C21} = V_{C22} = V_{C31} = V_{C32} = E$, each module can output four nonzero levels. As a result, the total port voltage u_{a1b3} of these cascaded three modules encompasses 13 levels, ranging from $-6E$ to $6E$ (including the level 0).

The port voltage (u_{a1b1}) of module 1 is denoted as x, the port voltage (u_{a2b2}) of module 2 as y, and the port voltage (u_{a3b3}) of module 3 as z. By defining $x = u_{a1b1}/E$, $y = u_{a2b2}/E$, and $z = u_{a3b3}/E$, with x, y, and z as integers, the port voltage vectors of these three modules can be expressed as (x, y, z). Each module, being composed of a three-level H-bridge, can output five levels ranging from $-2E$ to $2E$, as indicated in Table 4.1, specifically $x, y, z \in \{-2, -1, 0, 1, 2\}$. Consequently, the total number of output vectors available from the three-module three-level cascaded circuit is $5^3 = 125$. Given that there are actually only 13 output levels, this results in a significant number of redundant vectors, providing a wide array of optional vectors for voltage equalization modulation.

The total AC-side port voltage (u_{a1b3}) level is denoted as k. By defining $k = u_{a1b3}/E$, the following equation is established:

$$k = x + y + z \quad \left(k \in [-6,6], K \text{ as an integer}\right) \tag{4.65}$$

First, the range of level values for x, y, and z is analyzed while keeping the k-value constant. Taking x as an example, its minimum value is $k - 4$ when both y and z reach their maximum of 2, and its maximum value is $k + 4$ when y and z reach their minimum of -2. Considering $x \in \{-2, -1, 0, 1, 2\}$, the minimum value of x is determined as the larger of $k - 4$ and -2, while the maximum value of x is taken as the smaller of $k + 4$ and 2.

Similarly, the value ranges of x, y, and z can be determined using the resulting k-value as follows:

$$\max[k-4,-2] \le x, y, z \le \min[k+4,2] \tag{4.66}$$

The value ranges of y and z are determined in the sequence of modules x, y, and z.

After determining the value range of x, i.e., $\max[k - 4, -2] \le x \le \min[k + 4, 2]$, the value range of y can be obtained. Specifically, when z reaches its maximum of 2, y is taken as the minimum; conversely, when z reaches its minimum of -2, y is taken as the maximum. Based on Eq. (4.65), the minimum of y is calculated as $k - 2 - x$, and the maximum as $k + 2 - x$. Considering $y \in \{-2, -1, 0, 1, 2\}$, the value ranges for y and z can be determined as follows:

$$\max[k-2-x,-2] \le y, z \le \min[k+2-x,2] \tag{4.67}$$

Finally, the value of y is determined within the range of $\max[k - 2 - x, -2] \leq y \leq \min[k + 2 - x, 2]$. Once y has been established, the value of z can be calculated as follows:

$$z = k - x - y \tag{4.68}$$

For example, at $k = 4$, Eq. (4.66) results in $0 \leq x \leq 2$, that is, $x \in \{0, 1, 2\}$.

At $x = 0$, Eq. (4.67) results in $2 \leq y \leq 2$, that is, $y = 2$; then Eq. (4.68) results in $z = 2$, indicating that at $k = 4$ and $x = 0$, the only feasible vector is (0, 2, 2).

At $x = 1$, Eq. (4.67) results in $1 \leq y \leq 2$, that is, $y \in \{1, 2\}$. There are two possible values of $z = 2$ at $y = 1$ and $z = 1$ at $y = 2$, indicating two possible vectors (1, 2, 1) and (1, 1, 2).

At $x = 2$, Eq. (4.67) results in $0 \leq y \leq 2$, that is, $y \in \{0, 1, 2\}$. There are three possible values of $z = 2$ at $y = 0$, $z = 1$ at $y = 1$, and $z = 0$ at $y = 2$, indicating three possible vectors (2, 0, 2), (2, 1, 1), and (2, 2, 0).

In summary, there are six possible vectors at $k = 4$: (0, 2, 2), (1, 2, 1), (1, 1, 2), (2, 0, 2), (2, 1, 1), and (2, 2, 0), all of which are redundant vectors at the output level $k = 4$.

Similarly, all redundant vectors at different k-values can be obtained using Eqs. (4.66)–(4.68), leading to the one-dimensional vector spatial distribution of single-phase three-module cascaded converters as shown in Fig. 4.37.

In Fig. 4.37, all parallel redundant vectors are projected to the same point on the k-axis. The k-value represents a level value, and the number of redundant vectors corresponding to each k-value point is as follows: 1 vector at $k = -6$, 3 vectors at $k = -5$, 6 vectors at $k = -4$, 10 vectors at $k = -3$, 15 vectors at $k = -2$, 18 vectors at $k = -1$, and 19 vectors at $k = 0$. The number of redundant vectors at positive k-value points is symmetric with that at negative k-value points about the zero point (0).

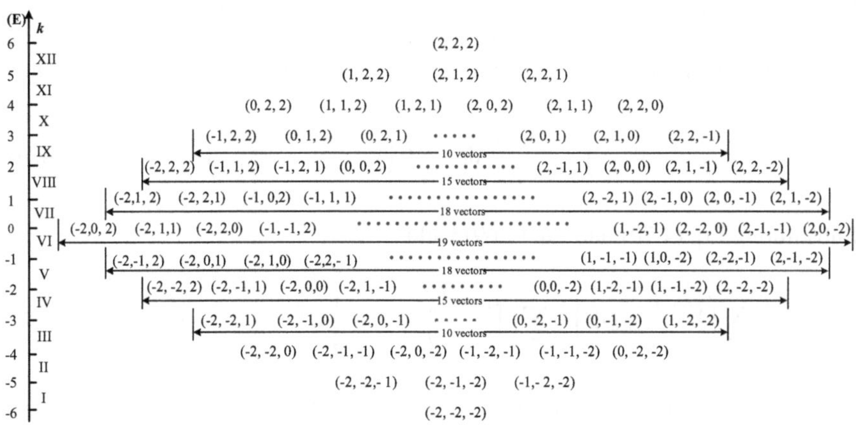

Fig. 4.37 One-dimensional space vector diagram of single-phase H-bridge three-module cascaded converter

Since there are 12 nonzero levels in the three-level cascade of three individual modules, the inclusion of level 0 results in a total of 13 levels, meaning that the total port voltage u_{a1b3} may assume values across these 13 levels ranging from $-6E$ to $6E$. Thus, the range of k-values is determined as $k \in [-6, 6]$. The vector space can be divided into 12 intervals based on k-values, denoted as I–XII. The reference vector can be positioned within any of these 12 intervals. At a given position, vectors corresponding to two adjacent k-values can be synthesized. Since each k-value corresponds to multiple redundant vectors, intra- and inter-module voltage equalization can be achieved by selecting appropriate redundant vectors. The SVPWM algorithm for voltage equalization of cascaded converters is discussed in the next section.

4.3.4 *SVPWM Algorithm for Voltage Equalization of Cascaded Converter*

4.3.4.1 Analysis of Voltage Equalization Principle

To balance the DC-side capacitance among modules within cascaded converters, it is essential to distribute the energy output from the AC-side power supply in accordance with the size of the DC loads served by each module. Therefore, the output energy from the AC-side power supply equals the sum of the energy absorbed by all the modules. This relationship can be expressed as follows, based on the three-module cascaded converters:

$$W = W_1 + W_2 + W_3 = \int_{t_0}^{t_0+\Delta t} u_{a1b1} \cdot i_s \mathrm{d}t + \int_{t_0}^{t_0+\Delta t} u_{a2b2} \cdot i_s \mathrm{d}t + \int_{t_0}^{t_0+\Delta t} u_{a3b3} \cdot i_s \mathrm{d}t \tag{4.69}$$

where W_1, W_2, and W_3 indicate the energy absorbed by the three modules. With the load of each module denoted as R_i and the DC-side capacitance expressed as $Ci1 = Ci2 = C$, where i denotes the number of modules (i = 1, 2, 3) and without considering the losses in the switching tubes, the following equation is established:

$$W_i = \int_{t_0}^{t_0+\Delta t} \frac{V_{\mathrm{d}ci(t)}^2}{R_i} \mathrm{d}t + \frac{1}{4} C \left(V_{\mathrm{d}ci(t_0+\Delta t)}^2 - V_{\mathrm{d}ci(t_0)}^2 \right) \tag{4.70}$$

Substituting Eq. (4.70) into Eq. (4.69) yields

$$\sum_{i=1}^{3} \left(\int_{t_0}^{t_0+\Delta t} u_{aibi} i_s \mathrm{d}t \right) = \sum_{i=1}^{3} \left[\int_{t_0}^{t_0+\Delta t} \frac{V_{\mathrm{d}ci(t)}^2}{R_n} \mathrm{d}t + \frac{1}{4} C \left(V_{\mathrm{d}ci(t_0+\Delta t)}^2 - V_{\mathrm{d}ci(t_0)}^2 \right) \right] \tag{4.71}$$

To analyze the voltage equalization of a DC-side voltage separately, Eq. (4.71) can be modified as follows:

Table 4.3 Voltage equalization patterns

Unbalanced condition	$i_s > 0$	$i_s < 0$
$V_{dci} > V_{ave}$	$u_{aibi}\downarrow$	$u_{aibi}\uparrow$
$V_{dci} < V_{ave}$	$u_{aibi}\uparrow$	$u_{aibi}\downarrow$

$$\int_{t_0}^{t_0+\Delta t} u_{aibi} i_s \mathrm{d}t = \int_{t_0}^{t_0+\Delta t} \frac{V_{dci(t)}^2}{R_n} \mathrm{d}t + \frac{1}{4} C\left(V_{dci(t_0+\Delta t)}^2 - V_{dci(t_0)}^2\right) \tag{4.72}$$

Based on Eq. (4.72), in the case of is greater than 0, when the AC-side port voltage (u_{aibi}) of the ith module decreases (or increases), its DC-side voltage (V_{dci}) also decreases (or increases). In the case of is below 0, when the AC-side port voltage (u_{aibi}) of the ith module decreases (or increases), its DC-side voltage (V_{dci}) increases (or decreases). The voltage equalization patterns under different conditions shown in Table 4.3 can be derived from these relationships as follows:

$$V_{ave} = \frac{1}{3}\left(V_{dc1} + V_{dc2} + V_{dc3}\right) \tag{4.73}$$

where V_{ave} denotes the DC average voltage of the three modules. To achieve voltage equalization, at is >0, in the case of $V_{dci} > V_{ave}$, V_{dci} needs to be reduced with the objective of minimizing the output level of u_{aibi}. In the case of $V_{dci} < V_{ave}$, V_{dci} needs to be increased, and the output level of u_{aibi} should be increased as much as possible. At is <0, the adjustments to u_{aibi} should be made in the opposite direction. From these voltage equalization patterns, it can be inferred that by distributing the AC port voltage (u_{aibi}) in each modulation cycle according to the direction of the AC current, the DC-side voltage of the modules can be adjusted toward equalization.

Since the three modules also need to meet the constraints given by Eq. (4.69), u_{aibi} cannot be changed arbitrarily. How to select a port voltage for voltage equalization modulation is further analyzed by introducing a priority order for voltage equalization below.

4.3.4.2 Analysis of Priority Order for Voltage Equalization

The priority order for voltage equalization is defined by the sequence of modules for selecting port levels, based on the magnitude of the deviations of each module's DC-side voltage (V_{dci}) from the average voltage (V_{ave}). Upon a vector change, the u_{aibi} value is adjusted first in the module with the highest priority for voltage equalization. A higher priority for voltage equalization indicates a faster rate of voltage equalization.

Letting $\Delta V_{dci} = |V_{dci} - V_{ave}|$, a larger ΔV_{dci} corresponds to a higher voltage equalization priority of the corresponding module. The priority order is shown in Table 4.4, where (Q_1, Q_2, Q_3) represent the vectors determined after establishing the voltage equalization priority.

Table 4.4 Priority order for voltage equalization

Condition	Priority order for voltage equalization	Condition	Priority order for voltage equalization
$\Delta V_{dc1} > \Delta V_{dc2} > \Delta V_{dc3}$	$Q_1 = x; Q_2 = y; Q_2 = z$	$\Delta V_{dc2} > \Delta V_{dc3} > \Delta V_{dc1}$	$Q_1 = y; Q_2 = z; Q_2 = x$
$\Delta V_{dc1} > \Delta V_{dc3} > \Delta V_{dc2}$	$Q_1 = x; Q_2 = z; Q_2 = y$	$\Delta V_{dc3} > \Delta V_{dc1} > \Delta V_{dc2}$	$Q_1 = z; Q_2 = x; Q_2 = y$
$\Delta V_{dc2} > \Delta V_{dc1} > \Delta V_{dc3}$	$Q_1 = y; Q_2 = x; Q_2 = z$	$\Delta V_{dc3} > \Delta V_{dc2} > \Delta V_{dc1}$	$Q_1 = z; Q_2 = y; Q_2 = x$

For example, with $(Q_1, Q_2, Q_3) = (x, y, z)$, the levels with the strongest voltage equalization capability for module 1 are selected first, followed by the levels for module 2, and finally the levels for module 3. Following this order of priority, in the case that a reference voltage change requires a vector adjustment from 6E to 5E, if x for the first module needs to change first, the vector changes from (2,2,2) to (1,2,2); if x remains unchanged while y is adjusted preferentially, the vector changes from (2,2,2) to (2,1,2); if neither x nor y needs to change, the vector reduces from (2,2,2) to (2,2,1). In practical applications, the relationship of ΔV_{dci} is constantly changing, and as a result, the voltage equalization priority may shift at any time.

4.3.4.3 Analysis of Rules of Selecting Adjacent Vectors

If vector selection only follows the above voltage equalization patterns and the established order of priority, skip-level jumping can occur at the module port when multiple redundant vectors are available for selection.

For example, with the reference voltage falling into interval XI at levels 5 and 4, as shown in Fig. 4.38a, assuming $(Q_1, Q_2, Q_3) = (x, y, z)$ and the voltage and current relationships as follows, is >0, $V_{dc1} > V_{ave}$, $V_{dc2} > V_{ave}$, and $V_{dc3} < V_{ave}$, if the current acting vector is (2,1,2), due to the highest priority in x, the expected voltage-equalizing vector is (0,2,2). As a result, the port level transitions through (1,1,2) and (1,2,2) before finally changing to (0,2,2), as shown by the curve in Fig. 4.38a, without any abrupt changes in the port level during this process.

Another case is illustrated in Fig. 4.38b. If the vector changes directly from (2,1,2) to (0,2,2), the vector for module 1 transitions directly from 2 to 0. While this selection allows for rapid voltage equalization for module 1, it results in a skip-level jump from 2E to 0, which can significantly impact the module and lead to a higher current change rate. To avoid this issue, the vector adjacent to the current vector should be selected instead. In the example shown in Fig. 4.38a, if the current vector is (2,1,2), the next acting vector can be chosen as (1,1,2). If the voltage and current conditions remain unchanged, the next vector can be selected as (1,2,2), followed by the final vector (0,2,2). Figure 4.38c shows the port level changes of the three modules after optimization, without level jumping greater than 1 for each module in adjacent cycles.

With the DC-side voltage of the module corresponding to Q_i denoted as V_{dc_Qi}, the correspondences are shown in Table 4.5, and Q_{i_step} indicates variation in Q_i.

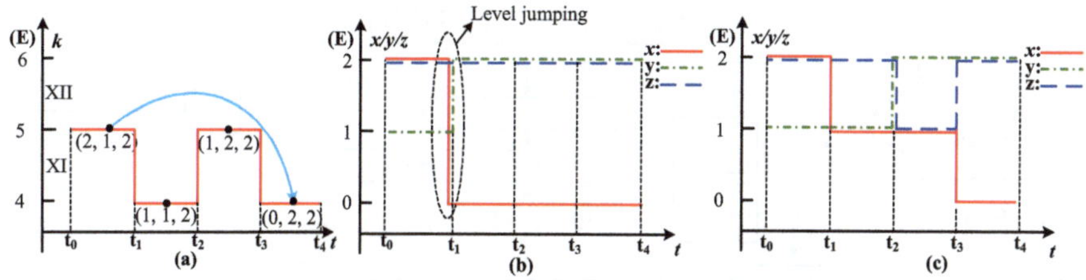

Fig. 4.38 Cases with and without level jumping greater than 1 in adjacent vectors

Table 4.5 DC-side voltage corresponding to V_{dc_Qi}

(Q_1,Q_2,Q_3)	$(V_{dc\_Q1},V_{dc\_Q2},V_{dc\_Q3})$	(Q_1,Q_2,Q_3)	$(V_{dc\_Q1},V_{dc\_Q2},V_{dc\_Q3})$
(x,y,z)	$(V_{dc1},V_{dc2},V_{dc3})$	(y,z,x)	$(V_{dc2},V_{dc3},V_{dc1})$
(x,z,y)	$(V_{dc1},V_{dc3},V_{dc2})$	(z,x,y)	$(V_{dc3},V_{dc1},V_{dc2})$
(y,x,z)	$(V_{dc2},V_{dc1},V_{dc3})$	(z,y,x)	$(V_{dc3},V_{dc2},V_{dc1})$

The rules of selecting adjacent vectors are applied to three-module systems, and two possibilities of level increase or decrease are considered:

1. Change in AC port level k to +1

   The current acting vector changes from $V_{\min}$ to $V_{\max}$. To ensure that skip-level jumping is avoided at the ports of all three modules—that is, the change in Q_i does not exceed 1—the value of Q_{i_step} is calculated using the following equation, considering the voltage equalization patterns:

$$Q_{i\_\text{step}} = \begin{cases} +1 & V_{\text{dc}\_Qi} \langle V_{\text{ave}}, i_{\text{s}} \rangle 0 \,||\, V_{\text{dc}\_Qi} > V_{\text{ave}}, i_{\text{s}} < 0 \\ 0 & V_{\text{dc}\_Qi} > V_{\text{ave}}, i_{\text{s}} > 0 \,||\, V_{\text{dc}\_Qi} < V_{\text{ave}}, i_{\text{s}} < 0 \end{cases} \tag{4.74}$$

2. Change in AC port level k to -1

   The current acting vector changes from $V_{\max}$ to $V_{\min}$. Similarly, to ensure that skip-level jumping is avoided at the ports of all three modules—that is, the change in Q_i does not exceed 1—the value of Q_{i_step} is calculated using the following equation, considering the voltage equalization patterns:

$$Q_{i\_\text{step}} = \begin{cases} 0 & V_{\text{dc}\_Qi} \langle V_{\text{ave}}, i_{\text{s}} \rangle 0 \,||\, V_{\text{dc}\_Qi} > V_{\text{ave}}, i_{\text{s}} < 0 \\ -1 & V_{\text{dc}\_Qi} > V_{\text{ave}}, i_{\text{s}} > 0 \,||\, V_{\text{dc}\_Qi} < V_{\text{ave}}, i_{\text{s}} < 0 \end{cases} \tag{4.75}$$

4.3.4.4 Algorithm Calculation Steps

Based on the voltage equalization patterns, the priority order for voltage equalization, and the selection of adjacent vectors discussed above, the SVPWM algorithm implementation process for three-module voltage equalization is shown in Fig. 4.39.

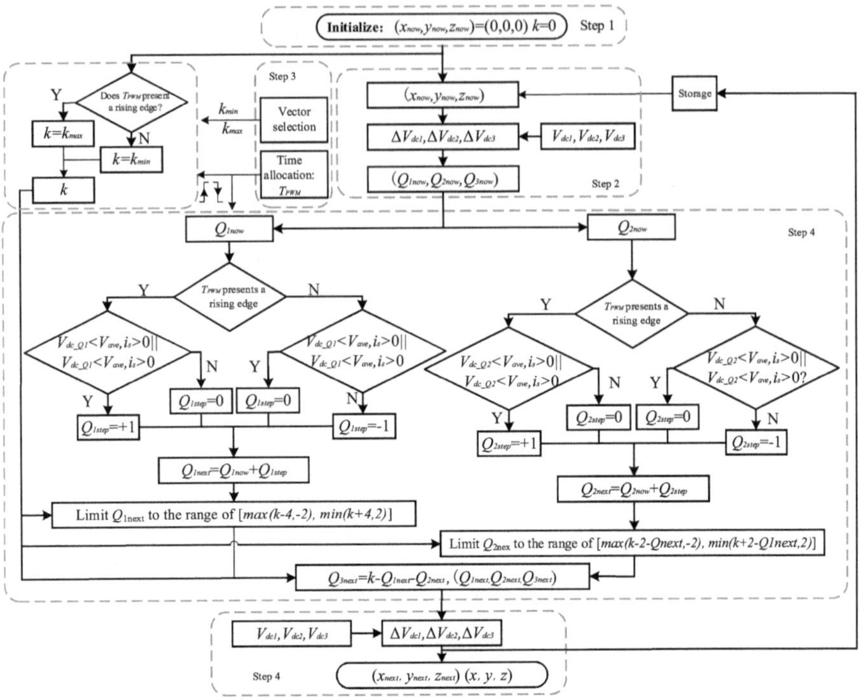

Fig. 4.39 Calculation process of three-module 1D-SVPWM algorithm

In this figure, $(x_{\text{now}}, y_{\text{now}}, z_{\text{now}})$ denote the current vectors, while $(x_{\text{next}}, y_{\text{next}}, z_{\text{next}})$ represent the vectors in the next cycle.

Step 1: The current acting vector $(x_{\text{now}}, y_{\text{now}}, z_{\text{now}})$ and the total port level k are initialized.

Step 2: The vectors $(Q_{1\text{now}}, Q_{2\text{now}}, Q_{3\text{now}})$ under the current state of voltage equalization are obtained according to the priority order for voltage equalization shown in Table 4.4.

Step 3: The number of levels $k_{\min}$ and $k_{\max}$ of the resultant vectors $V_{\min}$ and $V_{\max}$ is calculated using Eq. (4.61). Then, through time calculation and allocation, the action sequence (TPWM) of $V_{\min}$ and $V_{\max}$ is yielded using Eq. (4.64) and according to Fig. 4.40.

Step 4: The port level k-value when the current vector acts is obtained according to TPWM. First, $Q_{1\text{next}}$ is calculated among vectors $(Q_{1\text{next}}, Q_{2\text{next}}, Q_{3\text{next}})$, and the variation in Q_1 (Q_{1_step}) is yielded using Eqs. (4.74) and (4.75). Then, the Q_1-value $(Q_{1\text{next}})$ of the acting vector for the next cycle is calculated according to the current value $(Q_{1\text{now}})$ and the variation (Q_{1_step}):

$$Q_{1\text{next}} = Q_{1\text{now}} + Q_{1\text{step}} \tag{4.76}$$

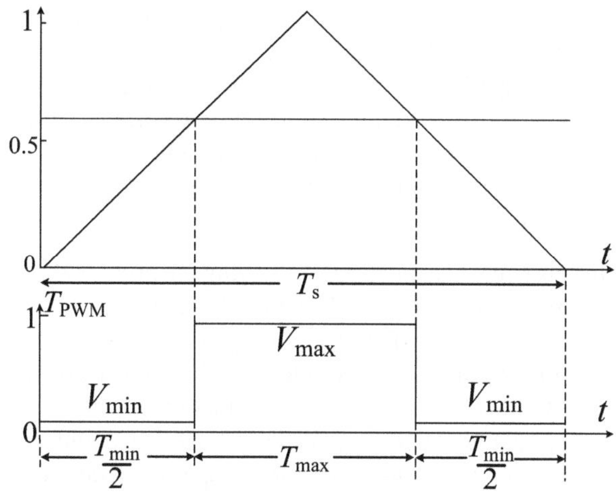

Fig. 4.40 Three-segment vector synthesis

The calculated Q_{1next} is constrained by the range given by Eq. (4.66) to ensure that the resulting Q_{1next} falls within this range.

Then, Q_{2next} is calculated. Similarly, variation in Q_2 (Q_{2step}) is calculated using Eqs. (4.74) and (4.75), and the Q_2-value (Q_{2next}) of the acting vector in the next cycle is calculated using the current value (Q_{2now}) and the variation (Q_{2step}):

$$Q_{2next} = Q_{2now} + Q_{2step} \tag{4.77}$$

The calculated Q_{2next} is constrained by the range given by Eq. (4.67) to ensure that the resulting Q_{2next} falls within this range.

Finally, substituting k, Q_{1next}, and Q_{2next} into Eq. (4.65) results in Q_{3next}, thereby completing the calculation of the acting vectors in the next cycle (Q_{1next}, Q_{2next}, Q_{3next}).

Step 5: According to Table 4.4, a reduction process transfers the vectors (Q_{1next}, Q_{2next}, Q_{3next}) into the vectors (x_{next}, y_{next}, z_{next}), which are used as the output vector (x, y, z) for the next cycle.

Taking the vector shown in Fig. 4.38 as an example, the calculation steps according to the process shown in Fig. 4.39 are as follows:

Step 1: Considering the current vector of (2, 1, 2), where $x_{now} = 2$, $y_{now} = 1$, and $z_{now} = 2$, the relationships (Q_1, Q_2, Q_3) = (x, y, z) lead to the establishment of $Q_{1now} = 2$, $Q_{2now} = 1$, and $Q_{3now} = 2$.

Step 2: With V^*_{ref} falling within interval XI, the calculation using Eq. (3.10) yields $k_{max} = 5$ and $k_{min} = 4$. As the vector changes from $5E$ to $4E$, TPWM presents a falling edge.

Step 3: Since TPWM presents a falling edge, $k = k_{min} = 4$ is established. First, Q_{1next} is calculated. Given that $V_{dc\_Q1} = V_{dc1} > V_{ave}$ and $i_s > 0$, $Q_{1step} = -1$ is obtained from Eq. (4.75). $Q_{1next} = 1$ is obtained by the calculation based on Eq. (4.76). Substituting $k = 4$ into Eq. (4.66) generates the range [0, 2]. As Q_{1next} falls within this range, $Q_{1next} = 1$ is established. Then, Q_{2next} is calculated. Given

$V_{dc\_Q2} = V_{dc2} > V_{ave}$ and $i_s > 0$, $Q_{2step} = -1$ is obtained from Eq. (4.75). $Q_{2next} = 0$ is obtained by the calculation based on Eq. (4.77). Substituting $k = 4$ and $Q_{1next} = 1$ into Eq. (4.66) generates the range (Zheng 2009; Xu 2013). Given $0 < 1$, $Q_{2next} = 1$ is established. Finally, substituting $k = 4$, $Q_{1next} = 1$, and $Q_{2next} = 1$ into Eq. (4.65) generates Q_{3next}, ultimately resulting in (Q_{1next}, Q_{2next}, Q_{3next}) = (1, 1, 2).

Step 4: The relationships (Q_1, Q_2, Q_3) = (x, y, z) lead to the establishment of (x, y, z) = (1, 1, 2).

In subsequent calculations, following the process shown in Fig. 4.39, the acting vector transitions from (1,1,2) to (1,2,2) and then to (0,2,2). This change process aligns with the red paths illustrated in Fig. 4.38. When voltage equalization is required for two or three modules, the selection should be made in accordance with their priorities. In the above example, since modules 1 and 2 are in the same conditions, the vector that effectively equalizes voltage for module 1 should be selected according to the default priority order for voltage equalization.

4.3.5 *SVPWM Algorithm for Voltage Equalization of* N-*Module Cascaded Converters*

4.3.5.1 One-Dimensional Vector Distribution for *N*-Module Cascaded Converters

Similarly, the vectors of single-phase n-module cascaded converters can be analyzed in the one-dimensional space. Letting the number of levels in the port voltage (u_{aibi}) of the ith module as $v_i = u_{aibi}/E$, the vector of nth module can be expressed as (v_1, v_2, v_3, …, v_n), where v_1, v_2, v_3, …, $v_n \in \{-2, -1, 0, 1, 2\}$, and the number of levels in the total port voltage (u_{a1bn}) is expressed as $k = u_{a1bn}/E$. The relationship is established as follows:

$$k = v_1 + v_2 + v_3 + \cdots + v_n \quad \left(k \in [-2n, 2n], k \text{ as an integer}\right) \tag{4.78}$$

First, the range of v_1 is determined, with $v_1 \in \{-2,-1,0, 1,2\}$, and the range can be expressed as follows:

$$v_1 \begin{cases} \geq \max\left[k - 2(n-1), -2\right] \\ \leq \min\left[k + 2(n-1), 2\right] \end{cases} \tag{4.79}$$

After v_1 is determined within the range given by Eq. (4.79), the range of v_2 can be expressed as follows:

$$v_2 \begin{cases} \geq \max\left[k - 2(n-2) - v_1, -2\right] \\ \leq \min\left[k + 2(n-2) - v_1, 2\right] \end{cases} \tag{4.80}$$

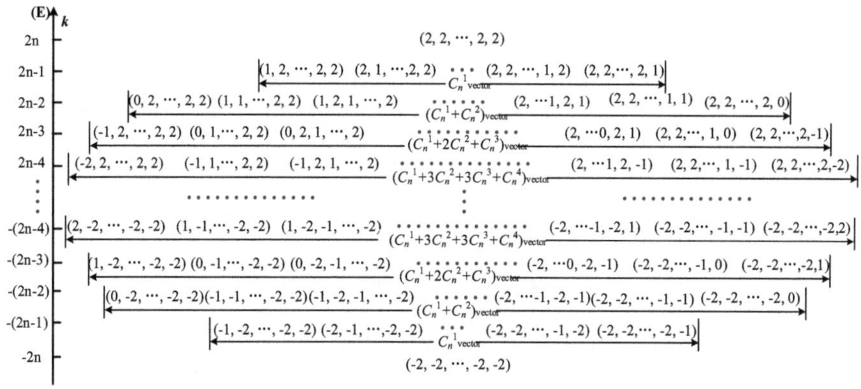

Fig. 4.41 One-dimensional space vector diagram of single-phase three-level H-bridge cascaded converter (n-module)

Subsequently, the range of v_i can be expressed as follows:

$$v_i \begin{cases} \geq \max\left[k-2(n-i)-v_1-v_2-\ldots-v_{i-1},-2\right] \\ \leq \min\left[k+2(n-i)-v_1-v_2-\ldots-v_{i-1},2\right] \end{cases} \tag{4.81}$$

Based on Eqs. (4.78)–(4.81), the one-dimensional space vector distribution of n-module cascaded converters can be generated, as shown in Fig. 4.41. Their vector distribution characteristics are similar to those of two-module and three-module configurations.

4.3.5.2 SVPWM Algorithm Implementation for Voltage Equalization

There are a total of $5n$ voltage vectors in cascaded converters with n three-level modules. Based on the voltage equalization principle, the priority order for voltage equalization, and the rules of selecting adjacent vectors, a cyclical calculation process yields the acting vector for module n, facilitating the DC-side voltage equalization of n-module cascaded converters. The entire calculation process, illustrated in Fig. 4.42, consists of five steps that are similar to the algorithm for the three-module configuration, where ($v_{1\text{now}}$, $v_{2\text{now}}$,..., $v_{n\text{now}}$) denote the current acting vectors and ($v_{1\text{next}}$, $v_{2\text{next}}$,..., $v_{n\text{next}}$) denote the acting vectors in the next cycle.

Step 1: The current acting vector ($v_{1\text{now}}$, $v_{2\text{now}}$, ..., $v_{n\text{now}}$) and port level k are initialized.

Step 2: The vectors ($Q_{1\text{now}}$, $Q_{2\text{now}}$, ..., $Q_{n\text{now}}$) under the current state of voltage equalization are obtained according to the priority order for voltage equalization shown in Table 4.4.

Step 3: The number of levels $k_{\min}$ and $k_{\max}$ of the resultant vectors $V_{\min}$ and $V_{\max}$ is calculated using Eq. (4.61). Then, through time calculation and allocation, the

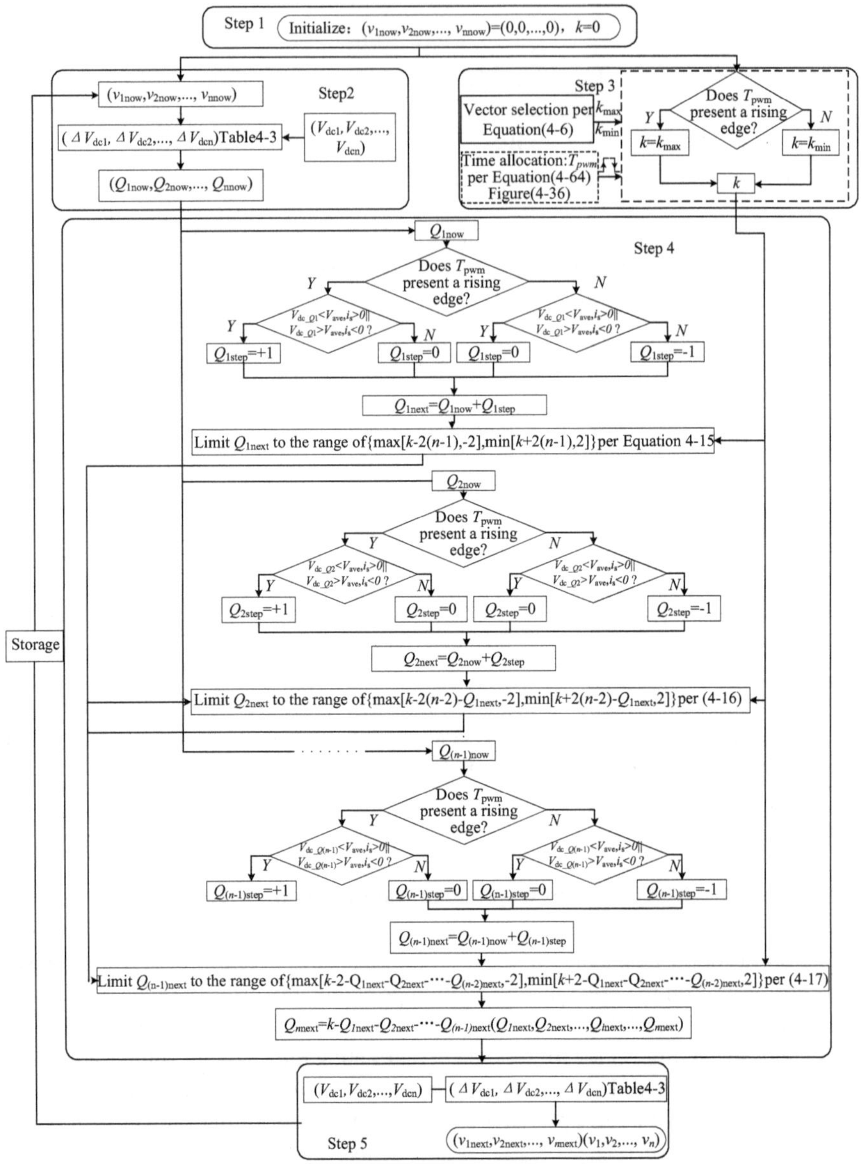

Fig. 4.42 Block diagram of 1D-SVPWM algorithm implementation for single-phase multi-module cascaded converter

action sequence (TPWM) of $V_{\min}$ and $V_{\max}$ is yielded using Eq. (4.64) and according to Fig. 4.36.

Step 4: The port level k-value when the current vector acts is obtained according to TPWM. First, $Q_{1\text{next}}$ is calculated among vectors ($Q_{1\text{next}}$, $Q_{2\text{next}}$, …, $Q_{n\text{next}}$), and the variation in Q_1 (Q_{1_step}) is yielded using Eqs. (4.74) and (4.75). Then, the Q_1-value

(Q_{1next}) of the acting vector for the next cycle is calculated according to the current value (Q_{1now}) and the variation (Q_{1_step}):

The calculated Q_{1next} is constrained by the range given by Eq. (4.79) to ensure that the resulting Q_{1next} falls within this range.

The values of Q_{1next}, Q_{2next}, ..., $Q_{(n-1)next}$ are calculated in a similar manner.

Finally, substituting k, Q_{1next}, Q_{2next}, ..., $Q_{(n-1)next}$ into Eq. (4.78) results in Q_{nnext}, thereby completing the calculation of the acting vectors in the next cycle (Q_{1next}, Q_{2next}, ..., Q_{nnext}).

Step 5: According to Table 4.4, a reduction process transfers the vectors (Q_{1next}, Q_{2next}, ..., Q_{3next}) into the vectors (v_{1next}, v_{2next}, ..., v_{nnext}), which are used as the output vector (v_1, v_2, v_3, ..., v_n) for the next cycle.

4.3.6 Simulation and Experimental Verification

4.3.6.1 Simulation and Experimental Verification of Intra-module Voltage Equalization

In the single-module configuration featuring a diode-clamped three-level circuit, redundant switch states are available in modes 2 and 4, among the operating states listed in Table 4.3. By selecting an appropriate switch state from these redundant options, dynamic equalization of capacitor voltage within the module can be achieved.

To verify the effectiveness of this intra-module voltage equalization strategy, loads of different sizes are connected to the two DC sub-capacitors C1 and C2 on the output side of the converter, for instance with $R_1 = 20\ \Omega$ and $R_2 = 50\ \Omega$. The simulation results are illustrated in Fig. 4.5.

As shown in Fig. 4.43, the simulation lasts for 0.5 s, during which the intra-module voltage equalization modulation strategy is not implemented until after 0.2 s. With the voltage reference for the DC side (U_{dc}) set at 100 V, the steady-state values of the two capacitor voltages U_{dc1} and U_{dc2} are 50 V when the sub-capacitor voltages are equalized. However, due to the absence of voltage equalization from $t = 0$ to $t = 0.2$ s, the voltage U_{dc1} of capacitor C1, which serves a larger load, gradually decreases to about 30 V, while the voltage U_{dc2} of capacitor C2, which serves a smaller load, gradually increases to about 70 V, with the total voltage stabilizing at 100 V. In this unbalanced state, the switching devices experience different voltage stresses. Additionally, the waveform of the port voltage u_{ab} on the input side is significantly distorted, as shown in Fig. 4.44a.

Capacitor voltage equalization modulation is implemented after 0.2 s. After approximately 0.05 s, the voltage values of both U_{dc1} and U_{dc2} return to balance at 50 V. Following this voltage equalization, the port voltage returns to its normal state, as shown in Fig. 4.44b.

The results of experiments conducted under the same conditions as the aforementioned simulation are shown in Fig. 4.45, where U_{dc} denotes the total output voltage on the DC side, U_{dc1} and U_{dc2} represent the voltages of the two sub-capacitors,

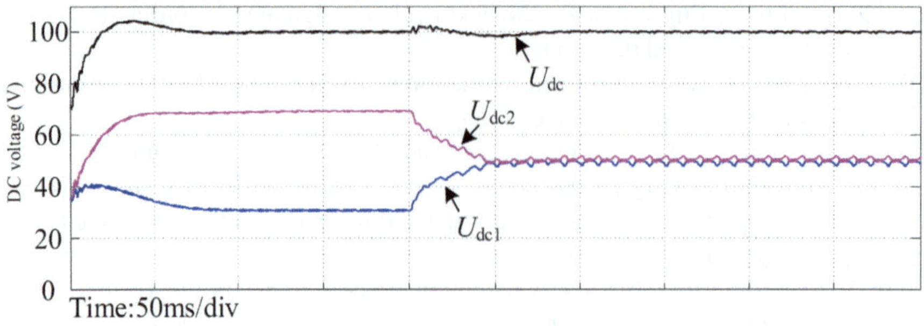

Fig. 4.43 Intra-module capacitor voltage equalization control

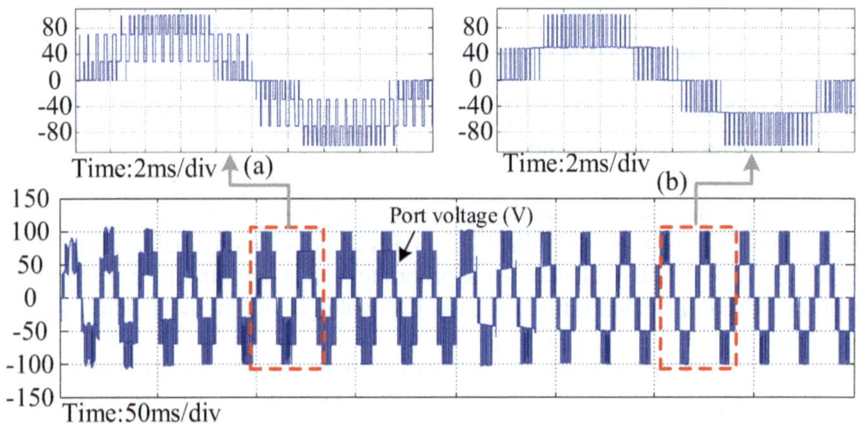

Fig. 4.44 Multilevels at AC input port

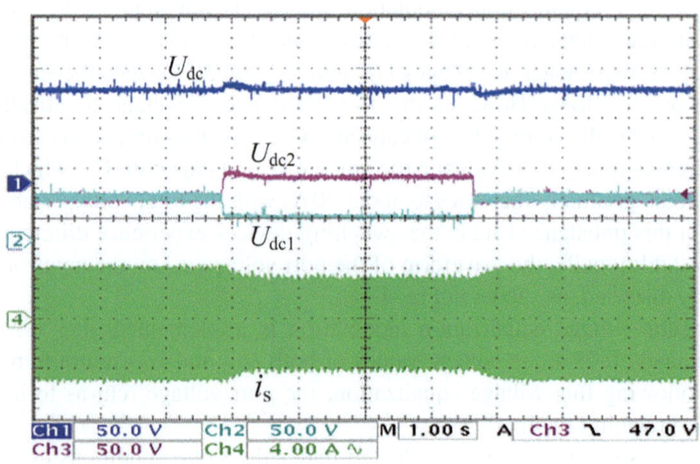

Fig. 4.45 Experiment of intra-module voltage equalization

and is indicates the current waveform on the AC grid side. After normal system operation for a period of time, the capacitor voltage equalization modulation is removed. Due to the larger load on capacitor C1, its voltage U_{dc1} decreases, while the voltage U_{dc2} increases due to the smaller load on capacitor C2. Following the restoration of voltage equalization modulation, the voltages on these two capacitors quickly return to a balanced state. The experimental results are consistent with the simulation results.

4.3.6.2 Simulation and Experiment of SVPWM Algorithm for Voltage Equalization

The SVPWM algorithm for voltage equalization of three-module cascaded converters, as discussed above, was simulated for verification using MATLAB/Simulink. The main simulation parameters are presented in Table 4.6.

To generate 13 levels from $-6E$ to $6E$, the total DC voltage reference is set at 189 V, the modulation degree is approximately 0.9, and each of the three modules is loaded with 20 Ω under the load equalization state. As shown in Fig. 4.46a, the AC voltage (u_s) and the AC current (i_s) remain aligned in phase. Meanwhile, the total port voltage (u_{a1b3}) exhibits 13 distinct levels, and the DC voltages across the 3 modules are maintained in balance at 63 V.

Table 4.6 Simulation parameters

Parameter	Value	Parameter	Value
AC-side power supply voltage, u_s/V	$120\sqrt{2}$	Inductance, L_s/mH	3
Frequency, f/HZ	50	Capacitance, C/μF	470*2
Switching frequency, f_s/Hz	4k	Unbalanced loads, R/Ω	50/20/20

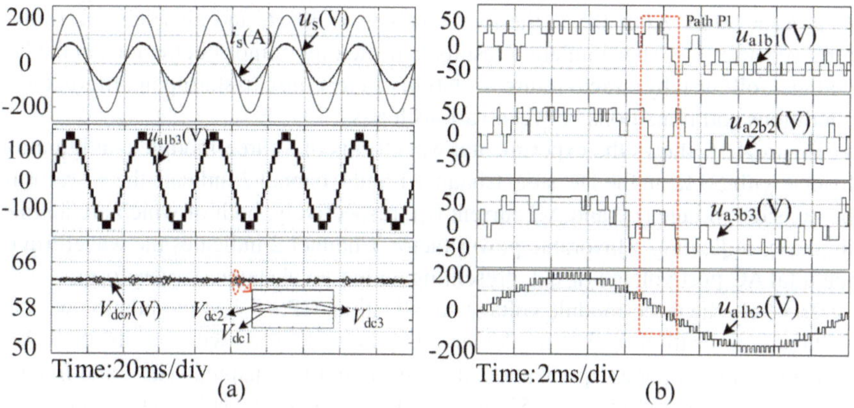

Fig. 4.46 Simulation of voltage equalization modulation for single-phase three-module cascaded converter. (**a**) the waveforms of AC voltage, AC current and DC voltages, (**b**) the waveforms of Modules 1, 2, and 3

Figure 4.46b shows the waveforms of modules 1, 2, and 3, as well as the total output port. The vector transition path P1 without jumps across the boxes in the figure is as follows: (1, 2, −1) → (1, 1, −1) → (2, 1, −1) → (2, 0, −1) → (2, 0, 0) → (2, −1, 0) → (2, −2, 0) → (2, −2, 1) → (1, −2, 1) → (2, −2, 1) → (1, −2, 1) → (1, −1, 1) → (1, −1, 0) → (0, − 1,0) → (0, 0, 0) → (−1, 0, 0) → (1, 1, 0) → (−2, 1, 0) → (−2, 2, −1). This indicates that all the module ports are free of level jumping, resulting in smooth transitions in the total output port levels, which closely aligns with sinusoidal waveforms.

Figure 4.47 displays the simulation waveforms of the converter when DC unbalanced loads ($R_1 = 50\ \Omega$, $R_2 = R_3 = 20\ \Omega$) are applied at a unity power factor. Here, M denotes the modulation degree, and λ represents the power factor. It is evident that, with varying modulation degrees and under different DC load conditions, the DC-side voltages of the three modules can be balanced without any skip-level jumping at the ports.

It is challenging to achieve voltage equalization in cases of significant load unbalance, particularly when one module is unloaded while the other modules are normally loaded. Figure 4.48a–c displays the simulation waveforms for the scenario of unbalanced loads, where module 1 is unloaded ($R_1 = \infty$), and modules 2 and 3 are loaded with $R_2 = R_3 = 20\ \Omega$.

Figure 4.48c presents significant fluctuations in the DC-side voltage of module 1 despite the possibility of voltage equalization at a modulation degree at $M = 0.8$. At a high modulation degree, such as $M = 0.9$, reducing the power factor to $\lambda = 0.80$ can achieve a better voltage equalization result, as shown in Fig. 4.48d, which indicates that reducing the power factor helps achieve DC voltage equalization.

Figure 4.49 illustrates the simulation results of the system's dynamic performance under various load conditions. The total DC-side voltage reference is set at 240 V, and the system operates at a unity power factor. The converter is initially started in an unloaded state and is then subjected to balanced loads after 0.5 s, specifically $R_1 = R_2 = R_3 = 50\ \Omega$, with the load R_1 being removed after 1 s. Figure 4.49b depicts the starting conditions in an unbalanced state, where $R_1 = \infty$ and $R_2 = R_3 = 50\ \Omega$. After 0.5 s, the load changes to 30 Ω, and all loads are further removed after 1 s. The simulation results indicate only minor fluctuations in the DC voltage, followed by a rapid return to stability in response to startup in different load conditions and transitions between various loads.

Figure 4.50 shows the experimental waveforms of a three-module configuration loaded with 50 Ω on the DC side. Channel 1 and Channel 2 indicate the waveforms of AC-side voltage (u_s) and AC current (i_s), respectively, both of which are aligned in phase, indicating a favorable power factor. Channel 2 indicates the waveform of the total AC port voltage (u_{a1b3}), presenting a nine-level stepwise waveform.

To verify the three-module voltage equalization, CPS-PWM modulation without voltage equalization is first implemented, followed by a switch to the SVPWM algorithm for voltage equalization discussed in this chapter. The experimental results are presented in Fig. 4.51. As shown, the three DC-side voltages are in an initially unbalanced state due to the use of CPS-PWM modulation. After implementing the SVPWM modulation, the module voltages quickly return to a balanced

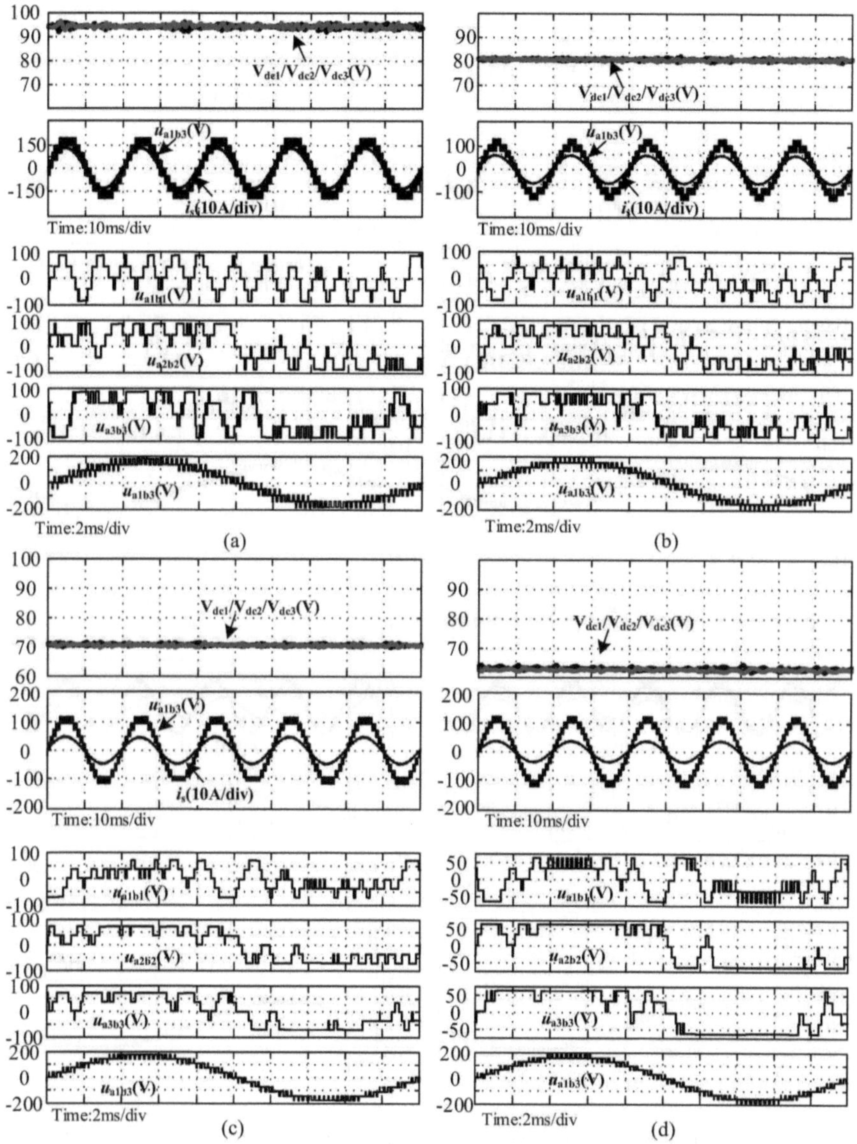

Fig. 4.47 Simulation waveforms under unbalanced loads ($R_1 = 50\ \Omega$, $R_2 = R_3 = 20\ \Omega$). (**a**) $M = 0.6$, $\lambda = 0.999$; (**b**) $M = 0.7$, $\lambda = 0.998$; (**c**) $M = 0.8$, $\lambda = 0.998$; (**d**) $M = 0.9$, $\lambda = 0.998$

state and remain stable. However, upon reapplying CPS-PWM modulation, the DC voltage becomes unbalanced again. These results demonstrate the effectiveness of the SVPWM algorithm in achieving voltage equalization.

Figure 4.52 displays the waveforms of port voltages for the individual modules and the total port voltage. These channels represent the port voltages of module 1, module 2, and module 3, denoted as u_{a1b1}, u_{a2b2}, and u_{a3b3}, respectively, along with

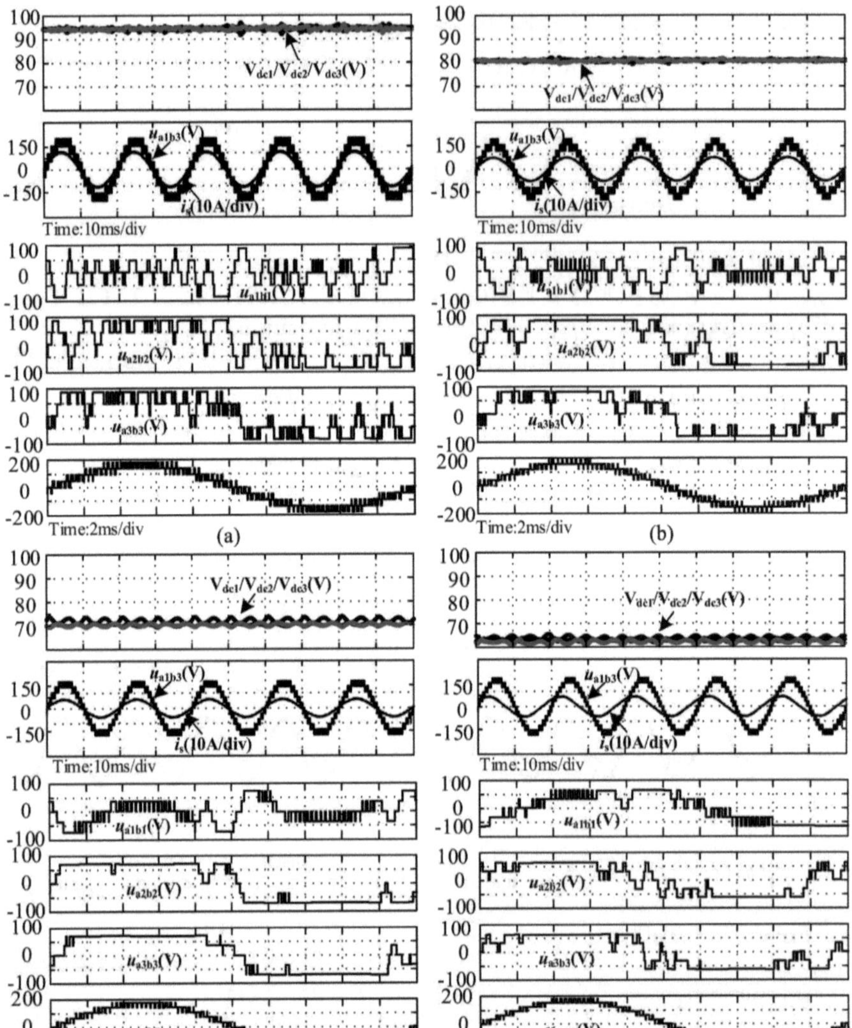

Fig. 4.48 Simulation waveforms under unbalanced loads ($R_1 = \infty$, $R_2 = R_3 = 20\ \Omega$). (**a**) $M = 0.6$, $\lambda = 0.998$; (**b**) $M = 0.7$, $\lambda = 0.997$; (**c**) $M = 0.8$, $\lambda = 0.995$; (**d**) $M = 0.9$, $\lambda = 0.80$

the total port voltage denoted as u_{a1b3}. As shown, the port levels of these modules exhibit different waveforms due to varying loads, with no instances of skip-level jumping. The total port level is synthesized into a sinusoidal waveform, which aligns with both theoretical and simulation results.

Figure 4.53a shows the experimental waveform of the system jumping from an unloaded state to a balanced load state and a two-load unbalanced state. Figure 4.53b presents the experimental waveforms of the system experiencing sudden changes from a two-load unbalanced state to a three-load unbalanced state and an unloaded

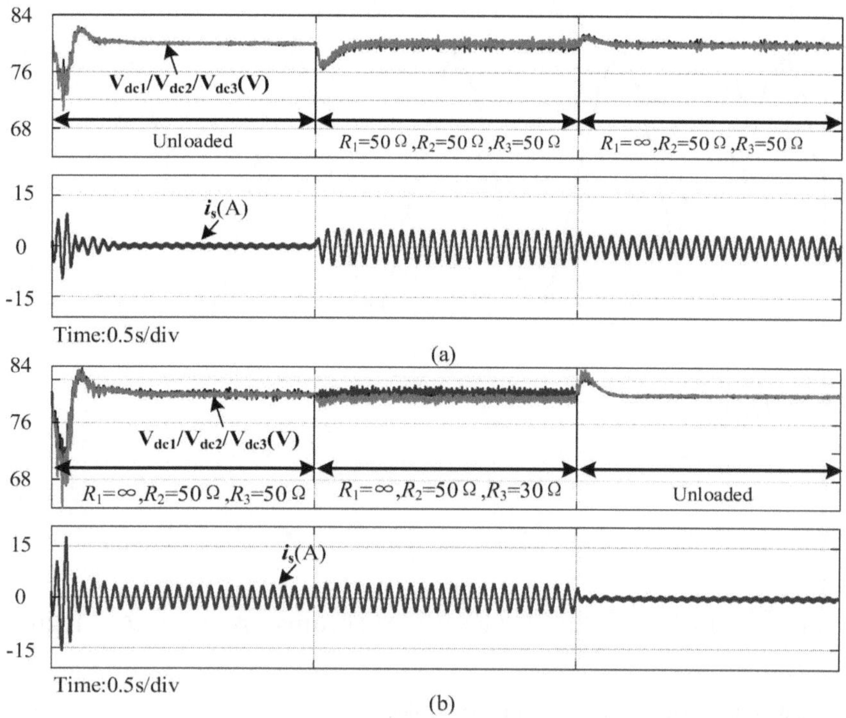

Fig. 4.49 Dynamic waveforms. (**a**) Waveform of system variations across unloaded startup, balanced loads, and unbalanced loads; (**b**) waveform of system variations across unbalanced startup, extremely unbalanced loads, and unloaded state

state. As indicated in Fig. 4.53, following sudden load changes, the DC voltage rapidly returns to a stable state, with the longest adjustment duration being approximately 0.4 s and the shortest about 0.2 s. The results demonstrate that the DC-side voltage can be equalized under various load conditions and that DC voltage equalization can be maintained even when the loads are unbalanced among the three modules.

4.3.6.3 Simulation of Six-Module Configuration

Figure 4.54 shows the simulation results of single-phase cascaded six-module converters. To present 25 levels ranging from $-12E$ to $12E$, the DC-side voltage reference is set at 590 V, the modulation degree M is taken as 0.91, and all the six modules are loaded with 20 Ω on the DC side. As shown in Fig. 4.54, rectification control ensures that the AC voltage (u_s) and AC current (i_s) are aligned in phase, allowing the system to operate at a unity power factor. The DC-side voltages among the six modules are equalized and maintained at a stable value of 98 V, resulting in a total port voltage (u_{a1b6}) with 25 distinct levels.

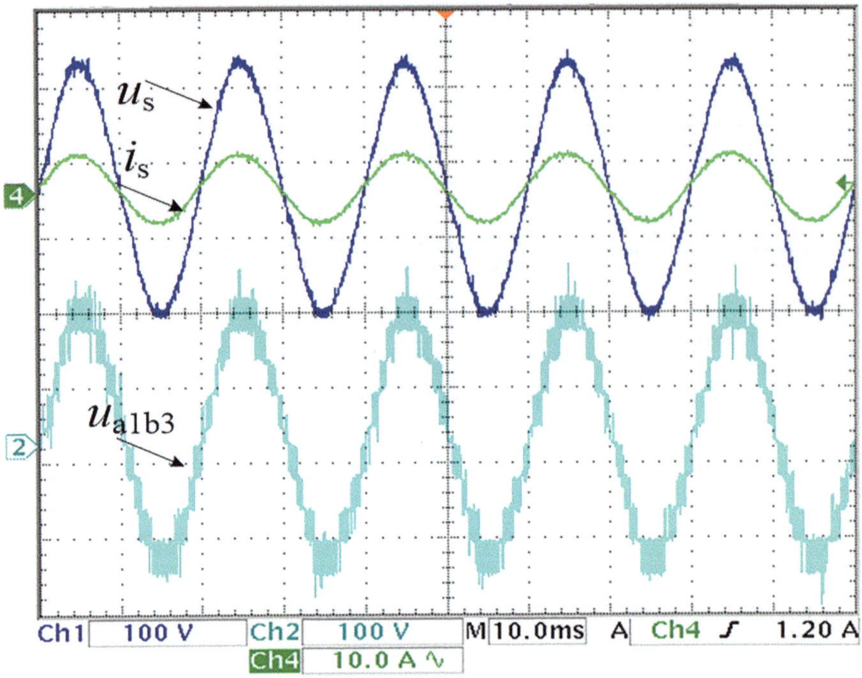

Fig. 4.50 Voltage and current waveforms on AC side

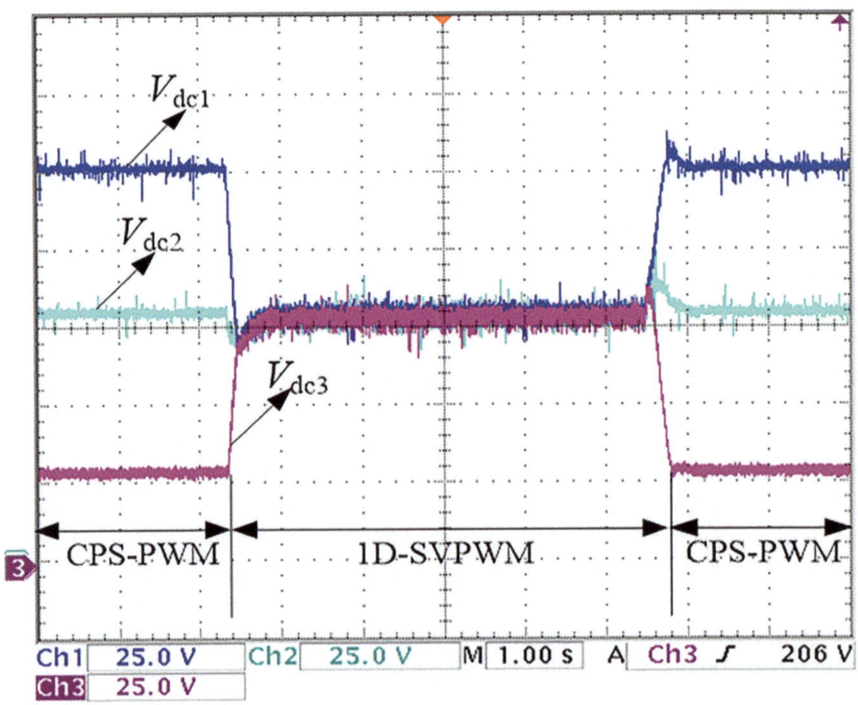

Fig. 4.51 DC-side voltage waveforms after switching modulation algorithms

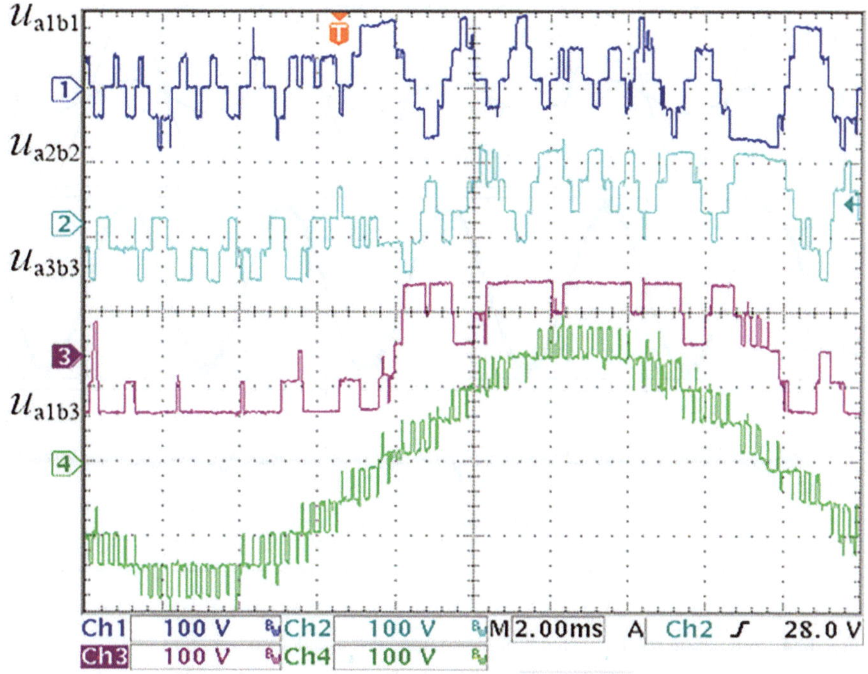

Fig. 4.52 Waveforms of port voltages for individual modules and total port voltage

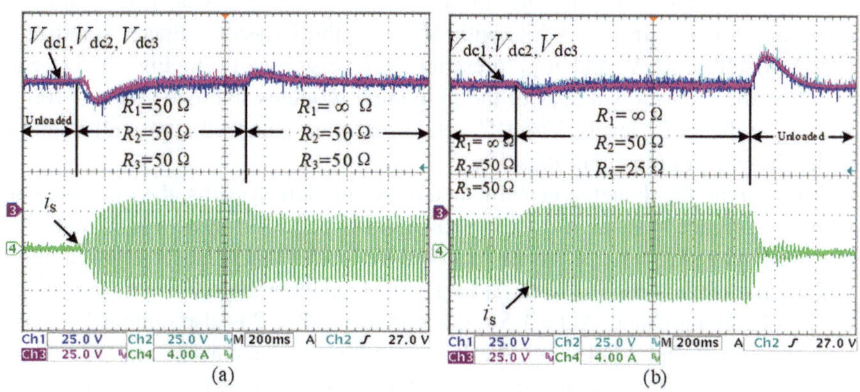

Fig. 4.53 Experimental waveform in the case of sudden load changes. (**a**) Unloaded state to balanced load state and then to unbalanced load state; (**b**) unbalanced load state to extremely unbalanced load state and then to unloaded state

Figure 4.55a–d presents simulation waveforms under unbalanced load conditions with modulation degrees (M) of 0.6, 0.7, 0.8, and 0.9, respectively. Specifically, module 1 is unloaded, while modules 2 through 6 are loaded with 20 Ω each. As shown, the DC-side voltages can be equalized despite the unbalanced load conditions. In Fig. 4.55a, with M = 0.6, the DC-side voltages of the individual modules

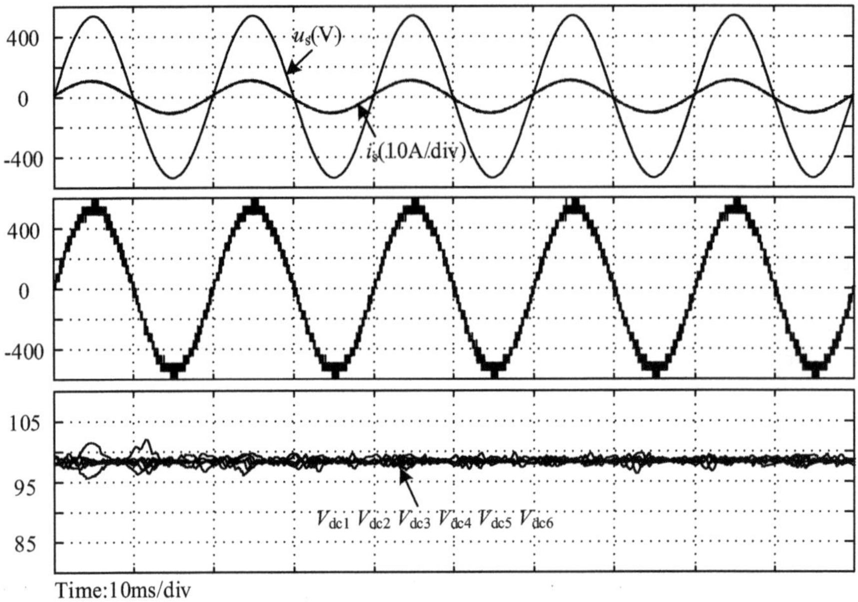

Fig. 4.54 Simulation results of single-phase six-module cascaded converter

stabilize at about 150 V, and the AC port voltage (u_{a1b6}) presents 17 levels. In Fig. 4.55b, with $M = 0.7$, the DC-side voltages stabilize at around 128 V, and u_{a1b6} presents 19 levels. In Fig. 4.55c, with $M = 0.8$, the DC-side voltages stabilize at approximately 112 V, and u_{a1b6} presents 21 levels. Finally, in Fig. 4.55d, with $M = 0.9$, the DC-side voltages of the individual modules stabilize at about 100 V, and u_{a1b6} reflects 23 levels. The DC voltages remain balanced across all modules under these operating conditions.

To simulate the dynamic performance of six-module systems, the total DC-side voltage reference is set at 760 V. As shown in Fig. 4.56, the system is started in an unloaded state. At $t = 0.5$ s, the six modules are loaded in a balanced configuration, with each module loaded at 20 Ω. At $t = 1$ s, the load on module 1 is removed, resulting in it being unloaded. At $t = 1.5$ s, the load on module 2 is also removed, leaving both modules 1 and 2 unloaded. As illustrated, the system current and DC voltage quickly return to stability. Specifically, it takes approximately 0.2 s to stabilize following the unloaded startup, about 0.1 s to stabilize after a sudden change from an unloaded state to a balanced load state, and around 0.05 s to stabilize following sudden changes from a balanced load state to an unbalanced load state, as well as from one unbalanced load state to another.

The six-module configuration has been simulated without further experimental research. Through the analysis of the two-module and three-module configurations, along with an extended analysis of N modules using six modules as an example, it can be concluded that inter-module voltage equalization of single-phase cascaded

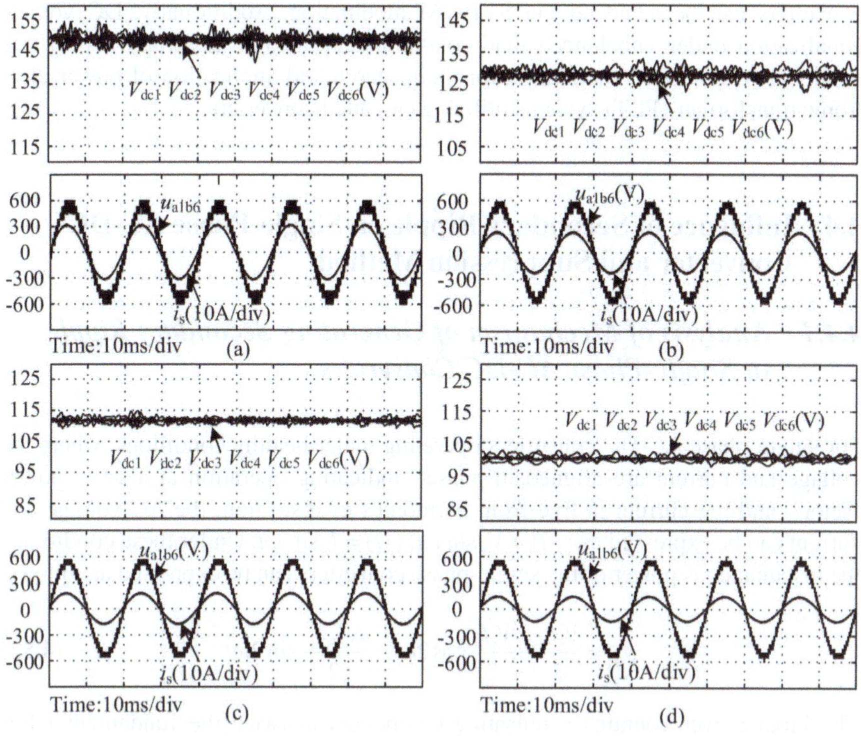

Fig. 4.55 Simulation waveforms under unbalanced load conditions ($R_1 = \infty$, $R_1 = R_2 = R_3 = R_4 = R_5 = 20\ \Omega$). (**a**) $M = 0.6$, $\lambda = 0.99$; (**b**) $M = 0.7$, $\lambda = 0.99$; (**c**) $M = 0.8$, $\lambda = 0.99$; (**d**) $M = 0.9$, $\lambda = 0.98$

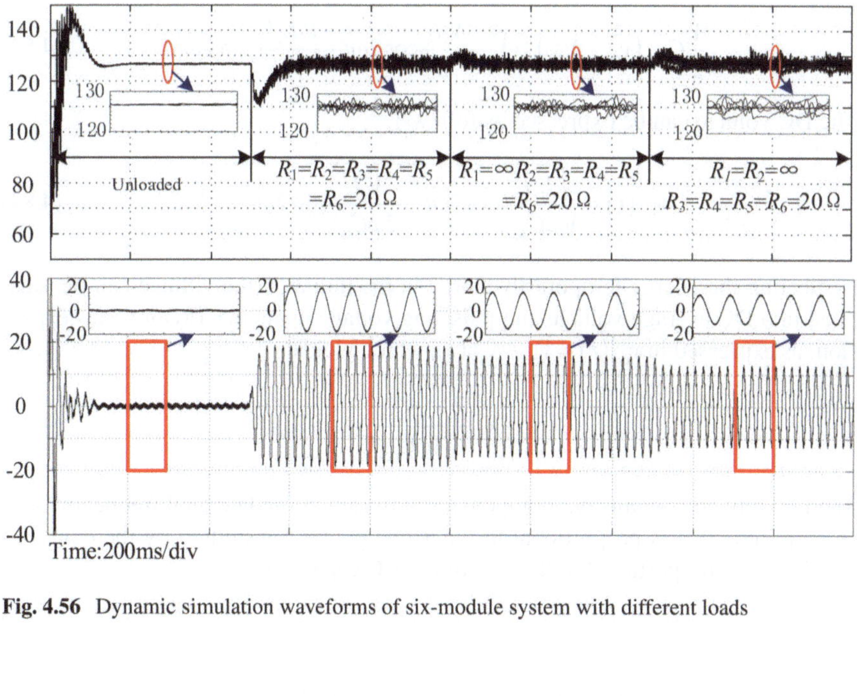

Fig. 4.56 Dynamic simulation waveforms of six-module system with different loads

converters can be achieved via SVPWM modulation. Additionally, DC voltage equalization under unbalanced load conditions can also be accomplished. These results provide a solid foundation for the research and application of power electronic transformer (PET) systems utilizing cascaded converters.

4.4 Influence of Secondary Ripple on Single-Phase AC-DC Converter and Suppression Method

4.4.1 Analysis of Mechanism of Generating Secondary Ripple in Single-Phase AC-DC Converters

For single-phase AC-DC converters operating under healthy conditions, where the voltage and current are aligned in phase—indicating operation at a unity power factor—and the current is free from harmonics or distortion, the AC voltage and current can be expressed as $v_s(t) = V_s \sin \omega t$, $i_s(t) = I_s \sin \omega t$. Under these conditions, the instantaneous power of the single-phase converters can be expressed as follows:

$$p_{\text{in}} = \frac{V_s I_s}{2} - \frac{V_s I_s}{2}\cos 2\omega t - \frac{\omega L_s I_s^2}{2}\sin 2\omega t \tag{4.82}$$

The input power contains a pulsating component at twice the fundamental frequency, affecting the ripple of the DC voltage. The magnitude of the ripple mainly depends on the AC values of the last two terms in Eq. (4.82). Under these conditions, the energy of the instantaneous ripple stored in the capacitor satisfies the following equation:

$$\frac{1}{2}C_{\text{dc1}}\left(v_{\text{dc}}^2 - V_{\text{dc}}^2\right) = \int_0^t -\frac{V_s I_s}{2}\cos 2\omega t - \frac{\omega L_s I_s^2}{2}\sin 2\omega t dt \tag{4.83}$$

The DC voltage can be expressed as follows:

$$v_{\text{dc}}(t) = \sqrt{\frac{L_s I_s^2}{2C_{\text{dc1}}}\cos 2\omega t - \frac{V_s I_s}{2\omega C_{\text{dc1}}}\sin 2\omega t + V_{\text{dc}}^2} \tag{4.84}$$

Neglecting the energy temporarily stored in the inductor, the voltage can be separated into DC average voltage and pulsating voltage using the Taylor series expansion, as expressed in $v_{\text{dc}}(t) = V_{\text{dc}} - \hat{v}_{\text{dc}}(t)$, which yields

$$v_{\text{dc}}(t) = V_{\text{dc}} - \frac{V_s I_s}{4\omega C_{\text{dc1}} V_{\text{dc}}}\sin 2\omega t \tag{4.85}$$

The DC voltage contains AC components at twice the fundamental frequency, with an amplitude that is proportional to the active power transmitted by the converter and inversely proportional to the size of the DC capacitor.

The flowing capacitive current satisfies the relational expression $i_c(t) = C_{dc}dv_{dc}(t)/dt$, which, when combined with Eq. (4.85), yields

$$i_c(t) = -\frac{V_s I_s}{2V_{dc}}\cos 2\omega t \tag{4.86}$$

It is evident that the capacitance of the DC capacitor does not affect the amplitude of the current flowing through it, and the capacitive current also contains AC components at twice the fundamental frequency.

To reduce the influence of secondary ripple, the general measures include the addition of LC secondary filter branches or capacitors with larger capacitance to reduce ripple amplitude. However, the addition of secondary filter branches features low frequencies, large operating currents, large volumes, and high hardware costs; increasing the capacitance also significantly increases the system volume and cost. Subsequent sections discuss suppressing the effects of ripple on AC ports of AC-DC converters through control and modulation while retaining DC ripple without increasing hardware cost.

4.4.2 Influence of DC Ripple on AC-DC Converter

The single-phase AC-DC converter can employ dual-loop control for both voltage and current, such as the classic transient current control illustrated in the block diagram in Fig. 4.19. The controller first compares the DC output voltage with the voltage reference and uses a proportional-integral controller to make the output voltage (v_{dc}) align with the reference voltage ($v_{dc}*$), thereby stabilizing the rectified output DC voltage. The controller then multiplies the output of the proportional-integral controller by the sinusoidal signal from the phase-locked loop (PLL) to generate the real-time reference current signal (i_s*) for the current control loop.

Based on the analysis in the previous section, when the DC capacitance is low or the transmission power is high, the DC ripple increases significantly. Focusing solely on the influence of the secondary ripple on the voltage outer loop, the following equation is established:

$$\begin{cases} e_v = v_{dc}^* - v_{dc} = \hat{v}_{dc}(t) \\ i_s^* = I_s^* \sin\omega t = \sin\omega t \times \left(k_p e_v + k_i \int e_v \mathrm{d}t\right) \end{cases} \tag{4.87}$$

Then, the reference current output by the controller PI can be expressed as follows:

$$I_s^* = \frac{k_i V_s I_s}{8\omega^2 C_{dc1} V_{dc}}\cos 2\omega t - \frac{k_p V_s I_s}{4\omega C_{dc1} V_{dc}}\sin 2\omega t \tag{4.88}$$

Based on Eq. (4.88), the influence of the proportional controller parameters on the reference current is more pronounced. By ignoring the influence of the integral controller, the current reference signal can be expressed as follows:

$$i_s^* = \frac{k_{pv} V_s I_s}{8\omega C_{dc1} V_{dc}} (\cos \omega t - \cos 3\omega t) \tag{4.89}$$

Under these conditions, the current contains both reactive current components at the fundamental frequency and third harmonic components, which significantly affect the power quality of the converter, including the AC-side power factor and harmonics. Furthermore, the third harmonic components of the AC current are transmitted through the AC-DC power of the converter, generating fourth harmonic voltage pulsations on the DC capacitor. These fourth harmonic pulsations then produce the fifth harmonic current on the AC side. Similarly, multiple higher-frequency harmonic voltages and currents are generated in succession.

Due to the mutual influence between voltage and current, effective control of current distortion helps to reduce voltage distortion. For instance, a trap filter can be added to the voltage control loop to eliminate harmonic components from the DC voltage. Assuming that all harmonic components are completely filtered out, the reference current will consist solely of fundamental components. Thus, if the current inner loop only tracks the reference current at the fundamental frequency, the modulation waveform will contain no third harmonics or other harmonics. By neglecting higher harmonic components, and with the modulation signal denoted as $u_{ab}^* \sin \omega t$, the port level can be expressed as follows:

$$\begin{aligned} u_{ab} &= u_{ab}^* \sin \omega t \left(V_{dc} - \frac{V_s I_s}{4\omega C_{dc} V_{dc}} \sin 2\omega t \right) \\ &= u_{ab}^* V_{dc} \sin \omega t + \frac{u_{ab}^* V_s I_s}{8\omega C_{dc} V_{dc}} (\cos \omega t - \cos 3\omega t) \end{aligned} \tag{4.90}$$

Applying Kirchhoff's voltage law (KVL), the grid-side current can be expressed as follows:

$$i_s = \frac{1}{L_s} \int_0^t V_s \sin \omega t - u_{ab} \mathrm{d}t \tag{4.91}$$

It is understood that the difference between the port voltage, as given by Eq. (4.90), and the grid-side voltage represents the voltage across both ends of the input inductor of single-phase converters. As indicated by Eq. (4.91), even if the harmonic components of the DC voltage are suppressed, the grid-side current still contains third harmonic components.

Therefore, the DC bus voltage of single-phase AC-DC converters exhibits low-frequency ripple, and the influence of harmonics cannot be completely eliminated by control methods such as trap filters. As a result, the grid-side current still contains third and other low-frequency harmonics, which generate reactive power components simultaneously and reduce the input power factor of the converters.

4.4.3 *Suppression of Low-Frequency Ripple Based on Trap Filter Method*

For single-phase converters configured with different DC capacitors or utilizing various control strategies, the DC voltage and controller modulation waveforms are presented in Fig. 4.57. When the DC capacitance is high, as illustrated in Fig. 4.57a, the DC voltage exhibits negligible ripples. Neglecting high-frequency components, the port voltage appears sinusoidal, and the total harmonic distortion (THD) of the input current remains low. However, when the DC capacitance is reduced, as shown in Fig. 4.57b, fluctuations in DC voltage ripple increase significantly, resulting in worse distortion of the port waveform and a substantial increase in harmonics within the input current. If the modulation waveform is dynamically adjusted in response to fluctuations in the DC voltage, it is possible to prevent distortion of the port voltage. The simulation results demonstrating this method are shown in Fig. 4.57c.

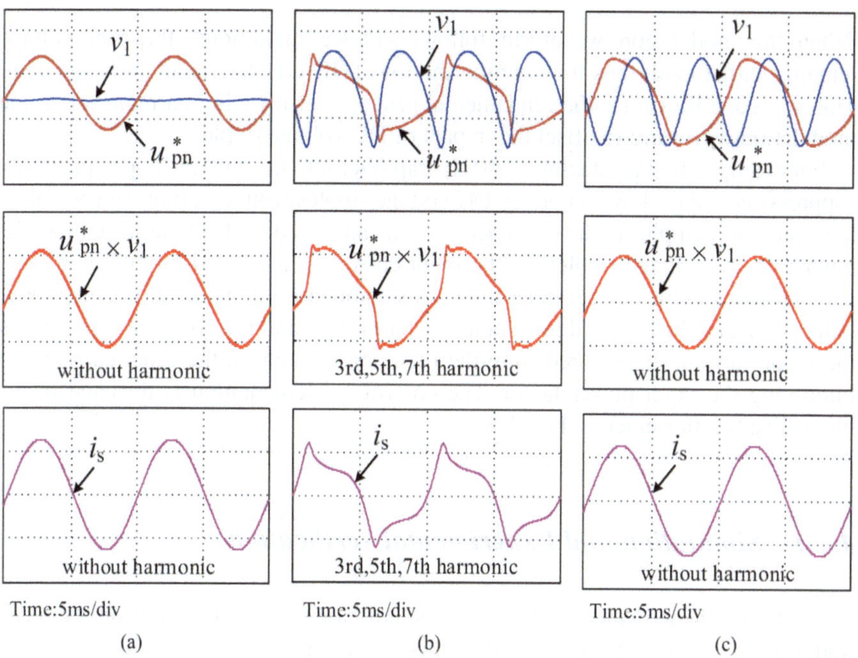

Fig. 4.57 Theoretical waveforms of single-phase converter under various conditions. (**a**) Conventional control under large capacitance conditions; (**b**) traditional control under small capacitance conditions with harmonic; (**c**) traditional control under small capacitance conditions without harmonic

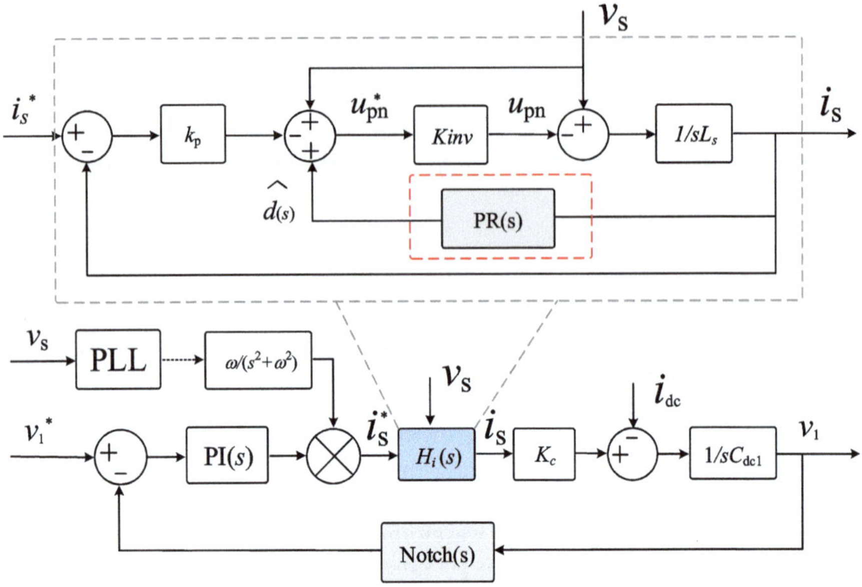

Fig. 4.58 Block diagram of trap filter method to suppress the influence of ripple

When the modulation waveform follows voltage distortions, the port voltage remains undistorted, thus maintaining the low THD characteristics of the current. This approach illustrates the fundamental principle of injecting harmonics into the modulation waveform to offset the impact of DC voltage ripple.

For example, by introducing a 100 Hz trap filter into the voltage loop to primarily suppress secondary low-frequency DC voltage ripples from entering the controller, low-frequency ripples in the reference current can be reduced. At the same time, the impact of DC voltage ripples on the input current is mitigated through harmonic compensation in the current inner loop. This compensation mode allows the modulation signals to be adjusted in response to harmonics and reactive power present in the input current, thereby reducing the harmonic content in the input current and improving the input power factor. The control implementation is depicted in the block diagram presented in Fig. 4.58.

4.4.4 Simulation and Experimental Verification

In small DC capacitance conditions, the simulation results are compared between traditional control with the control method mentioned in the previous section, as shown in Fig. 4.59. Under conventional control, the total harmonic distortion (THD) in the grid current reaches 35.69%, with third harmonic components as high as 33.29%, while other odd harmonics decrease with increasing orders, as illustrated

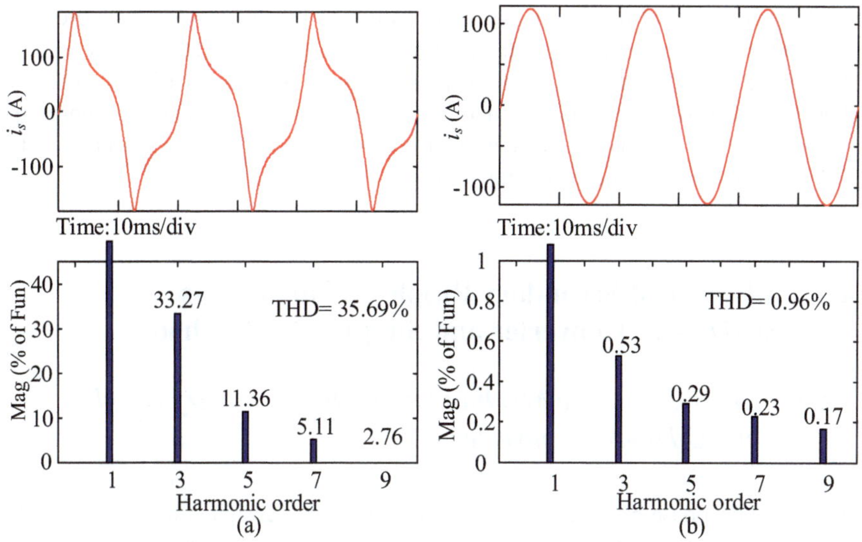

Fig. 4.59 FFT analysis of grid-side current under two control strategies: (**a**) traditional control, (**b**) proposed control

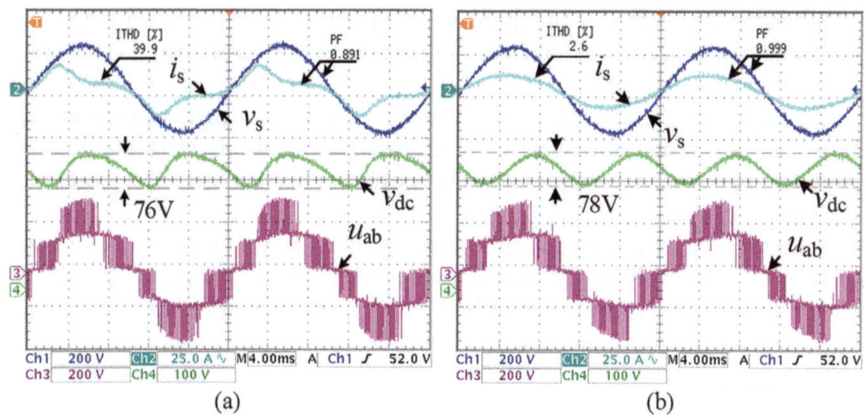

Fig. 4.60 Experimental waveforms of single-phase converter: (**a**) traditional control strategy, (**b**) proposed control strategy

in Fig. 4.59a. In contrast, the implementation of the optimization control described in the previous section resulted in a reduction of THD in the grid current to 0.96%, demonstrating a substantial improvement.

The results of the comparative experiments are shown in Fig. 4.60. When the DC voltage is set to 300 V, the DC voltage ripple under traditional control measures 76 V. Due to this ripple, the THD in the grid-side current reaches 39.9%, and the power factor (PF) is only 0.891. This indicates that the DC voltage ripples significantly impact the input power quality of the single-phase converter.

The experimental results of the optimization control proposed to suppress the ripple effect are shown in Fig. 4.60b. With DC voltage ripples at 78 V, the THD in the grid-side current decreases to 2.6%, while the power factor (PF) increases to 0.999. These results demonstrate that the proposed control method for suppressing the ripple effect can significantly mitigate the impact of DC voltage ripple on the DC input power of single-phase converters.

4.5 Influence of Secondary Ripple on Single-Phase AC-DC-AC Converter and Suppression Method

4.5.1 *Cross-Influence Mechanism of DC Secondary Ripple in AC-DC-AC Converter*

Advanced co-phase traction power supply systems are designed to convert the AC input from the grid into AC voltage for the traction network. The AC-DC-AC converter topology facilitates the connection of dual AC ports. Based on the two-level AC-DC-AC converter circuit topology, the connection mode is illustrated in Fig. 4.61.

Similar to rectifiers, this circuit topology represents a single-phase AC-DC conversion and is subject to the influence of secondary ripple, as discussed in the previous section. Additionally, this circuit topology functions as a bidirectional single-phase AC-DC converter, with both ports acting as single-phase rectifiers. When the voltage and current phases of the two ports are inconsistent, the effects of their dual-port secondary ripples may either offset or overlap each other, necessitating further analysis.

Assuming that phase A input can operate at a unity power factor when connected to the grid, where voltage and current are aligned in phase, converter is considered ideally loss-free, and harmonics of both voltage and current on the grid side are neglected, the following equation is established:

$$\begin{cases} v_{s\alpha} = V_s \sin \omega t \\ i_{s\alpha} = I_{sd} \sin \omega t \end{cases} \tag{4.92}$$

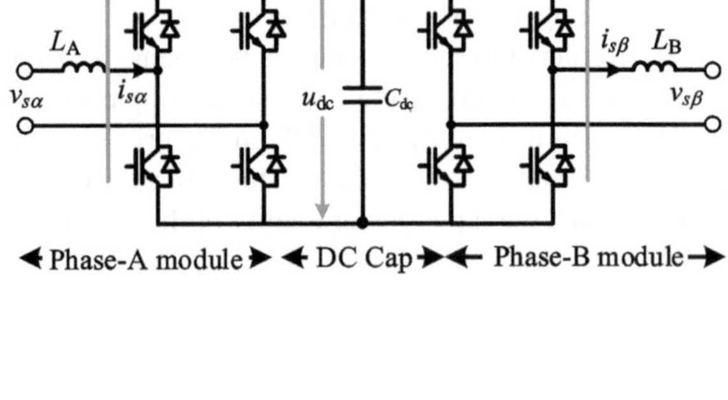

Fig. 4.61 Single-module topology of back-to-back two-level full-bridges for co-phase power supply

The instantaneous power of inductor A can be expressed as follows:

$$P_{\text{LA}} = \frac{\omega L_{\text{A}} I_{sd}^2 \sin 2\omega t}{2} \tag{4.93}$$

The input power of the converter port can be expressed as follows:

$$p_{\text{in}} = V_{\text{s}} I_{sd} \frac{1 - \cos 2\omega t}{2} - P_{\text{LA}} \tag{4.94}$$

Similarly, the output power of phase B can be expressed as follows:

$$\begin{cases} v_{s\beta} = V_{\text{s}} \cos \omega t \\ i_{s\beta} = I_{sd} \cos \omega t + I_{sq} \sin \omega t \end{cases} \tag{4.95}$$

The instantaneous power on inductor B can be expressed as follows:

$$P_{\text{LB}} = \frac{\omega L_{\text{B}} \sin 2\omega t}{2}\left(I_{sq}^2 - I_{sd}^2\right) \tag{4.96}$$

The output power of the converter port can be expressed as follows:

$$p_{\text{out}} = V_{\text{s}} I_{sd} \frac{1 + \cos 2\omega t}{2} + \frac{V_{\text{s}} I_{sq}}{2} \sin 2\omega t + P_{\text{LB}} \tag{4.97}$$

If the inductance at both ends is the same, the difference in power between the two AC ports can be expressed as follows:

$$p_{\text{in}} - p_{\text{out}} = \frac{V_{\text{s}} - \omega L_{\text{B}} I_{sq}}{2} I_{sq} \sin 2\omega t - V_{\text{s}} I_{sd} \cos 2\omega t \tag{4.98}$$

Based on Eq. (4.98), the power difference between the two AC ports is reflected in the DC capacitor voltage, with its pulsating power primarily influenced by secondary power. The voltage fluctuation range on the DC side satisfies the following equation:

$$\frac{1}{2} C_{\text{dc}} \left(v_{\text{dc}}^2 - V_{\text{dc}}^2\right) = \int_0^t p_{\text{in}} - p_{\text{out}} \mathrm{d}t \tag{4.99}$$

Derivation from Eq. (4.99) yields

$$v_{\text{dc}}(t) = \sqrt{V_{\text{dc}}^2 - \frac{V_{\text{s}} I_{\text{sd}}}{\omega C_{\text{dc}}} \sin 2\omega t + \left(1 - \cos 2\omega t\right) \frac{V_{\text{s}} - \omega L_{\text{B}} I_{\text{sq}}}{2\omega C_{\text{dc}}} I_{\text{sq}}} \tag{4.100}$$

The DC voltage waveform in response to the different powers transmitted, as derived from the above equation, is shown in Fig. 4.62.

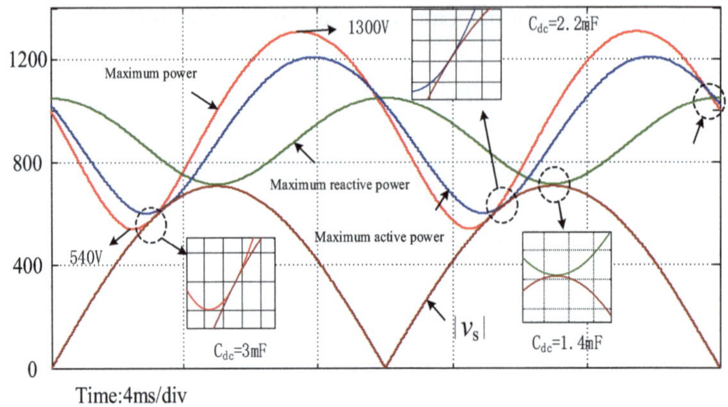

Fig. 4.62 Fluctuation in DC bus voltage under various operating conditions

With harmonic compensation provided by the converter, the power stored in the DC capacitor due to the fluctuating voltage can be expressed as follows:

$$p_{dch} = -V_s \sin \omega t \sum_{k=3}^{2n+1} i_{hn} \tag{4.101}$$

The resulting DC voltage fluctuation can be expressed as follows:

$$\hat{v}_{dch} = \frac{V_s I_{ck}}{2\omega V_{dc} C_{dc}} \sum_{k=3}^{2n+1} \frac{\cos\left[(k+1)\omega t + \varphi_{k1}\right]}{k+1} - \frac{\cos\left[(k-1)\omega t + \varphi_{k2}\right]}{k-1} \tag{4.102}$$

The fluctuation of DC voltage during harmonic compensation is depicted in Fig. 4.63. As the order of harmonic compensations increases, the distortion of the DC voltage becomes more pronounced.

When both the input and output contain harmonics, the generation mechanism of DC voltage ripple is similar to the analysis provided above and is not repeated in this book.

4.5.2 Low-Frequency Ripple Effect Suppression Strategy Based on Compensation of Harmonics at Characteristic Orders

As shown in Fig. 4.59 and other figures, the converter current harmonics exhibit typical distribution characteristics at characteristic orders, such as the third, fifth, and seventh harmonics. Based on the principle of individual compensation for

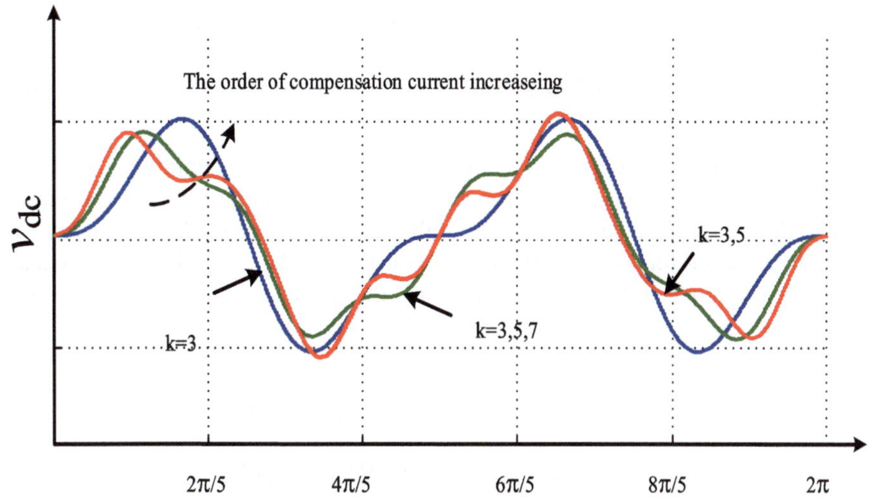

Fig. 4.63 DC voltage ripple with increasing harmonic order under sole harmonic compensation

harmonics at these characteristic orders, *k*th harmonic closed-loop control can be employed to generate *k*th compensation voltages (u_{hk}) for compensating *k*th harmonic currents. The superposition of compensation voltages at all characteristic orders effectively suppresses the influence of low-frequency ripple on current distortion.

First, a single-phase phase-locked loop (PLL) is used to detect the real-time phase information of phases A and B, ranging from the fundamental frequency to *k* times the fundamental frequency, respectively. Based on a concept similar to active and reactive decoupling control, the *k*th harmonic is converted into a *d-q* DC component at the *k*th harmonic frequency.

Second, the DC components at each harmonic order are aligned with a reference value of 0, and closed-loop control is achieved using a PI controller.

Finally, the harmonic reference voltage in the *k*th *d-q* coordinate system is converted into the instantaneous voltage reference (u_{hk}) in the *abc* coordinate system using the inverse transformation at the *k*th frequency.

Similar to the active and reactive decoupling analysis conducted at the fundamental frequency, the mathematical model for the *k*th harmonics can be expressed as follows:

$$\begin{cases} V_{kd} = -L\dfrac{\mathrm{d}i_{kd}}{\mathrm{d}t} - Ri_{kd} + k\omega Li_{kq} \\ V_{kq} = -L\dfrac{\mathrm{d}i_{kq}}{\mathrm{d}t} - Ri_{kq} + k\omega Li_{kd} \end{cases} \tag{4.103}$$

With $i_{kd} = |i_{kd}|e^{jk\omega t}$, $L\mathrm{d}i_{kd}/\mathrm{d}t$ can be converted into the following equation:

$$L\frac{\mathrm{d}i_{kd}}{\mathrm{d}t} = jk\omega L\left|i_{kd}\right|\mathrm{e}^{jk\omega t} = jk\omega L i_{kd} \tag{4.104}$$

Assume that there is a coefficient relationship (m_1) between the amplitudes of i_{kd} and i_{kq}, and its value varies across different system states, and let $|i_{kd}| = m_1|i_{kq}|$; meanwhile, due to the i_{kd} delay by $\pi/2$ of i_{kq}, the relationship of $jk\omega Li_{kd} = k\omega Lm_1 i_{kq}$ is established. Therefore, Eq. (4.104) can be converted to $L\mathrm{d}i_{kd}/\mathrm{d}t = k\omega Lm_1 i_{kq}$. To sum up, the error between the real-time value and the reference value can be expressed as follows:

$$\begin{cases} -L\dfrac{\mathrm{d}\left(i_{kd} - i_{kd\mathrm{ref}}\right)}{dt} = k\omega Lm_1\left(i_{kq\mathrm{ref}} - i_{kq}\right) \\ -R\left(i_{kd} - i_{kd\mathrm{ref}}\right) = k\omega Lm_2 \int\left(i_{kq\mathrm{ref}} - i_{kq}\right)\mathrm{d}t \end{cases} \tag{4.105}$$

In the Eq. (4.105), m_2 is also a conversion coefficient. Following the transformation method described above, the d-axis reference (V_{kd}) of the kth harmonic plane can be obtained for the d-axis component of the kth harmonics by integrating the q-axis error:

$$\begin{aligned} V_{kd} &= k\omega Lm_1\left(i_{kq\mathrm{ref}} - i_{kq}\right) + k\omega Lm_2\int\left(i_{kq\mathrm{ref}} - i_{kq}\right)\mathrm{d}t + n\omega L\left(i_{kq} - i_{kq\mathrm{ref}}\right) \\ &= k\omega L\left[k_\mathrm{p}\left(i_{kq\mathrm{ref}} - i_{kq}\right) + k_\mathrm{I}\int\left(i_{kq\mathrm{ref}} - i_{kq}\right)\mathrm{d}t\right] \end{aligned} \tag{4.106}$$

Similarly, the q-axis reference (V_{kq}) of the kth harmonic plane can be obtained. Thus, Eq. (4.103) is transformed as follows:

$$\begin{cases} V_{kd} = k\omega L\left[k_\mathrm{p}\left(i_{kq\mathrm{ref}} - i_{kq}\right) + k_\mathrm{I}\int\left(i_{kq\mathrm{ref}} - i_{kq}\right)\mathrm{d}t\right] \\ V_{kq} = -k\omega L\left[k_\mathrm{p}\left(i_{kd\mathrm{ref}} - i_{kd}\right) + k_\mathrm{I}\int\left(i_{kd\mathrm{ref}} - i_{kd}\right)\mathrm{d}t\right] \end{cases} \tag{4.107}$$

The block diagram for control based on harmonics at specific orders to suppress the influence of ripple voltage can be designed according to Eq. (4.107), as illustrated in Fig. 4.64.

The suppression strategy for harmonics at characteristic orders is similar to the active and reactive decoupling control strategy implemented at the fundamental frequency. Independent controllers are designed for the $2k \pm 1$ harmonics, such as the third, fifth, and seventh harmonics, to generate harmonic compensation voltages that suppress the corresponding harmonic currents. The cumulative summation of these compensation voltages at all characteristic orders, along with the reference voltage for active and reactive power control at the fundamental frequency, results in the modulation waveform for the converters.

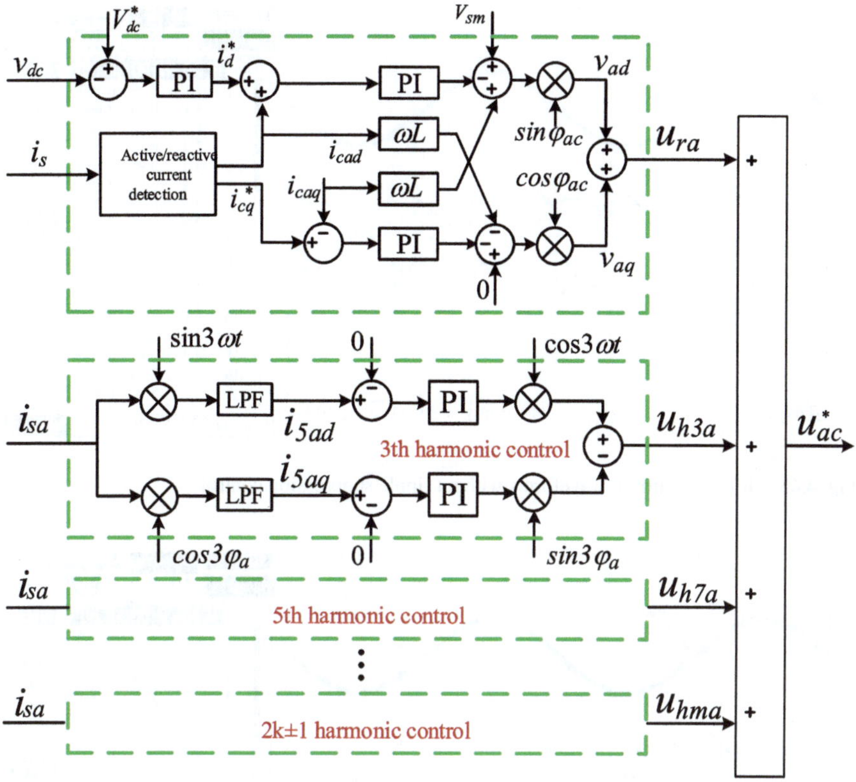

Fig. 4.64 Ripple effect suppression strategy based on compensation of harmonics at specific orders

4.5.3 Experimental Verification

Based on the analysis, even-order ripples impact the performance of the converters in multiple ways, leading to the generation of $2k \pm 1$ harmonics, primarily dominated by the third harmonics. The load configuration is as follows: a 10 Ω resistor connected between phases A and C, and a 25 Ω resistor connected between phases B and C.

Without any control measures to suppress the influence of ripple, the experimental results are shown in Fig. 4.65. The distortion in the grid-side current is evident, with the THD in the three-phase current being approximately 7%. The harmonic currents for phase C are as follows: 0.12 A for the third harmonic, 0.05 A for the fifth harmonic, and 0.03 A for the seventh harmonic.

The grid-side current waveforms with third harmonic suppression, derived from the experimental results, are shown in Fig. 4.66. The current distortion is significantly reduced, with the THD in the currents of phases A, B4–34, and C decreasing to 2.8%, 2.0%, and 3.0%, respectively. Notably, the third harmonic current decreases from 0.12 A to 0.01 A. However, the control of the third harmonic generated by the

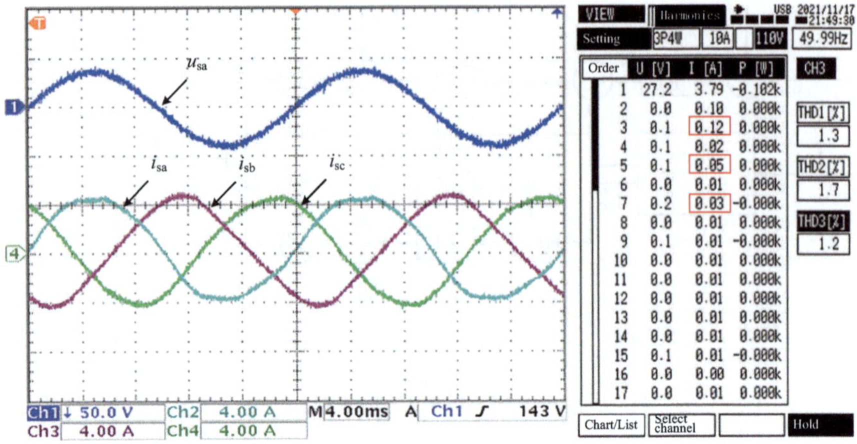

Fig. 4.65 Grid-side current waveforms without ripple suppression control

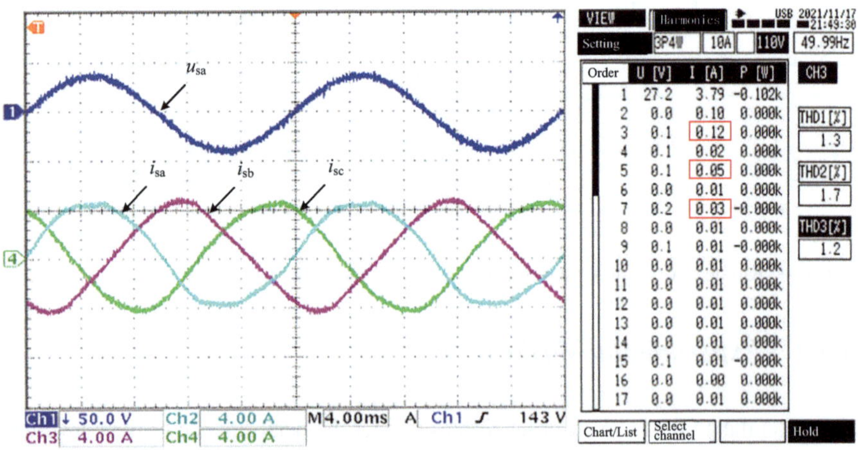

Fig. 4.66 Grid-side current waveform with trap filter control and third harmonic suppression

converter causes the DC capacitor to produce second and fourth ripples, which in turn generate third and fifth harmonics. Consequently, when the third harmonic is suppressed, the fifth harmonic will also decrease accordingly. As indicated, the fifth harmonic decreases from 0.05 A to 0.04 A without specific control for the fifth harmonic.

The current waveforms on the grid side, with the suppression of both the third and fifth harmonics, derived from the experimental results, are shown in Fig. 4.67. The current waveforms appear more sinusoidal, with the THD for the currents in phases A, B, and C decreasing to 2.3%, 1.9%, and 2.9%, respectively. Notably, the fifth harmonic decreases from 0.5 A to 0.02 A.

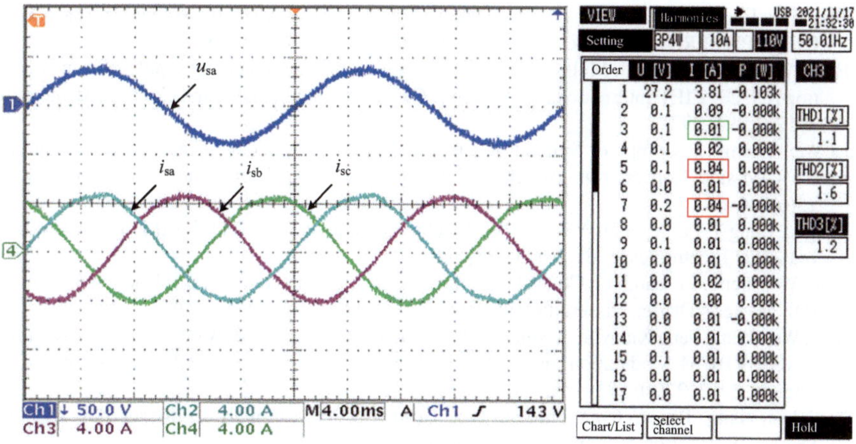

Fig. 4.67 Grid-side current waveform with trap filter control and third and fifth harmonic suppression

Based on the experimental results presented, it can be concluded that DC voltage ripple has a minimal effect on the output power quality of grid-connected converters. The control strategy for harmonics at specific orders proposed in this book effectively mitigates the harmonic influences generated by the converter due to capacitor voltage ripple.

4.6 Summary

This chapter primarily analyzes the operating principles and mathematical models of single-phase cascaded H-bridge converters in both two-level and three-level configurations. It elaborates on several control methods suitable for single-phase PWM converters, with a particular focus on the implementation of the SVPWM algorithm for voltage equalization in single-phase converters. Further studies address the generation mechanism, impact, and suppression of secondary ripple in single-phase converters. The feasibility of the proposed algorithm is verified through experiments, demonstrating its effectiveness in mitigating the influence of DC secondary ripple on both input and output characteristics.

References

Borghetti G, Carpaneto M, Marchesoni M, et al. Energy efficiency improvement in ac/dc/ac diode-clamped multilevel conversion systems [C]. Proc. Int. Symp. Power Electron., Electr. Drives, Autom. Motion, 2008:798–803.

Chen Yao, Tong Yibin, Jin Xinmin. A New Algorithm for SVPWM Harmonic Analysis Based on PWM Rectifier [J]. Proceedings of the CSEE, 2007, 27(13): 76–80.

Gao Jian, Liu Zhen. Research on Two Direct Current Control Methods for Single-Phase Half-Bridge PWM Rectifier [J]. Electrical Application, 2006, 25(6): 26–35.
He X, Wang Y, Shu Z, et al. Multiple AC-DC-AC substations traction power supply system simulation [C]. 2012 IET International Conference On Information Science and Control Engineering, 2012, 1:294–297.
Li Wei, Zhang Li. Computer Simulation Implementation of Predictive Current Control Method for Grid-side Converter in AC-DC-AC Drive System [J]. China Railway Science, 2002, 25(6): 46–63.
Peng L, Kang Y, Pei X, et al. A novel PWM technique in digital control [J]. IEEE Transactions on Industrial Electronics, 2007, 54(1):338–346.
Song Wensheng. Research on Control and Modulation Algorithms of Electric Traction Converters [D]. Master's Degree Dissertation of Southwest Jiaotong University, 2008.
Song Wensheng, Feng Xiaoyun. A Study on the Intrinsic Relationship Between Single-Phase Three-Level SVPWM Modulation and Carrier SPWM [J]. Transactions of China Electrotechnical Society, 2012, 27(6): 131–138.
Xu Zhengxiang. Research on Through-type Co-phase Power Supply Converter [D]. Master's Degree Dissertation of Southwest Jiaotong University, 2013.
Zhang Xing. Research on PWM Rectifier and Its Control Strategies [D]. Master's Degree Dissertation of Hefei University of Technology, 2003.
Zheng Huaxi. Simulation Analysis and Parameter Study on Traction Auxiliary System of 350 km/h EMU [D]. Master's Degree Dissertation of Beijing Jiaotong University, 2009.

Chapter 5
Diode-Clamped Multilevel Three-Phase-to-Single-Phase Converter

5.1 Advanced Co-phase Traction Power Supply System Based on Three-Phase-to-Single-Phase Converter

If all traction substations within a traction power supply system are equipped with AC-DC-AC converters, continuous power supply can be achieved along the entire railway through interconnected substations, eliminating segmentation between them. In this configuration, the AC-DC-AC converters at these substations act as loads connected to the electric power system, with their single-phase sides interconnected via the traction power supply network. This setup forms a mutually supported traction power supply microgrid that spans the entire railway. Consequently, the three-phase side of the AC-DC-AC converters in this advanced co-phase traction power supply system is linked to the electric power system, while the single-phase side connects to the traction network. Typically, this is arranged in a three-phase-to-single-phase converter topology, with AC ports on both sides connected to the active power supply network. The AC ports on either side of the converters operate in a four-quadrant mode, enabling active interconnection and control of both active and reactive power.

The controllable frequency, amplitude, and phase of the port voltages on both the three-phase and single-phase sides of these converters enable connection to the three-phase electric power system and the single-phase traction network. Chapters 3 and 4 discuss three-phase and single-phase converters, respectively. In this chapter, further discussion involves the topology, characteristics, and control of three-phase-to-single-phase converters for advanced co-phase traction power supply applications.

To align the voltage of the electric power system and the traction network, and the operating voltage level of the converters, matching transformers should be installed on both sides of the three-phase-to-single-phase converters for continuous power supply. The three-phase side connects to the three-phase electric power system via a three-phase transformer, while the single-phase side is linked to the single-phase traction network through a single-phase transformer.

Z. Shu, *Power Electronic Converters in Advanced Co-Phase Traction Power Supply System*, https://doi.org/10.1007/978-981-95-3276-6_5

To facilitate power transmission at traction substations, three-phase-to-single-phase converters for continuous power supply are typically configured with multiple modules connected in parallel or arranged in a multilevel circuit topology. For simplicity in analysis, the initial discussion concentrates on the single-module three-phase-to-single-phase converter topology illustrated in Fig. 5.1. This substation model primarily consists of a three-phase-to-single-phase PWM converter (which includes a three-phase PWM converter, a DC link, and a single-phase PWM converter), as well as a three-phase step-down transformer and a single-phase step-up transformer.

In the case of a traction load in the traction condition, active power is transferred from the three-phase grid to the single-phase traction network. Therefore, the active power on the three-phase AC side is transmitted to the DC capacitor side through the three-phase PWM converter, followed by inversion and output to the single-phase traction network through the single-phase PWM converter, thereby providing active power support for the traction load.

In the case of a traction load in the braking condition, active power is transferred from the single-phase traction side to the three-phase grid. The single-phase PWM converter operates in rectification mode, and the active power is fed back to the three-phase power grid through the three-phase PWM converter after passing through the DC capacitor. In addition, reactive power and harmonics are independently controlled by the three-phase PWM converter and the single-phase PWM converter, without an intermediate DC link for transmission.

To sum up, the exchange of only active power occurs between the three-phase electric power system and the traction network. The advanced co-phase traction power supply system operates with purely resistive characteristics in relation to the three-phase power grid, exhibiting a high power factor, reduced harmonic current distortion, and favorable load characteristics. The three-phase-to-single-phase PWM converter achieves balance across the three phases by equalizing active

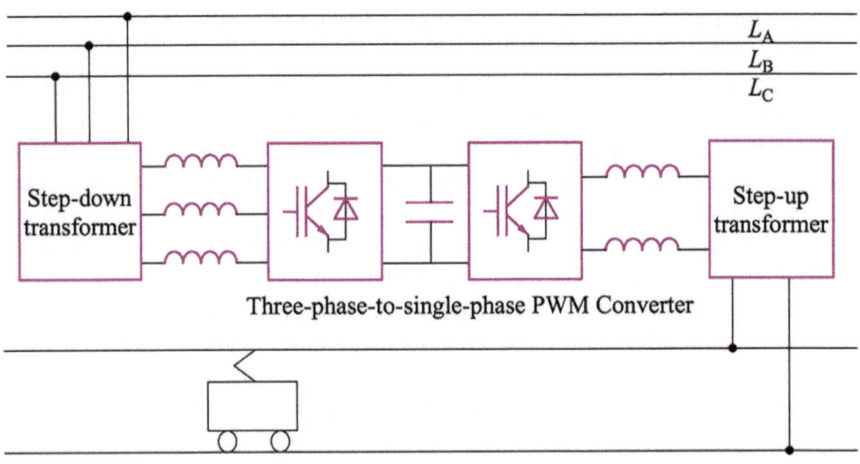

Fig. 5.1 Substation model based on three-phase-to-single-phase PWM converter

power, thereby mitigating the negative sequence that is prevalent in current traction loads and significantly enhancing the power quality of the three-phase grid. Furthermore, the single-phase-side converters of multiple substations facilitate a continuous co-phase power supply throughout the entire railway, eliminating the need for any neutral section passing devices. This effectively addresses the challenges associated with neutral sections in traditional traction power supply systems. Additionally, traction loads are serviced by all substations along the railway, fostering mutual active power support among them. This arrangement reduces the capacity and reliability demands placed on individual substations. With interconnections established among the substations, both ends of the traction network are connected to these substations, which increases the voltage at the midpoints and significantly extends the power supply distance.

5.1.1 Brief Introduction to Two-Level Three-Phase-to-Single-Phase Converter

Figure 5.2 shows a three-phase-to-single-phase converter in a two-level circuit topology. The system consists of a three-phase-side PWM rectifier, a single-phase-side PWM rectifier, a three-phase step-down transformer (T3), and a single-phase step-up transformer (T2).

When the advanced co-phase traction power supply system needs to transmit active power from the three-phase side to the single-phase side, the three-phase-side PWM rectifier supplies active power and maintains voltage stability at the DC capacitor. Meanwhile, the single-phase-side PWM rectifier facilitates interconnection on the single-phase traction side and outputs active power while also compensating for reactive power and harmonic currents on the traction network side. Conversely, when the system needs to transmit active power from the single-phase side to the three-phase side, the two converters simply need to reverse the direction of the active power reference.

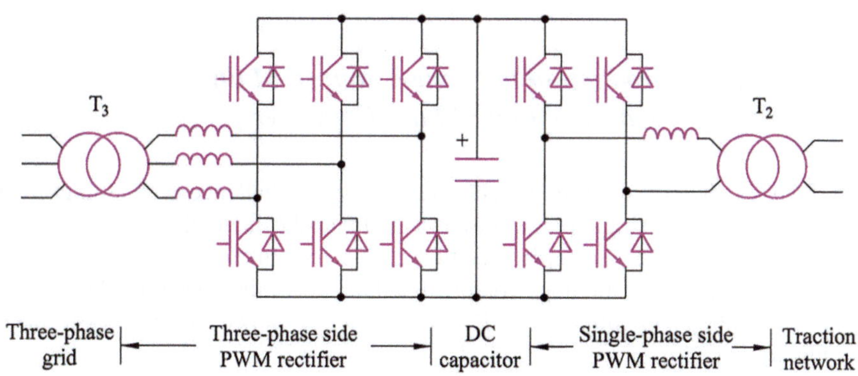

Fig. 5.2 Three-phase-to-single-phase PWM converter based on two-level circuit

Due to the significant output voltage jumps (i.e., d*u*/d*t*) produced by two-level converters, the output port voltage exhibits a high total harmonic distortion (THD). Furthermore, the limitations imposed by the voltage withstand capacities of existing power electronic switching devices make it challenging for the voltage and capacity levels of two-level converters to directly align with the requirements of the electric power system and traction network. As a result, the design must incorporate multiple modules connected in parallel, along with transformers that feature a high transformation ratio.

The implementation of advanced co-phase traction power supply systems based on three-phase-to-single-phase converters still faces the following two severe challenges:

(1) Existing individual converters with low capacities struggle to meet the large-capacity requirements of traction substation systems. Due to the limitations of the rated parameters of power electronic switching devices, such as IGBTs and IGCTs, their interface voltage and capacity levels often fall short of the requirements for substations. While it is possible to match the system voltage and power through the use of transformers and by connecting multiple modules in parallel, these approaches significantly increase both the cost and complexity of the system.
(2) Advanced co-phase traction power supply systems necessitate energy transmission between the three-phase electric power system and the single-phase traction power supply system, imposing stringent requirements for power quality on the three-phase side and for interconnection adjustment capacity on the single-phase side. The converters on the three-phase side focus on addressing power quality issues such as three-phase unbalance, low power factors, and high harmonics while also ensuring stability of the DC voltage. Meanwhile, the single-phase-side converters facilitate the interconnection of substations through the traction network. This not only requires these converters to supply or absorb active power in both forward and reverse directions to and from traction loads but also demands their capability to compensate for reactive power and harmonic currents to meet the needs of traction loads.

5.1.2 Brief Introduction to Diode-Clamped Multilevel Three-Phase-to-Single-Phase Converter

Multilevel converters present a viable solution to the challenge discussed in item (1). Three-phase-to-single-phase converters utilizing the diode-clamped three-level circuit topology offer an implementation option for multilevel circuits, as illustrated in Fig. 5.3. This topology comprises a three-phase three-level PWM rectifier and a single-phase three-level PWM rectifier arranged in a back-to-back configuration.

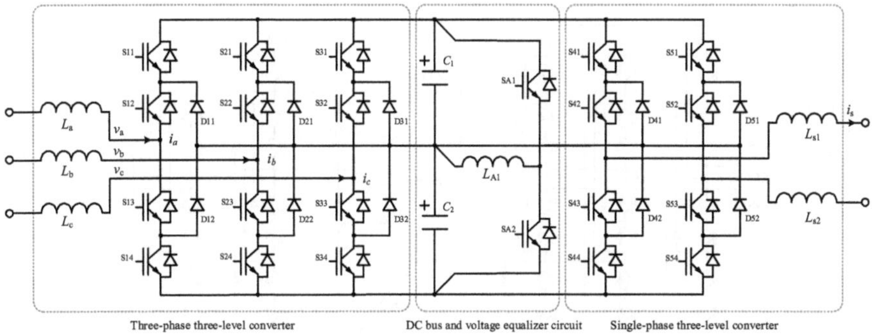

Fig. 5.3 Topology of three-level three-phase-to-single-phase converter

In this converter topology, each bridge arm of the diode-clamped three-level circuit consists of four switching devices and two clamping diodes. C1 and C2 represent the two DC-link capacitors, for which voltage equalization must be addressed within these multilevel circuits. As illustrated in Fig. 5.3, the intermediate inductor voltage equalizer circuit ensures that the capacitor voltages are equalized. This circuit can also be replaced with modulation voltage equalization or other methods. Generally, the two capacitor voltages on the DC side are evenly divided to $V_{dc}/2$, thereby reducing the withstand voltage requirement for the switching devices to $V_{dc}/2$. Consequently, the diode-clamped three-level circuit topology can significantly increase the rated operating voltage and power levels compared to the two-level circuit topology. This diode-clamped three-level circuit can achieve a port three-level output through a specific combination of switches. Existing literature and publications (Pelicer et al. 2015; Jie 2006) have analyzed the specific operating paths and modulation methods in detail, so these concepts are not covered further in this book.

5.2 Diode-Clamped Five-Level Three-Phase-to-Single-Phase Converter

Two-level and three-level circuits are suitable for medium- to low-voltage and medium- to small-power applications, whereas diode-clamped multilevel circuits can further enhance the voltage and power levels of converters. To address the high-voltage and high-power requirements of advanced co-phase traction power supply systems, this chapter focuses on the diode-clamped five-level three-phase-to-single-phase circuit shown in Fig. 5.4.

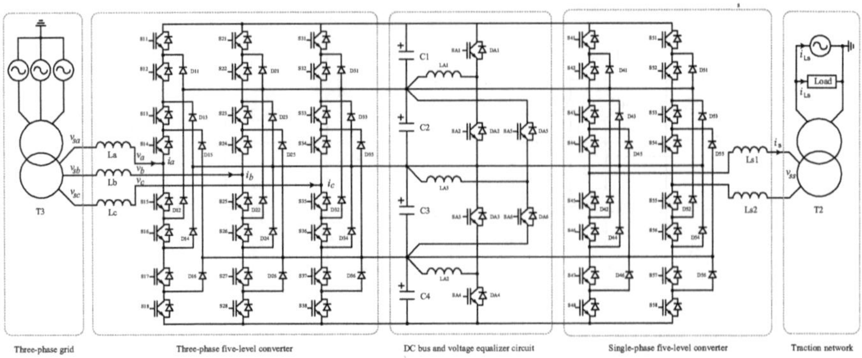

Fig. 5.4 Diode-clamped five-level three-phase-to-single-phase circuit topology

Table 5.1 Relationships between switching states and output voltages of five-level diode-clamped circuit

	Switch drive state (0: off; 1: on)							
Output port voltage level, Vo	S11	S12	S13	S14	S15	S16	S17	S18
$V_0 = 0$	0	0	0	0	1	1	1	1
$V_1 = 1/4V_{dc}$	0	0	0	1	1	1	1	0
$V_2 = 1/2V_{dc}$	0	0	1	1	1	1	0	0
$V_3 = 3/4V_{dc}$	0	1	1	1	1	0	0	0
$V_4 = V_{dc}$	1	1	1	1	0	0	0	0

5.2.1 Brief Introduction to Circuit Operating Principle

In the circuit shown in Fig. 5.4, each bridge arm is a diode-clamped five-level configuration. Assuming that the capacitance and voltage of the four DC capacitors are equal, the total voltage on the DC side is divided by these four capacitors into five levels: V_0, V_1, V_2, V_3, and V_4.

The switching tubes (S11, S15), (S12, S16), (S13, S17), and (S14, S18) form complementary switch pairs. When one switch is turned on, the other is turned off, which can be represented using a binary function: S1i = 1 indicates the on state, while S1i = 0 (i = 1–8) indicates the off state. The relationships between the bridge arm output levels and the drive states of the switching tubes are illustrated in Table 5.1, which outlines the five output levels for this circuit. Each output level can support both positive and negative current directions, resulting in a total of ten current operating paths.

5.2.2 Analysis of Current Operating Path

The current flowing from the DC terminal to the AC terminal (VO) is defined as positive, while the reverse direction is considered negative. The switching sequence is represented by the states S11, S12, S13, S14, S15, S16, S17, and S18. For instance,

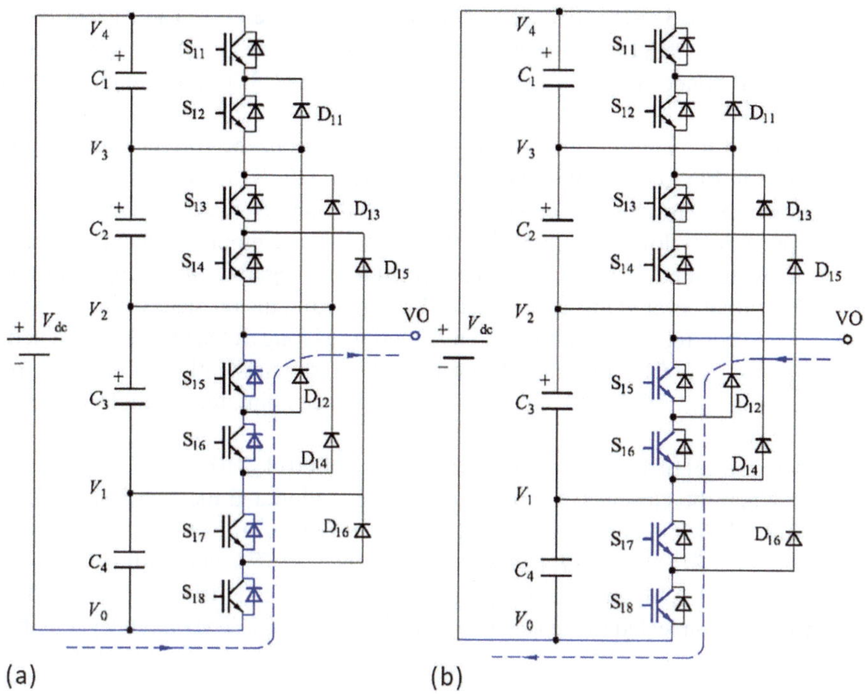

Fig. 5.5 Flow path at output level of V_0: (**a**) positive current, (**b**) negative current

a switching state of 0000_1111 indicates that switches S11, S12, S13, and S14 are turned off, while switches S15, S16, S17, and S18 are turned on.

(1) Output level of $V_0 = 0$ at switching state of 0000_1111

When the current flows in the positive direction and the switching sequence outputs are in the 0000_1111 state, the current path is illustrated in Fig. 5.5a. In this state, switches S11, S12, S13, and S14 are all turned off. The current flows through the body diodes of switches S15, S16, S17, and S18 to reach the AC terminal (VO), which is connected to the DC negative terminal. The output level in this case is V_0.

When the current flows in the negative direction and the switching sequence outputs are in the 0000_1111 state, the current path is illustrated in Fig. 5.5b. In this state, switches S11, S12, S13, and S14 are all turned off, while switches S15, S16, S17, and S18 are all turned on. Consequently, the input current flows through switches S15, S16, S17, and S18. The AC terminal (VO) remains connected to the negative DC terminal, and the output level remains V_0.

In short, regardless of the current direction, as long as the bridge arm switching state is 0000_1111, the AC terminal (VO) is consistently connected to the terminal V_0. With $V_0 = 0$ regarded as the reference point, the output level of VO is 0.

(2) Output level of $V_1 = 1/4V_{dc}$ at switching state of 0001_1110

When the current flows in the positive direction and the switching state is 0001_1110, the current path is illustrated in Fig. 5.6a. In this state, switches

S11, S12, S13, S15, S16, S17, and S18 are all turned off. The current flows through switch S14 and the clamping diode D15 to reach the AC terminal (VO), which is connected to the negative end of capacitor C3. The output level in this case is V_1.

When the current flows in the negative direction, the current path is illustrated in Fig. 5.6b. In this state, switches S11, S12, S13, S14, and S18 are all turned off. However, since switches S15, S16, and S17 are all turned on, the input current flows through S15, S16, S17, and the clamping diode D16 to reach the AC terminal (VO), which is connected to the negative end of capacitor C3. The output level in this case remains V_1, neglecting the on-state voltage drop.

In short, regardless of the current direction, as long as the bridge arm switching state is 0001_1110, the AC terminal (VO) is consistently connected to the terminal V_1. With $V_0 = 0$ regarded as the reference point, the output level of VO is $1/4V_{dc}$.

(3) Output level of $V_2 = 1/2V_{dc}$ at switching state of 0011_1100

When the current flows in the positive direction and the switching state is 0011_1100, the current path is illustrated in Fig. 5.7a. In this state, switches S11, S12, S15, S16, S17, and S18 are all turned off. The current flows through switch S13, S14, and the clamping diode D13 to reach the AC terminal (VO), which is connected to the negative end of capacitor C2. The output level in this case is V_2.

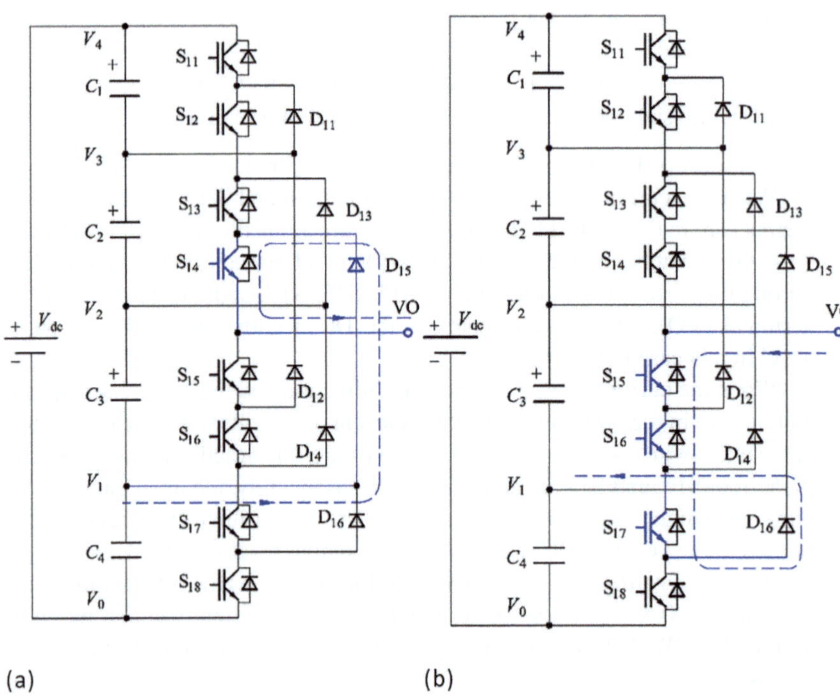

Fig. 5.6 Flow path at output level of V_1

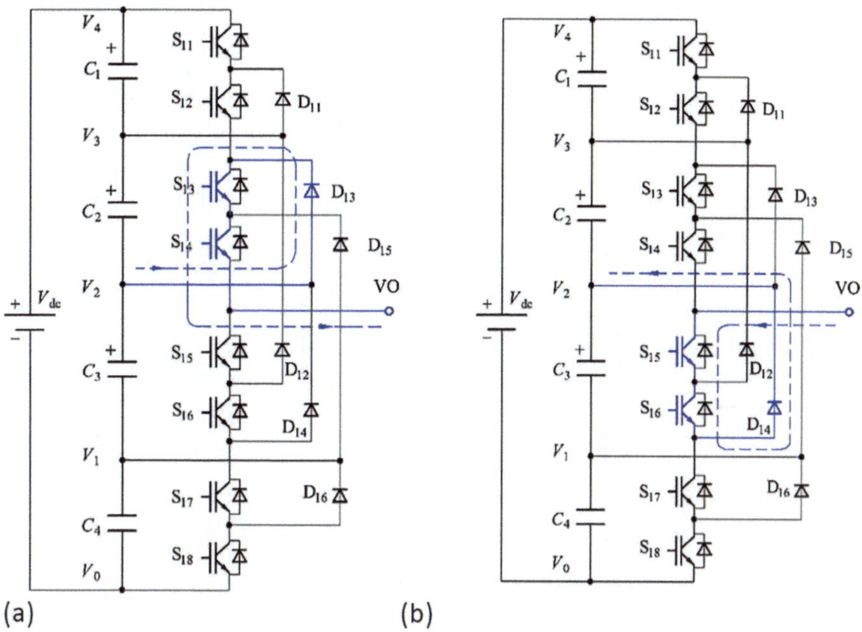

Fig. 5.7 Flow path at output level of V_2

When the current flows in the negative direction, the current path is illustrated in Fig. 5.7b. In this state, switches S11, S12, S13, S14, S17, and S18 are all turned off. However, since switches S15 and S16 are both turned on, the input current flows through S15, S16, and the clamping diode D14 to reach the AC terminal (VO), which is connected to the negative end of capacitor C2. The output level in this case remains V_2, neglecting the on-state voltage drop.

In short, regardless of the current direction, as long as the bridge arm switching state is 0011_1100, the AC terminal (VO) is consistently connected to the terminal V_2. With $V_0 = 0$ regarded as the reference point, the output level of VO is $1/2V_{dc}$.

(4) Output Level of $V_3 = 3/4V_{dc}$ at Switching State of 0111_1000

When the current flows in the positive direction and the switching state is 0111_1000, the current path is illustrated in Fig. 5.8a. In this state, switches S11, S15, S16, S17, and S18 are all turned off. The input current flows through switches S12, S13, and S14 and the clamping diode D11 to reach the AC terminal (VO), which is connected to the negative end of capacitor C1. The output level in this case is V_3.

When the current flows in the negative direction, the current path is illustrated in Figure 5.8(b). In this state, switches S11, S12, S13, S14, S16, S17, and S18 are all turned off. However, since switch S15 is turned on, the input current flows through S15 and the clamping diode D12 to reach the AC terminal (VO), which is connected to the negative end of capacitor C1. The output level in this case remains V_3, neglecting the on-state voltage drop.

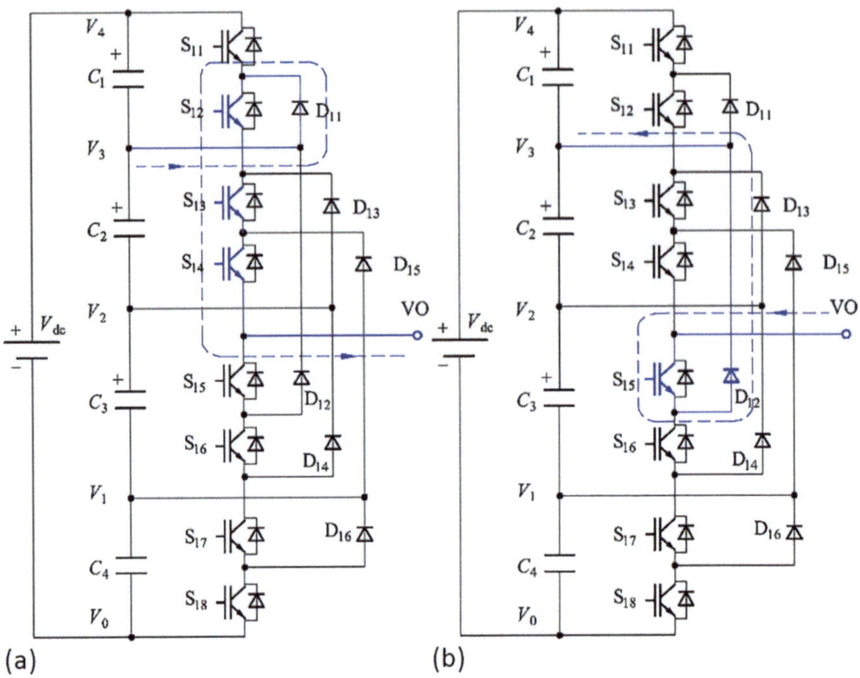

Fig. 5.8 Flow path at output level of V_3

In short, regardless of the current direction, as long as the bridge arm switching state is 0111_1000, the AC terminal (VO) is consistently connected to the terminal V_3. With $V_0 = 0$ regarded as the reference point, the output level of VO is $3/4V_{dc}$.

(5) Output Level of $V_4 = V_{dc}$ at Switching State of 1111_0000

When the current flows in the positive direction and the switching state is 1111_0000, the current path is illustrated in Fig. 5.9a. In this state, switches S15, S16, S17, and S18 are all turned off. The input current flows through switches S11, S12, S13, and S14 to reach the AC terminal (VO), which is connected to the positive end of capacitor C1. The output level in this case is V_4.

When the current flows in the negative direction, the current path is illustrated in Fig. 5.9(b). In this state, switches S15, S16, S17, and S18 are all turned off. However, the input current flows through the body diodes of S11, S12, S13, and S14 to reach the AC terminal (VO), which is connected to the positive end of capacitor C1. The output level in this case remains V_4, neglecting the on-state voltage drop.

In short, regardless of the current direction, as long as the bridge arm switching state is 1111_0000, the AC terminal (VO) is consistently connected to the terminal V_4. With $V_0 = 0$ regarded as the reference point, the output level of VO is V_{dc}.

To sum up, for the specified states of the eight switches, the output level remains consistent regardless of the current direction.

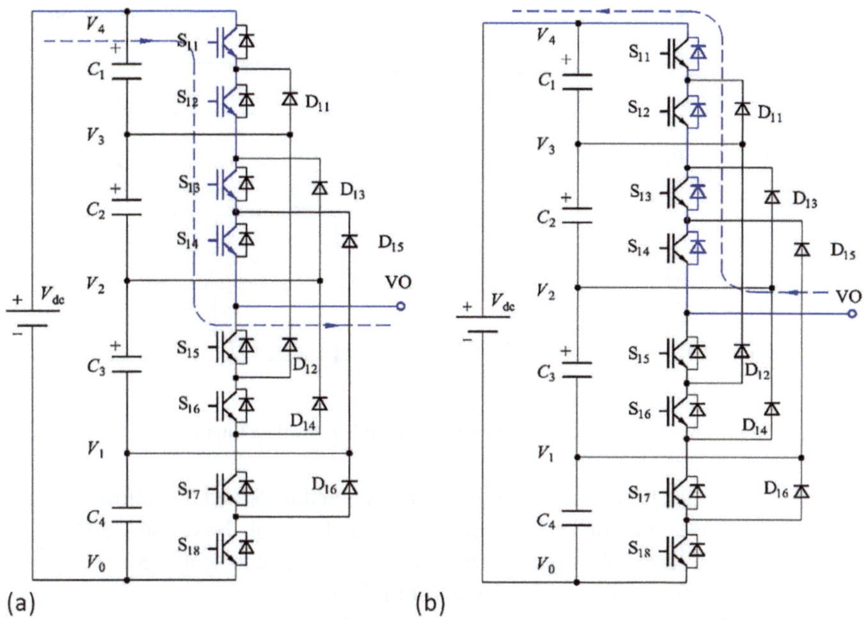

Fig. 5.9 Flow path at output level of V_4

5.2.3 Ideal Switch Model and Control Strategy for Five-Level Circuits

The previous section introduces the basic operating path of diode-clamped five-level circuits. The analysis indicates that, regardless of the current direction at the port, the five switching states correspond uniquely to five output levels, neglecting factors like the on-state voltage drop. Consequently, the controller can control the converter port level by outputting specific switching states, thereby controlling the amplitude and phase of the converter's output voltage. This facilitates the control of operational stability for the converter, including managing current amplitude, power level, and power factor (Pelicer et al. 2015; Jie 2006). The specific control method is described as follows:

For a circuit assumed to utilize ideal switches, based on the analysis in the previous section, the switching function (S_a) of the A-phase bridge arm can be defined as follows:

$$S_a = \begin{cases} 1 & S_{11}S_{12}S_{13}S_{14} \text{ turned on} \left(1111\_0000\right) \\ 3/4 & S_{12}S_{13}S_{14}S_{15} \text{ turned on} \left(011\_1000\right) \\ 2/4 & S_{13}S_{14}S_{15}S_{16} \text{ turned on} \left(0011\_1100\right) \\ 1/4 & S_{14}S_{15}S_{16}S_{17} \text{ turned on} \left(0001\_1110\right) \\ 0 & S_{15}S_{16}S_{17}S_{18} \text{ turned on} \left(0000\_1111\right) \end{cases} \tag{5.1}$$

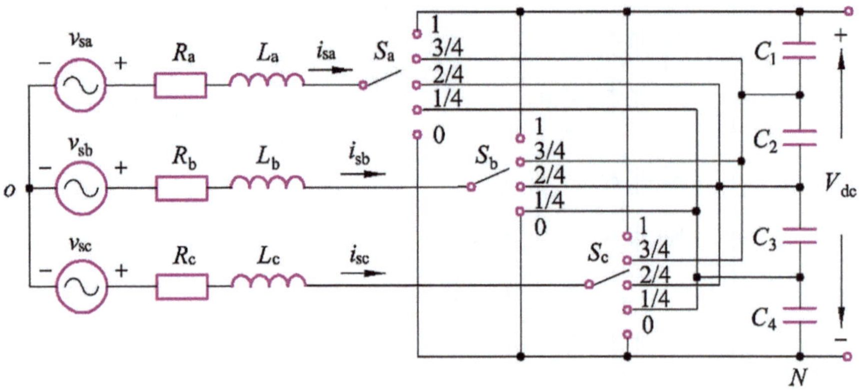

Fig. 5.10 Equivalent ideal switch model of three-phase five-level rectifier circuit

The switching functions for phases B and C, denoted as S_b and S_c, follow a similar form to the above equation and are not repeated in this book. Based on Eq. (5.1), the three-phase five-level PWM rectifier main circuit can be represented equivalently as an ideal switching circuit, as illustrated in Fig. 5.10.

Each diode-clamped five-level bridge arm in the figure is equivalent to a single-pole five-throw switch, connecting the AC terminal of each phase to five level points on the DC side, namely 0, 1/4 V_{dc}, 1/2 V_{dc}, 3/4V_{dc}, and V_{dc}. Consequently, the voltage (vmN) from the AC port of each bridge arm to the negative terminal on the DC side can be expressed as

$$v_{mN}(t) = S_m \times V_{dc} \tag{5.2}$$

where S_m ={0, 1/4, 1/2, 3/4, 1}, and m = a, b, c.

The control strategy for diode-clamped five-level three-phase circuits is fundamentally similar to that of two-level circuits. For instance, the active and reactive decoupling control outlined in Sect. 3.2 can be implemented for diode-clamped five-level three-phase rectifiers, as illustrated in Fig. 5.11. The specific details are not repeated in this section.

It should be noted that the modulation method shown in the figure requires adaptation for a diode-clamped five-level circuit. For instance, carrier disposition or carrier phase-shifted multilevel PWM modulation can be utilized to achieve the conversion between the modulation waves (V_{fa}, V_{fb}, and V_{fc}) and the corresponding five-level switching states discussed in Sect. 5.2.2.

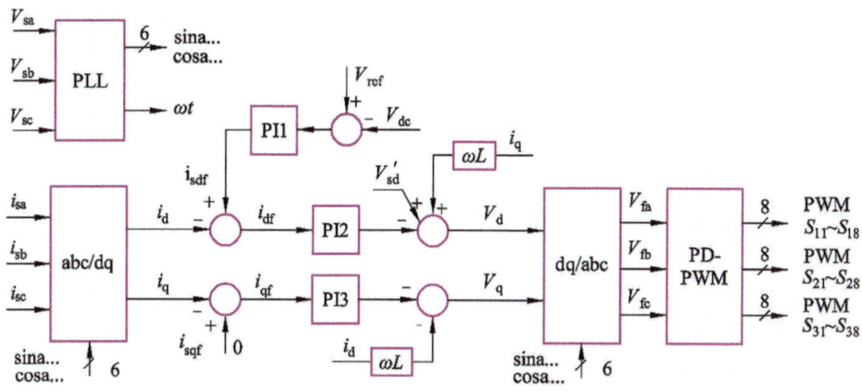

Fig. 5.11 Block diagram of active and reactive decoupling control strategy for five-level rectifier

5.3 Clamping Auxiliary Voltage Equalizer Circuit and Its Control

5.3.1 Voltage Unbalance Among DC Capacitors

The analysis of multilevel converter circuits in the previous sections is based on stable and balanced voltages across multiple DC capacitors. However, system control methods, modulation techniques, or unbalanced DC loads can cause deviations in the DC capacitor voltages from the equilibrium value, leading to voltage imbalances among them. Such unbalances result in deviations in the output port level from the ideal value, causing distortions in both the port voltage and the AC current. In cases of extreme voltage imbalance, such as scenarios where the voltage of certain DC capacitors drops to zero, the multilevel converter may function as a low-level converter, disrupting the equal distribution of voltage stress. As a result, switching devices that experience high voltage stress are susceptible to overvoltage failures, potentially leading to damage to devices, circuits, and systems. Therefore, equalization of DC-side capacitor voltages is deemed a precondition for ensuring the safe operation and high output performance of multilevel converters (Liu et al. 2016; Li et al. 2016; He and Chen 2006).

The capacitor voltages in a multilevel converter can remain balanced when the phase difference between the output voltage and current is 90°, meaning that voltage balance is maintained during the output of pure reactive power. In cases where the converter transmits partial active power, the balance zone of the DC voltages is influenced by the modulation degree. Figure 5.12 illustrates the balance zone of the capacitor voltages on the DC side of a diode-clamped multilevel converter (Shu et al. 2013a), highlighting the unstable zone as the white area and the stable balance zone as the gray area. As shown, a higher number of levels (indicating a higher modulation degree) correlates with a higher power factor, resulting in increased difficulty in maintaining voltage balance. In limit conditions where the power factor is

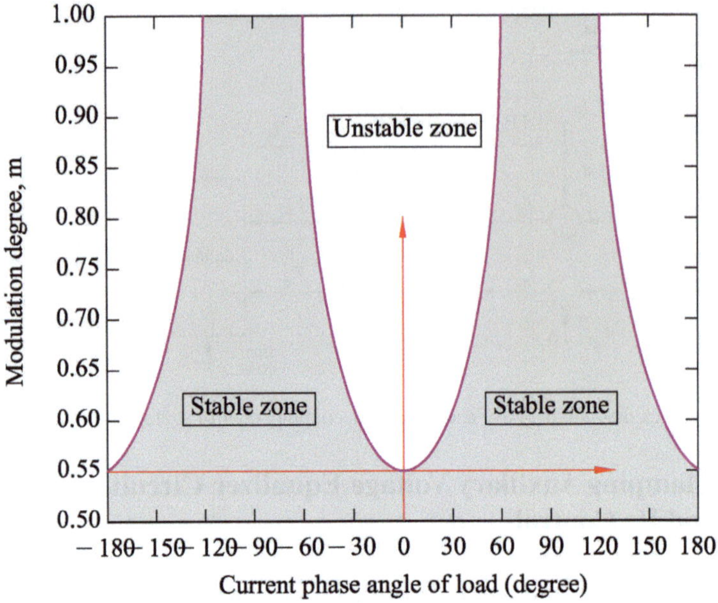

Fig. 5.12 Balance zone of DC-side capacitor voltage in multilevel converter relative to current phase angle and modulation degree

equal to 1, the maximum modulation degree is only 0.55; any modulation degree below this maximum further complicates efforts to sustain voltage balance.

Unbalance among DC-side capacitor voltages must be solved for multilevel converters. The following are the two primary methods to achieve this:

(1) Voltage-equalizing control strategy using space vector pulse width modulation (Shu et al. 2007, 2013b)

This method utilizes a combination of redundant switching states to control capacitor voltage, achieving voltage equalization without the need for additional hardware circuits, resulting in low costs. However, the capacity and range of voltage balancing are influenced by the specific voltage-balancing strategy and the operating state of the system.

(2) Voltage-equalizing control strategy using auxiliary voltage equalizer circuit (Shu et al. 2013c; Dixon and Boon-Teck 1988)

The capacitor voltage balance in multilevel circuits is maintained through a special auxiliary voltage equalizer circuit. Typically, an auxiliary voltage equalizer circuit operates independently of the main circuit, remaining unaffected by the main circuit's operating parameters, such as load power factor and modulation degree. Two types of auxiliary voltage equalizer circuits are discussed below: capacitive and inductive.

5.3.2 *Capacitor-Assisted Voltage Equalizer Circuit*

Capacitor-assisted voltage equalizer circuits encompass various configurations, including general capacitor, primary capacitor, single capacitor, and simplified single capacitor-assisted voltage equalizer circuits (Zheng 2011). These circuits are organized into a basic unit, represented by the capacitor-assisted voltage equalizer circuit shown in Fig. 5.13. This unit consists of four switching devices labeled Sa1–Sa4, each with body diodes, two DC-side capacitors (C1 and C2), and an auxiliary

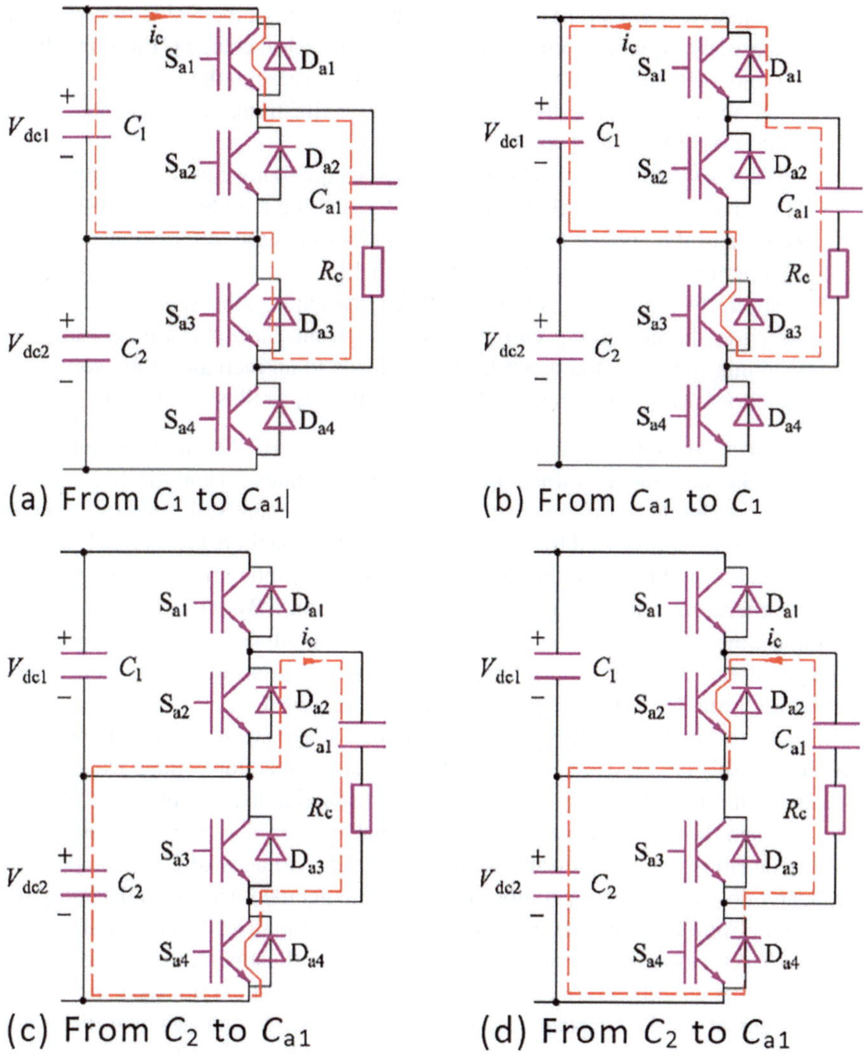

Fig. 5.13 Operating process of basic unit in capacitor-assisted voltage equalizer circuit. (**a**) From C1 to Ca1. (**b**) From Ca1 to C1. (**c**) From C2 to Ca1. (**d**) From C2 to Ca1

equalizing capacitor (Ca1). To mitigate the impulse tendency during capacitor charging, the peak value of the equalizing current (ic) can be limited by connecting a resistor in series (Rc) or employing other methods. The operating principle of the circuit is as follows:

(1) In the case of VC1 > VCa1

When switching tube Sa1 is turned on, the equalizing current (ic) flows from C1 to charge Ca1 through switching tube Sa1 and diode Da3. This results in a decrease in the voltage of capacitor C1 and an increase in the voltage of Ca1. The operating circuit is illustrated in Fig. 5.13a.

Similarly, when VC1 is less than VCa1, switching tube Sa3 is turned on. In this case, the equalizing current (ic) flows from Ca1 to charge C1 through diode Da1 and switching tube Sa3. This leads to an increase in the voltage of capacitor C1 and a decrease in the voltage of Ca1, as illustrated in Fig. 5.13b.

(2) In the case of VC2 > VCa1

When switching tube Sa4 is turned on, the equalizing current (ic) flows from C2 to charge Ca1 through diode Da2 and switching tube Sa4. This results in a decrease in the voltage of capacitor C2 and an increase in the voltage of Ca1. The operating circuit is illustrated in Fig. 5.13c.

Similarly, when VC2 is less than VCa1, switching tube Sa2 is turned on. In this case, the equalizing current (ic) flows from Ca1 to charge C2 through switching tube Sa2 and diode Da4. This leads to an increase in the voltage of capacitor C2 and a decrease in the voltage of Ca1, as illustrated in Fig. 5.13d.

As shown, when the circuit operates in states (a) and (b), Sa1 and Sa3 are turned on, while Sa2 and Sa4 are turned off. Conversely, in states (c) and (d), Sa2 and Sa4 are turned on, while Sa1 and Sa3 are turned off. Consequently, the energy stored in the capacitors is transferred back and forth between capacitors C1 and Ca1, as well as between C2 and Ca1, similar to the motion of a ping-pong ball. This operational mode is therefore referred to as "ping-pong operation." By repeating this process multiple times, the voltages of the two capacitors gradually converge toward the voltage of the auxiliary capacitor, resulting in a decreasing voltage difference between C1 and C2 until voltage equalization is achieved.

By controlling the switching on and off of the voltage equalizer circuit, the effectiveness of this circuit can be verified. The DC-side capacitor voltage waveform is illustrated in Fig. 5.14 during the operating states of enabled, disabled, and enabled again. When the voltage equalizer circuit is enabled, the capacitor voltages remain balanced. However, upon disabling the circuit, the voltages of all the capacitors gradually deviate from the equilibrium value. The detailed waveform depicting the transition from disabling to enabling the voltage equalizer circuit is shown in Fig. 5.15. As shown, after approximately 400 ms of adjustment, the voltages of all the capacitors quickly return to a balanced state. Curve (Ch1) represents the operating current of the DC capacitor (Ca3), with the positive and negative directions indicating capacitor charging and discharging, respectively. The experimental results confirm the effectiveness of the capacitor-assisted voltage equalizer circuit shown in Fig. 5.13.

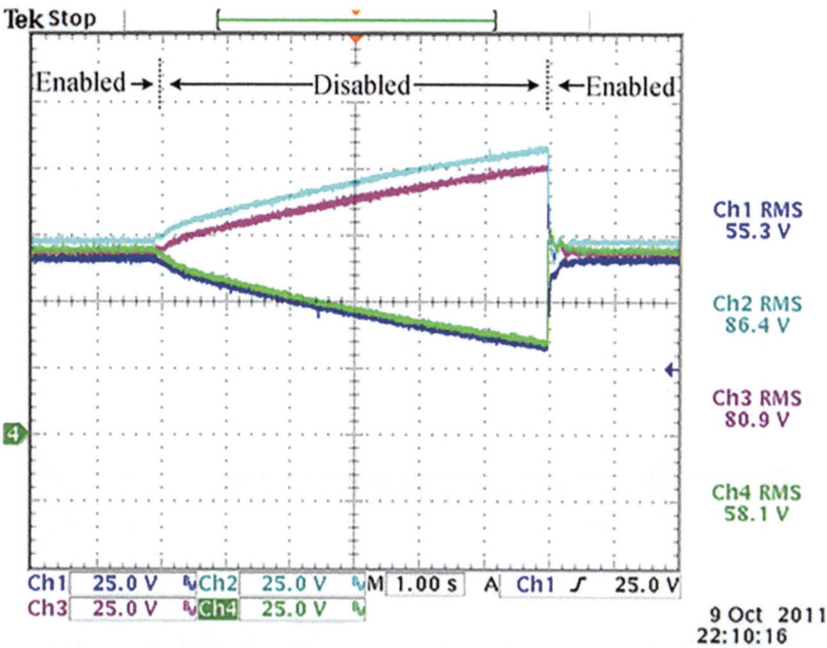

Fig. 5.14 Waveforms of DC capacitor voltages (vC1–vC4) during transitions of voltage equalizer circuit from enabled to disabled and back to enabledCh1: current on Ca3, Ch2–4: voltage on C2–C4

However, the voltage-balancing capability of capacitor-assisted voltage equalizer circuits is significantly influenced by the current-limiting resistor (Rc). A high resistance value restricts the voltage-balancing capacity, while a low resistance value may not adequately limit the equalizing current peak. Additionally, the current-limiting resistor contributes to increased losses in the voltage equalizer circuit, which poses a considerable limitation to the practical application of capacitive voltage equalizer circuits.

5.3.3 Inductor-assisted Voltage Equalizer Circuit

For diode-clamped five-level circuits with four capacitors on the DC side, the three-inductor-assisted voltage equalizer circuit depicted in Fig. 5.16 can be employed to achieve voltage balancing among the four capacitors. The circuit topology illustrated in the figure consists of three single-inductor voltage equalizer units, each designed to balance the voltages between pairs of capacitors: C1 and C2, C3 and C4, and C2 and C3. Utilizing the same control method, these voltage equalizer units operate independently from one another.

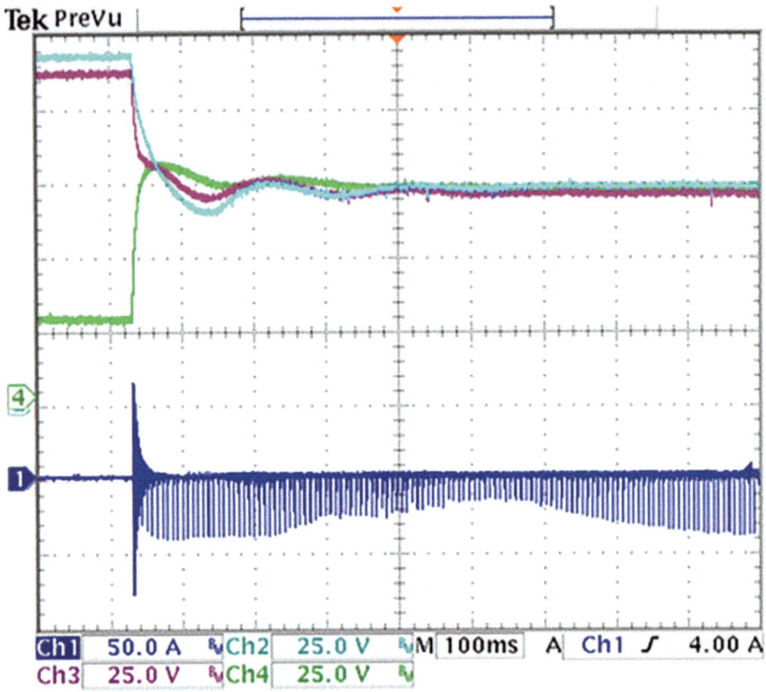

Fig. 5.15 Dynamic waveforms during transition of voltage equalizer circuit from disabled to enabled

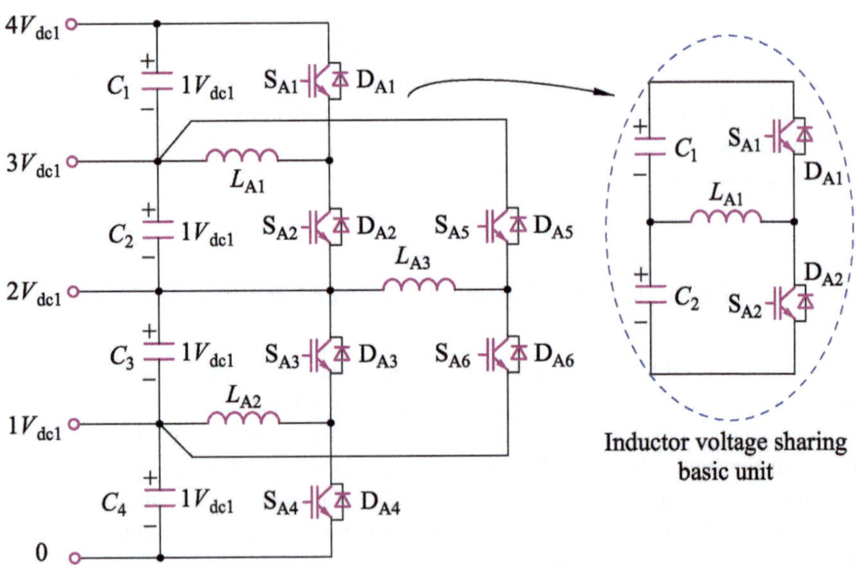

Fig. 5.16 Three-inductor voltage equalizer circuit and inductor voltage-sharing basic unit

The analysis of these single-inductor voltage equalizer units is based on the first unit. This unit consists of two series-connected switching tubes (SA1 and SA2) accompanied by body diodes (DA1 and DA2), as well as an intermediate energy storage voltage-sharing inductor (LA1). The function of inductor LA1 is to prevent short-circuiting of the DC capacitor through either a switching tube or a body diode.

To achieve voltage equalization, the auxiliary voltage equalizer circuit should meet the following requirements:

(1) The two series-connected switching tubes within a basic voltage-balancing unit should form a complementary switch pair, meaning that they should not be turned on simultaneously. Specifically, when one switch is activated, the other must remain off.
(2) During a switching cycle, the switching tube with a higher capacitor voltage is turned on first. This ensures that energy in an equilibrium state is transferred from the discharging capacitor to the voltage-sharing inductor. Subsequently, the switching tube with a lower capacitor voltage is turned on, allowing energy to flow from the voltage-sharing inductor to the capacitor that needs to be charged.

To simplify the analysis, the following assumptions are made during a switching cycle of a voltage equalizer circuit:

(1) The switching frequency is sufficiently high that the current increases and decreases in an almost linear manner.
(2) The main circuit is regarded as being solely related to the auxiliary voltage equalizer circuit, with any influence on the capacitor voltage considered negligible.

Based on the above assumptions, the voltage equalization process within a switching cycle can be divided into the following three stages:

In the first stage, the switching tube connected to the capacitor with a higher voltage is turned on first, causing the capacitor with a higher voltage to discharge, and the electrical energy is stored in the inductor.

In the second stage, both switching tubes are turned off, the capacitor with a lower voltage is charged, and the current in the voltage equalizer inductor is reduced to zero.

In the third stage, once the inductor current reaches zero, the equalizing current remains at zero until the next equalization cycle begins.

The specific operating process is shown in Fig. 5.17, where V_{dc1} and V_{dc2} indicate the voltages of DC capacitors C1 and C2, respectively.

When V_{dc1} is higher than V_{dc2}, the operating process is illustrated in Fig. 5.17a, b. During the initial part of the cycle, switching tube SA1 is turned on, causing capacitor C1 to discharge and resulting in a decrease in its voltage (V_{dc1}) while electric energy is stored in inductor LA1, as shown in Fig. 5.17a. During the latter part of the cycle, both switching tubes SA1 and SA2 are turned off, allowing inductor LA1 to discharge through diode DA2. This process charges capacitor C2, leading to an increase in its voltage (V_{dc2}).

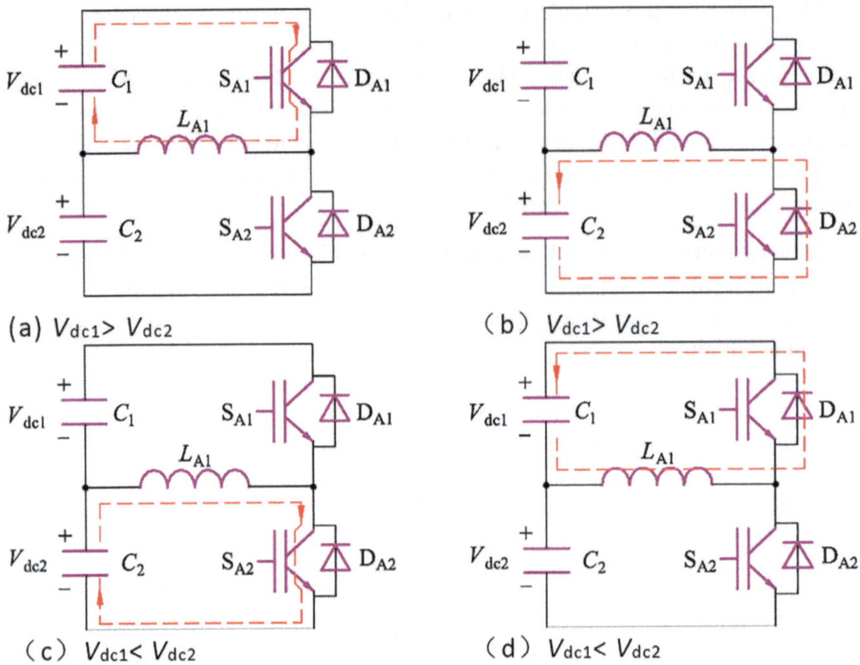

Fig. 5.17 Basic operating process of basic unit of inductor-assisted voltage equalizer circuit. (**a**) $V_{dc1} > V_{dc2}$. (**b**) $V_{dc1} > V_{dc2}$. (**c**) $V_{dc1} < V_{dc2}$. (**d**) $V_{dc1} < V_{dc2}$

Similarly, when V_{dc1} is less than V_{dc2}, the operating process is illustrated in Fig. 5.17c, d. During the initial part of the cycle, switch SA2 is turned on, causing capacitor C2 to discharge, leading to a decrease in its voltage (V_{dc2}), while energy is stored in inductor LA1. In the latter part of the cycle, both SA1 and SA2 are turned off, allowing inductor LA1 to discharge through diode DA1, which charges capacitor C1, resulting in an increase in its voltage (V_{dc1}).

To verify the effectiveness of the aforementioned voltage equalizer circuit, a five-level PWM rectifier equipped with an auxiliary voltage equalizer circuit was designed for simulations. In the diode-clamped five-level main circuit depicted in Fig. 5.4, the parameters are as follows: input inductance is 1000 μH, capacitor capacitance is 18 mF, initial capacitor voltage is 240 V, voltage reference for the capacitor is 250 V, and load inductance and resistance are 500 μH and 2 Ω, respectively. Figure 5.16 also includes the voltage equalizer circuit shown in Fig. 5.4, with a voltage-balancing inductance of 10 μH.

The first half of Fig. 5.18 displays the DC voltage waveform without a voltage equalizer circuit, while the second half illustrates the waveform with inductor-assisted voltage equalization. When the converter operates under a 100 kW DC resistive load without a voltage equalizer circuit, the four DC voltages (Vc1–Vc4) remain stable but fail to achieve complete balance. Following a sudden load change, the DC voltage significantly deviates. However, after 0.04 s, the inductor-assisted

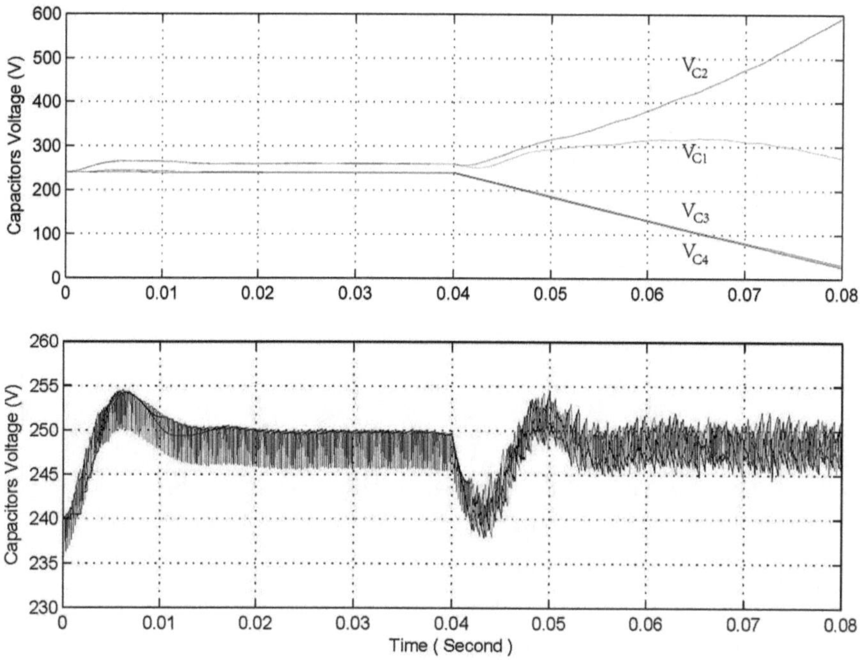

Fig. 5.18 Waveform diagram of system startup process ($t = 0$ s ~ $t = 0.04$ s) and on-load operating process ($t = 0.04$ s ~ $t = 0.08$ s). Above: DC voltage waveform without voltage equalizer circuit; below: DC voltage waveform with single inductor-assisted equalizer circuit

voltage equalizer circuit is activated, resulting in improved balance among the DC voltages. Even with the DC resistive loads connected, the multilevel PWM rectifier continues to operate normally and stably at a unity power factor and suitably high modulation conditions.

Figure 5.19 illustrates the output line voltage and input current waveforms for a converter without an equalizer circuit compared to one with a single inductor-assisted equalizer circuit. In the absence of an auxiliary equalizer circuit on the DC side, the DC voltage remains unstable, as shown in Fig. 5.21 (above), where the terminal voltage waveform output by the PWM rectifier transitions from seven levels to three levels following a sudden load connection. In contrast, when a single inductor-assisted voltage equalizer circuit is implemented, the PWM rectifier can effectively respond to sudden load changes. Consequently, the DC voltage stabilizes and balances, while the AC output terminal voltage maintains its output levels. Additionally, the smooth input current allows for a rapid transition into a stable state. Similar results can also be achieved with a well-designed two-stage voltage equalizer circuit.

The experimental circuit, which is consistent with the simulation circuit, also verified the effectiveness of inductor-assisted voltage equalizer circuits. The experimental waveforms are presented in Fig. 5.20, which illustrates only three DC voltages and the equalizing current due to channel limitations. During the experiment,

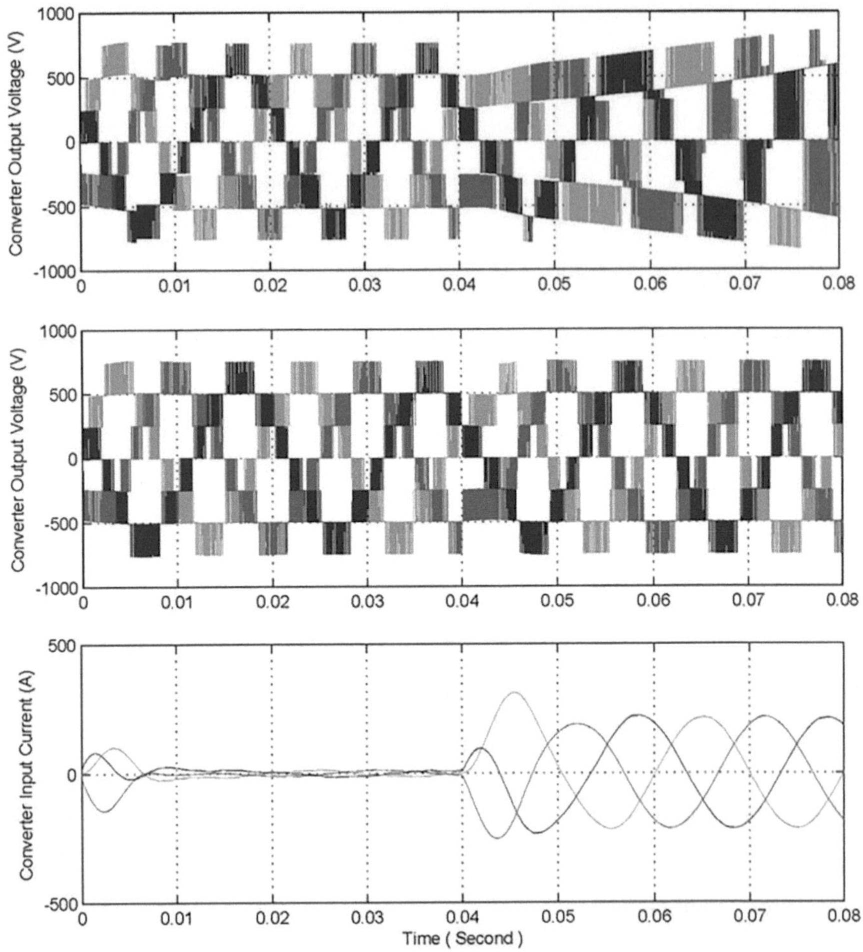

Fig. 5.19 Above: VSI output line voltage without voltage equalizer circuit; middle: VSI output line voltage with single-inductor voltage equalizer circuit; below: VSI input current with single-inductor voltage equalizer circuit

the auxiliary voltage equalizer circuit was disabled for the initial 40 ms, resulting in significant differences in capacitor voltages and serious unbalance. After the activation of the auxiliary voltage equalizer circuit at 40 ms, the severely unbalanced capacitor voltages from the early stage were rapidly adjusted and gradually approached a balanced state. The equalizing current also decreased progressively, with its peak effectively limited to within 20 A.

Inductor-assisted voltage equalizer circuits can limit the equalizing current while incurring less loss, offering several advantages over capacitor-assisted voltage equalizer circuits.

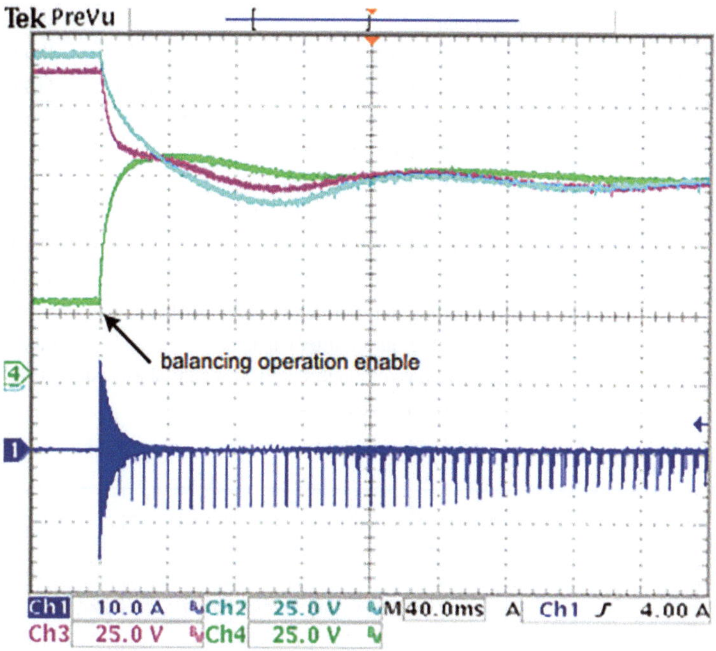

Fig. 5.20 Dynamic waveforms with voltage equalizer circuit enabled (Ch1: equalizing current, Ch2–4: DC voltage)

5.4 Simulation and Experimental Verification of Three-Phase-to-Single-Phase System

To verify the circuit control and modulation strategy after integrating the front and rear ends of the five-level diode-clamped three-phase-to-single-phase converter previously discussed with voltage equalizer circuits, as well as their feasibility in advanced co-phase traction power supply systems, this chapter employs Simulink to conduct simulation research. The focus is on power flow control in the following configurations: the three-phase PWM rectifier, single-phase PWM rectifier, three-inductor voltage equalizer circuit, and three-phase-to-single-phase converter.

A three-phase DQ decoupling control strategy is implemented on the three-phase side, while a single-phase DQ control strategy is employed on the single-phase side. The converters on both the three-phase and single-phase sides are controlled independently. The three-phase side manages the DC voltage, and the DC side is equipped with a three-inductor-assisted voltage equalizer circuit. The simulation parameters are detailed in Table 5.2.

Table 5.2 Simulation parameters of three-phase-to-single-phase converter system

Parameter	Value
Line voltage of three-phase grid	110 kV
Transformation ratio of step-down transformer	110 kV/400 V
AC input inductance on three-phase side	5 mH
Single DC-side capacitor	2.35 mF
Secondary filter circuit inductor	1 mH
Secondary filter circuit capacitor	2.2 mF
Total DC-side voltage (Vdc)	700 V
Single-phase-side AC voltage peak (Vss)	$400\times\sqrt{2}$ V
Single-phase output filter inductor (L1/L2)	5 mH/0.1 mH
Single-phase output filter capacitor	0.05 mF
Carrier frequency on three-phase/single-phase side	10 kHz
Simulation step size	1 μs

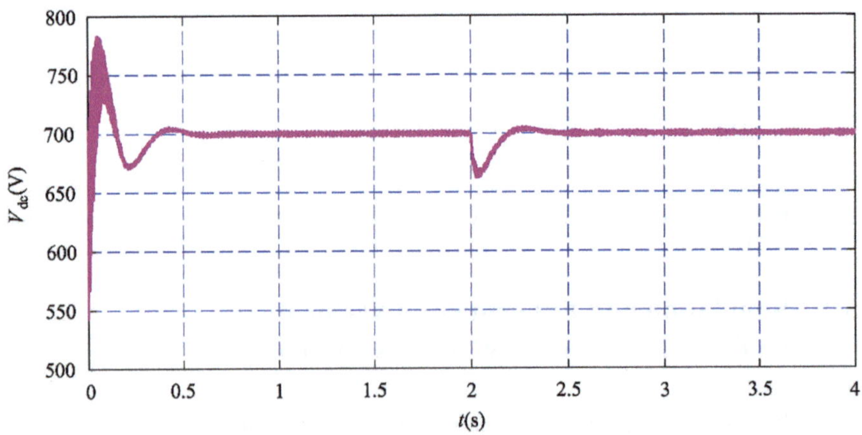

Fig. 5.21 Waveform of total DC voltage (Vdc) on three-phase rectifier side

5.4.1 Simulation of Three-Phase Five-Level Rectifier

Figure 5.21 displays the simulation waveform of the total voltage on the DC side during the process where the three-phase rectifier is unloaded from $t = 0$ to $t = 2$ s and then connected to a 100 Ω resistive load at $t = 2$ s. As shown, the total DC voltage quickly reaches and stabilizes at 700 V at $t = 0.6$ s after a brief adjustment, exhibiting an overshoot of approximately 10%. When the 100 Ω load is connected, the DC capacitor voltage briefly drops. After about 0.3 s, the total DC voltage returns to and stabilizes at a new steady value following an adjustment process, with a voltage overshoot of about 5.7%. Therefore, the three-phase DQ decoupling control strategy proves effective in maintaining the DC voltage stability of the rectifier at the reference level.

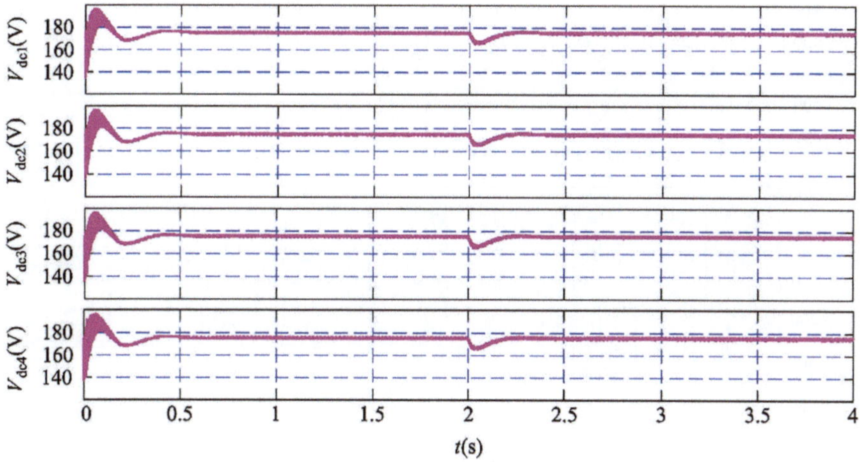

Fig. 5.22 Simulation waveforms of four sub-capacitor voltages

Figure 5.22 displays the simulation waveforms of the four sub-capacitor voltages. Under both unloaded and loaded conditions, the voltages of the four sub-capacitors on the DC side stabilize at approximately 175 V. After sudden load changes, the voltages return to a stable value within 0.3 s of adjustment. The simulation results demonstrate the effectiveness of the three-inductor-assisted voltage equalizer circuits in maintaining voltage balance across the four DC capacitors in both unloaded and loaded conditions.

Figure 5.23 displays the simulation waveforms of the three-phase AC port line voltage. During the interval $t = 3 - 3.1$ s, the three-phase-side port line voltage of the three-phase five-level PWM rectifier exhibits a nine-level stepwise waveform. Figure 5.23b presents the AC-side input current. The instantaneous impulse current for startup before $t = 0.3$ s remains below 60 A, while the current amplitude during the unloaded stage is close to 0 A. Upon load connection at $t = 2.0$ s, the current increases rapidly and stabilizes at approximately 12 A after about 0.1 s, demonstrating good dynamic performance. As highlighted in the detail, during the period from $t = 3$–3.1 s, the three-phase currents are sinusoidal, featuring a low total harmonic distortion (THD) and a mutual phase angle difference of 120°.

Figure 5.24 displays the simulation results of the reference currents (I_d, I_q) and the reference voltages (V_d, V_q) in the three-phase rectifier control system. Before the load is connected, I_d, I_q, and V_q all stabilize near the reference value of 0, while V_d maintains a stable level. After the 100 Ω load is connected, V_d and V_q experience slight changes, whereas the current components I_d and I_q undergo significant variations. They return to new stable values within 0.5 s.

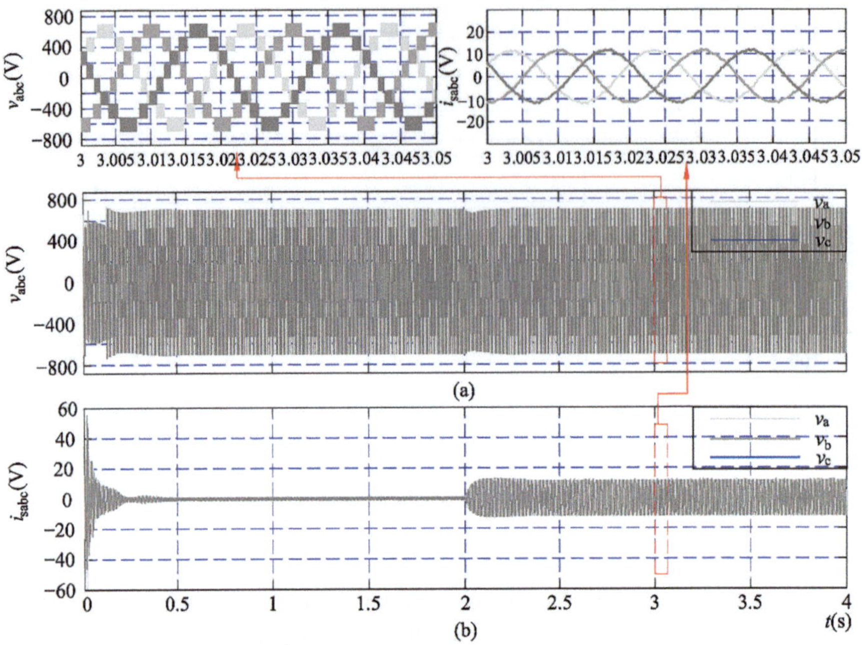

Fig. 5.23 AC-side port line voltage (vabc) and input current (isabc) of three-phase rectifier. (**a**) AC-side port line voltage (vabc) of three-phase rectifier; (**b**) AC-side input current (isabc) of three-phase rectifier

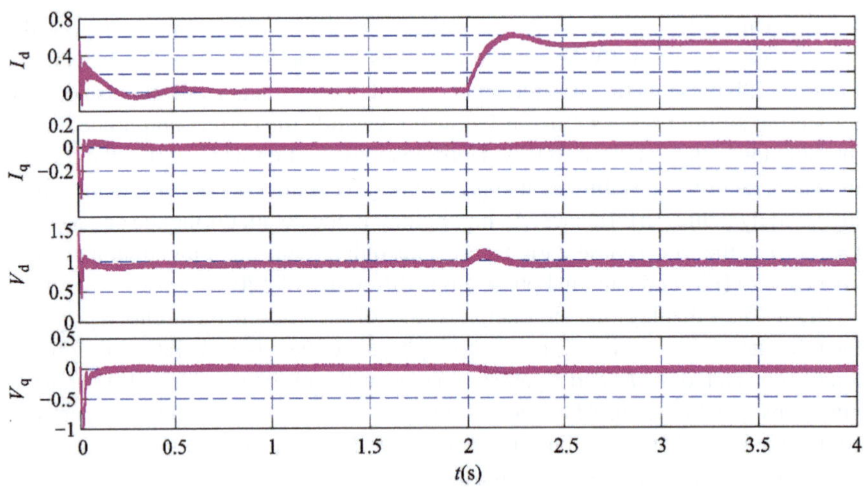

Fig. 5.24 Simulation waveforms of normalized values of reference currents (I_d and I_q) and reference voltages (V_d and V_q) of three-phase converter

5.4.2 Simulation of Single-Phase Five-Level Rectifier

Figure 5.25 displays the normalized values of active, reactive, and DC bias currents generated by decomposing single-phase AC currents when the single-phase DQ decomposition control strategy is applied. As shown, during the unloaded stage from $t = 0$ to $t = 2$ s, the active, reactive, and DC bias currents remain around 0 A after adjustments. When a resistive load is connected from $t = 2 - 4$ s, the active current rises rapidly and stabilizes quickly, while the reactive and DC bias currents return to 0 A after adjustments.

Figure 5.26 displays the simulation waveforms of the active voltage component (V_{sp}), reactive voltage component (V_{sq}), and DC bias voltage component (V_{sz}) in a

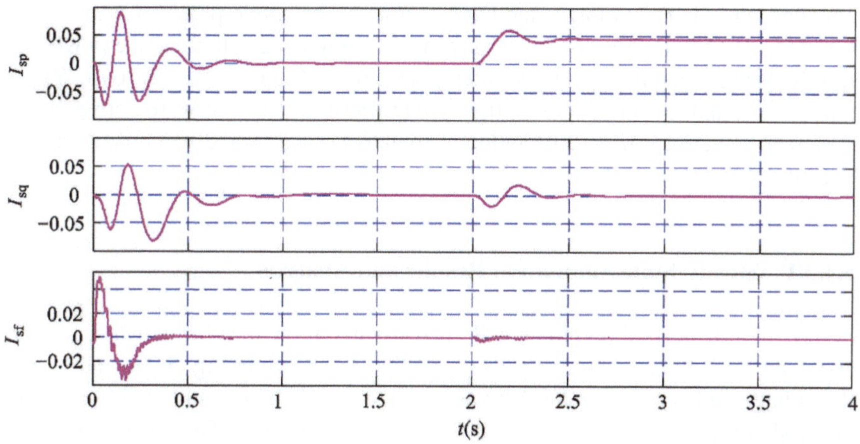

Fig. 5.25 Normalized values of active current (I_{sp}), reactive current (I_{sq}), and DC bias current (I_{sf}) of single-phase AC currents

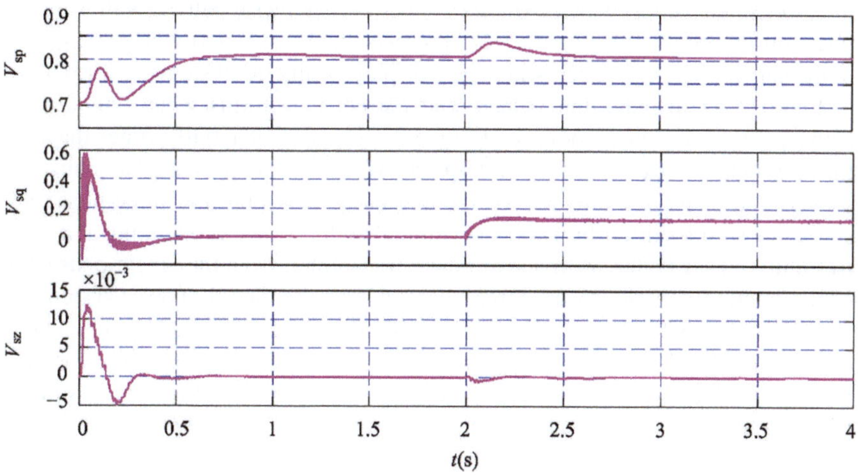

Fig. 5.26 Simulation waveforms of normalized reference voltages (V_{sp}, V_{sq}, and V_{sz}) of single-phase rectifier controller

single-phase PWM rectifier control system. As shown, under unloaded conditions, V_{sp} stabilizes at a normalized value of 0.8, while both V_{sq} and V_{sz} stabilize at 0 after approximately 0.7 s of adjustment. Under loaded conditions, all three voltage components respond to load variations and reach a new stable state within about 0.5 s.

5.4.3 Co-simulation of Three-Phase-to-Single-Phase Converter System

Co-simulations were conducted for three-phase-to-single-phase converter systems, with a simulation duration from $t = 0$ to $t = 10$ s. At $t = 0$ s, the three-phase rectifier operates, and the DC voltage stabilizes at 700 V. The single-phase inverter is activated at $t = 0.5$ s. During the interval from $t = 0.5$ to $t = 2$ s, both active power and reactive power remain at 0. In the intervals from $t = 2$ to $t = 4$ s, $t = 4$ to $t = 6$ s, $t = 6$ to $t = 8$ s, and $t = 8$ to $t = 10$ s, the system transmits 20% forward active power, 20% reverse active power, 20% inductive reactive power, and 20% capacitive reactive power under control.

5.4.3.1 Steady-State Simulation Results and Analysis

Figure 5.27 displays the voltage and current simulation waveforms when forward active power is injected from $t = 3$ to $t = 3.2$ s. As shown in Fig. 5.27a and b, both the voltage and current on the three-phase side and the single-phase side maintain phase alignment, indicating operation in a resistive load state. The successful implementation of system control at a unity power factor demonstrates that three-phase-to-single-phase converter systems can reliably transmit pure active power to traction networks. Figure 5.27c illustrates the three-phase port voltage waveform of the three-phase converter, which exhibits a nine-level stepwise pattern with a mutual phase difference of 120°. Figure 5.27d presents the AC-side port voltage waveform of the single-phase converter, which also displays a nine-level stepwise format.

Figure 5.28 displays the voltage and current waveforms when only reverse active power is injected from $t = 5$ to $t = 5.2$ s. Figure 5.28a, b illustrates that both the three-phase current and the single-phase AC current are presented as sinusoidal waves. The phases of the AC voltage and current on both the three-phase and single-phase sides differ by 180°, indicating that the system is operating at a negative unity power factor, with active power transmitted from the single-phase side of the traction network to the three-phase side of the electric power system. As shown in Fig. 5.28c and d, the three-phase and single-phase AC port voltages exhibit nine-level stepwise waveforms, indicating that the traction network can also feedback active power to the three-phase grid. This functionality is suitable for application scenarios such as energy recovery of braking power by the traction network.

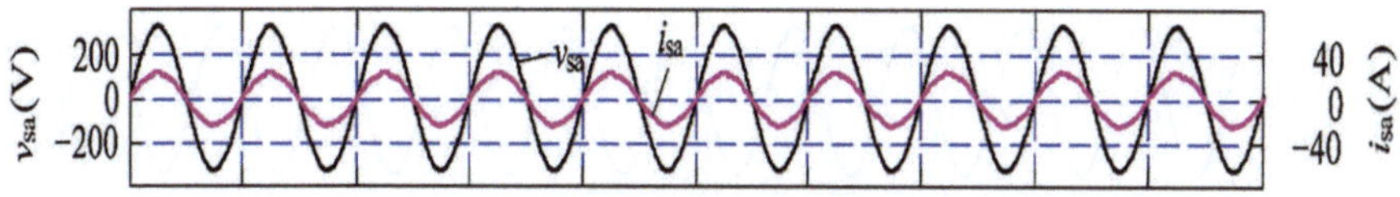

(a) A-phase AC Voltage (v_{sa}) and A-phase AC Current (i_{sa})

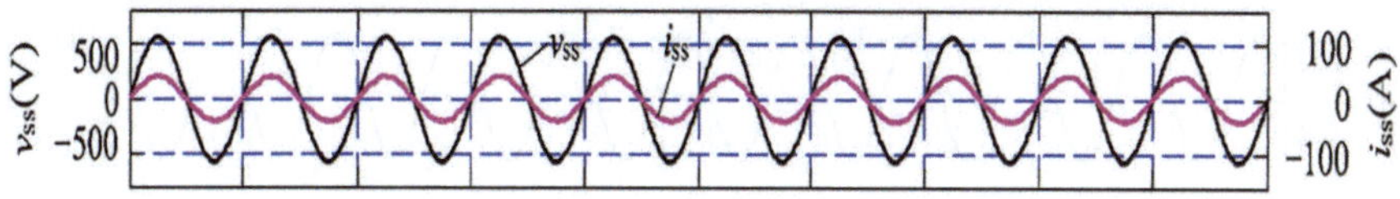

(b) Single-phase AC Voltage (v_{ss}) and current (i_{ss})

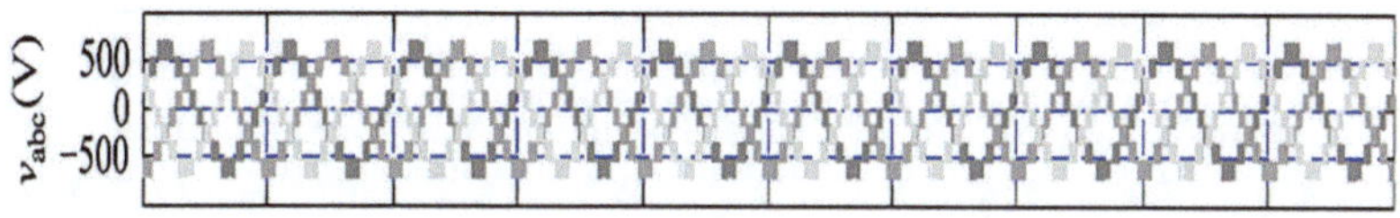

(c) Three-phase AC Port Voltage (v_{abc})

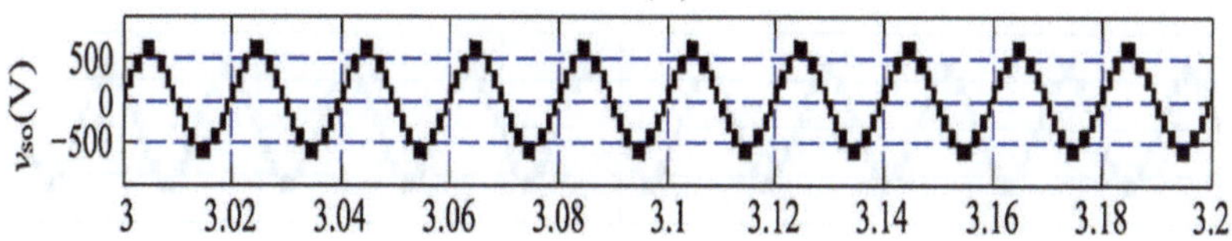

(d) Single-phase AC Port Voltage (v_{so})

Fig. 5.27 Voltage and current simulation waveforms during forward active power injection. (**a**) A-phase AC voltage (v_{sa}) and A-phase AC current (i_{sa}). (**b**) Single-phase AC voltage (v_{ss}) and current (i_{ss}). (**c**) Three-phase AC port voltage (v_{abc}). (**d**) Single-phase AC port voltage (v_{so})

Figure 5.29 displays the voltage and current simulation waveforms when only inductive reactive power is injected from $t = 7$ to $t = 7.2$ s. As shown in Fig. 5.29a, the A-phase current on the three-phase side is approximately 0 A, with reactive power flowing solely on the single-phase side. The absence of reactive power output on the three-phase side and the phase lag of the single-phase current behind the voltage by 90° indicate that the single-phase-side converter is operating in an inductive state. Figure 5.29c and d illustrates the three-phase AC port voltage and the single-phase AC port voltage, both of which exhibit nine-level stepwise waveforms.

Similarly, Figure 5.30 displays the voltage and current simulation waveforms when capacitive reactive power is injected from $t = 9$ to $t = 9.2$ seconds. As shown in Figure 5.30 (a) and (b), the three-phase current reads 0 A, while the single-phase current leads the voltage by 90° in phase, indicating a capacitive operating state. Figure 5.30 (c) and (d) depicts that the three-phase port voltage presents a nine-level stepwise waveform, while the single-phase AC port voltage exhibits a seven-level stepwise waveform. The decrease in the number of steps is associated with the reduction in the modulation degree of the single-phase converter.

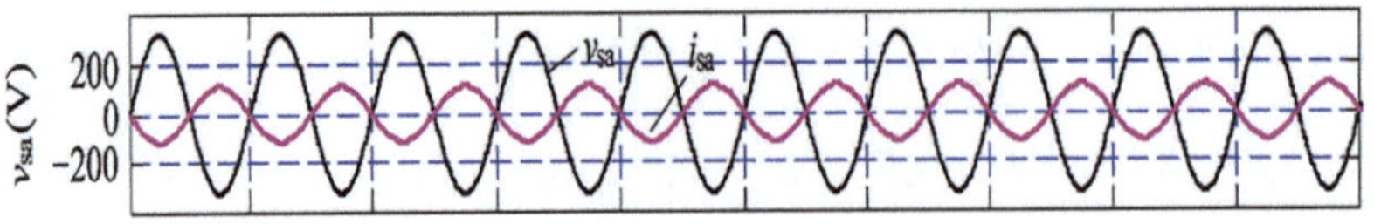

(a) A-phase AC Voltage (v_{sa}) and A-phase AC Current (i_{sa})

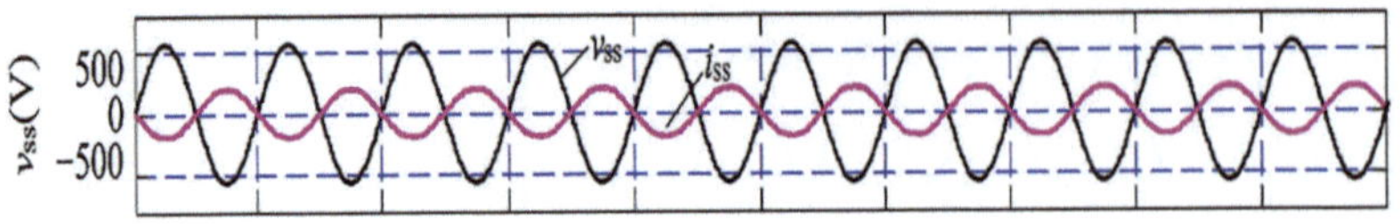

(b) Single-phase AC Voltage (v_{ss}) and current (i_{ss})

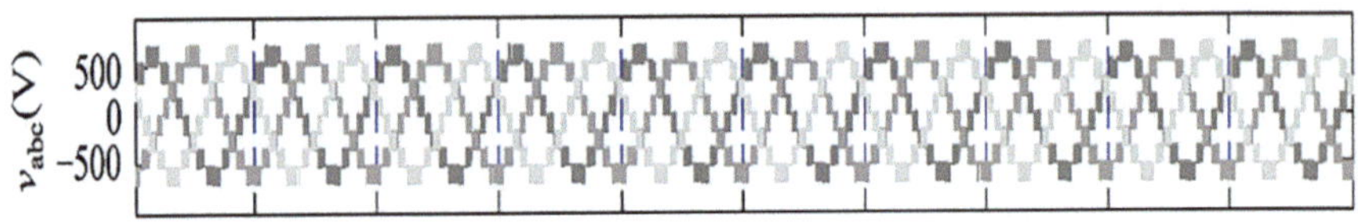

(c) Three-phase AC Port Voltage (v_{abc})

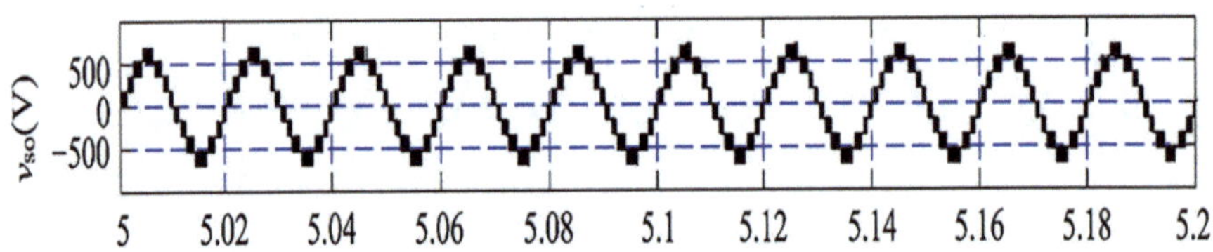

(d) Single-phase AC Port Voltage (v_s)

Fig. 5.28 Voltage and current simulation waveforms during reverse active power injection. (**a**) A-phase AC voltage (v_{sa}) and A-phase AC current (i_{sa}). (**b**) Single-phase AC voltage (v_{ss}) and current (i_{ss}). (**c**) Three-phase AC port voltage (v_{abc}). (**d**) Single-phase AC port voltage (v_s)

5.4.3.2 Dynamic Simulation Results and Analysis

As shown in Fig. 5.31a, the DC voltage remains at the reference level of 700 V during power fluctuations, with an adjustment duration of approximately 0.2 s, demonstrating a fast dynamic response. Figure 5.31b displays the A-phase AC current on the three-phase side, which also undergoes a dynamic adjustment process within 0.2 s. Additionally, when forward active power and reverse active power are injected, the A-phase current increases and then stabilizes. In contrast, when inductive reactive power and capacitive reactive power are injected, the A-phase current reads 0 A. Figure 5.31c illustrates the simulation waveform of the single-phase port voltage (v_{so}). As shown, when the active or reactive power changes dynamically, the port voltage reaches stability within 0.5 s, indicating good dynamic response performance, with its amplitude also varying following changes in modulation degree. Similarly, the single-phase AC current depicted in Fig. 5.31d demonstrates satisfactory dynamic performance. A comparison of Fig. 5.31b and d verifies that active

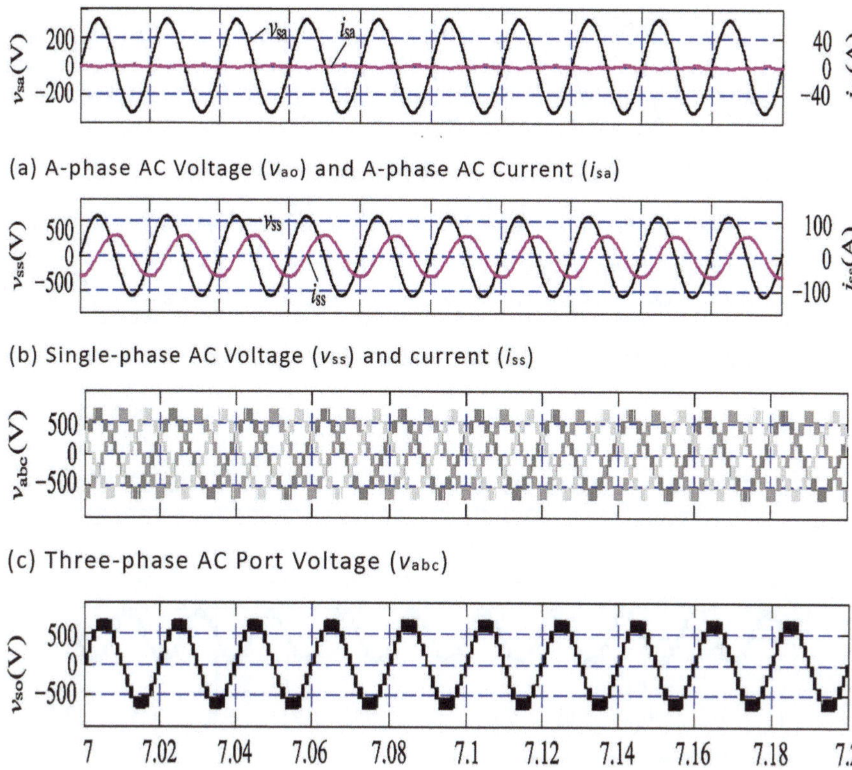

Fig. 5.29 Voltage and current simulation waveforms during inductive reactive power injection. (**a**) A-phase AC voltage (v_{ao}) and A-phase AC current (i_{sa}). (**b**) Single-phase AC voltage (v_{ss}) and current (i_{ss}). (**c**) Three-phase AC port voltage (v_{abc}). (**d**) Single-phase AC port voltage (v_{so})

power is alternately transferred between the three-phase and single-phase sides, while reactive power is present only at the single-phase port.

Figures 5.29, 5.30, and 5.31 verify that three-phase-to-single-phase converters can either feed back active power from the traction network to the three-phase grid or absorb active power, while the reactive power output on the single-phase side does not disrupt the normal operation of the three-phase side. The following conclusions can be drawn: The DQ control strategy on the single-phase side prevents the three-phase grid from supplying reactive power while facilitating efficient power transmission. Furthermore, the amplitude, phase, and frequency of the output voltage on the single-phase side are adjustable, providing a solid foundation for multisource grid connections.

Figure 5.32 displays the dynamic simulation waveforms of active current (I_d), reactive current (I_q), and port voltages (V_d and V_q) when the single-phase traction network outputs forward and reverse active power, as well as inductive and capacitive reactive power. When the active power changes, V_d and V_q reach new stable values after short adjustments, accompanied by corresponding changes in the active

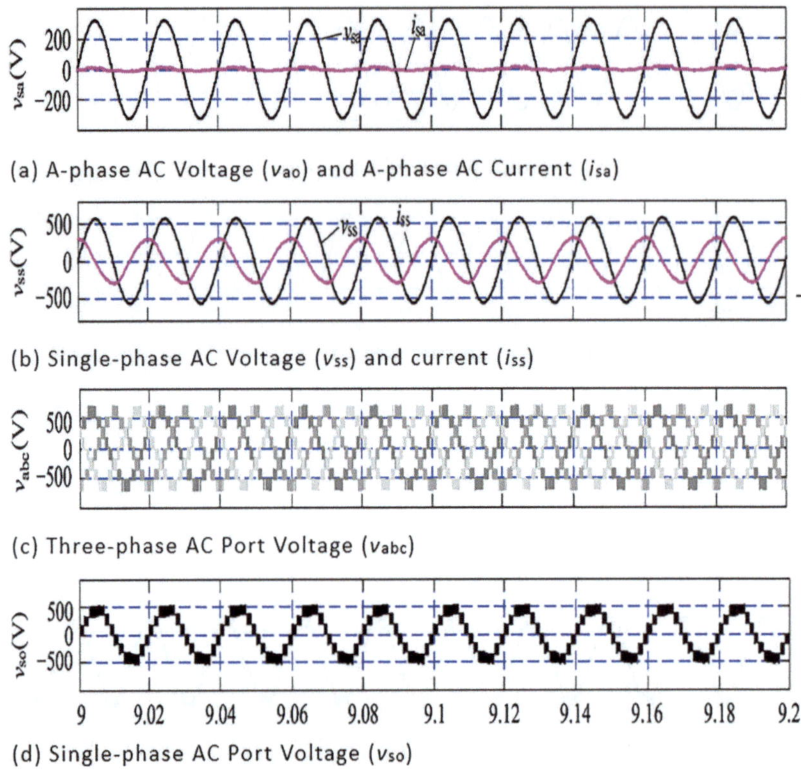

Fig. 5.30 Voltage and current simulation waveforms during capacitive reactive power injection. (**a**) A-phase AC voltage (v_{ao}) and A-phase AC current (i_{sa}). (**b**) Single-phase AC voltage (v_{ss}) and current (i_{ss}). (**c**) Three-phase AC port voltage (v_{abc}). (**d**) Single-phase AC port voltage (v_{so})

and reactive current outputs, demonstrating a fast response. When the reactive power on the single-phase side changes, I_d and I_q return to 0 after approximately 0.2 s of adjustment, aligning with the three-phase side, which does not output reactive power.

Similarly, Figure 5.33 presents the simulation waveforms of active, reactive, and DC bias reference voltages in the controller on the single-phase side. As shown, the port voltages (V_{sp} and V_{sq}) respond to changes in active and reactive power. When power changes occur at $t = 2$, $t = 4$, $t = 6$, and $t = 8$ seconds, the control curves return to a new steady state after approximately 0.4 s of adjustments.

Figure 5.34 displays the simulation waveforms of active, reactive, and DC bias currents on the single-phase side. As shown, after the injection of forward and reverse active power, stability is achieved at normalized stable values of 0.2 and −0.2, respectively, after a short adjustment period, while the reactive current stabilizes at 0. Following the injection of inductive and capacitive reactive power, the active current stabilizes around 0, while the reactive current stabilizes at −0.2 and 0.2, respectively, after a brief period of adjustment. The DC bias current

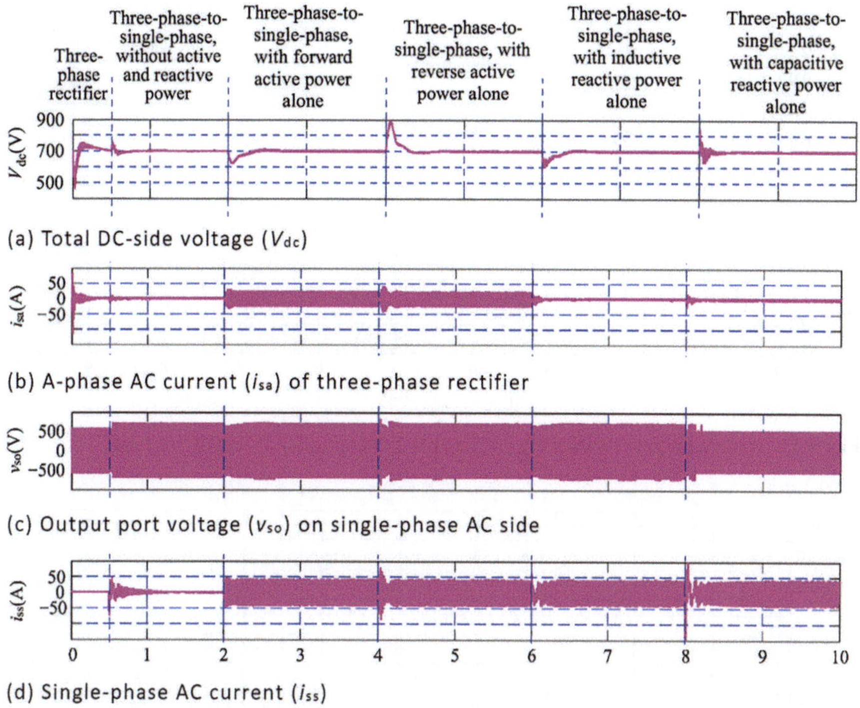

Fig. 5.31 Dynamic waveforms of voltage and current in grid-connected three-phase-to-single-phase converter during injection of forward active power, reverse active power, inductive reactive power, and capacitive reactive power. (**a**) Total DC-side voltage (V_{dc}). (**b**) A-phase AC current (i_{sa}) of three-phase rectifier. (**c**) Output port voltage (v_{so}) on single-phase AC side. (**d**) Single-phase AC current (i_{ss})

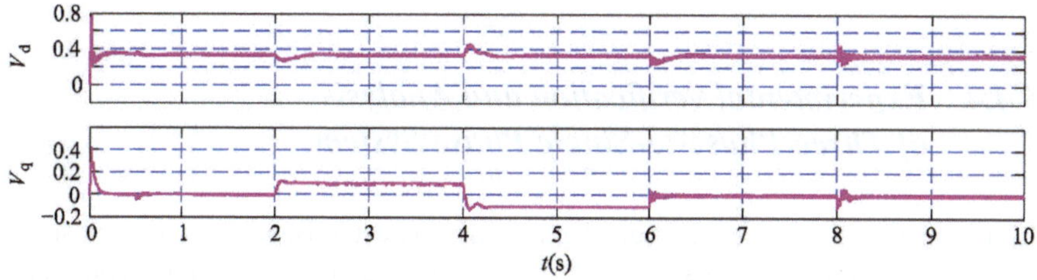

Fig. 5.32 Active current (I_d), reactive current (I_q), reference adjustment voltage (V_d and V_q) of controller at three-phase side

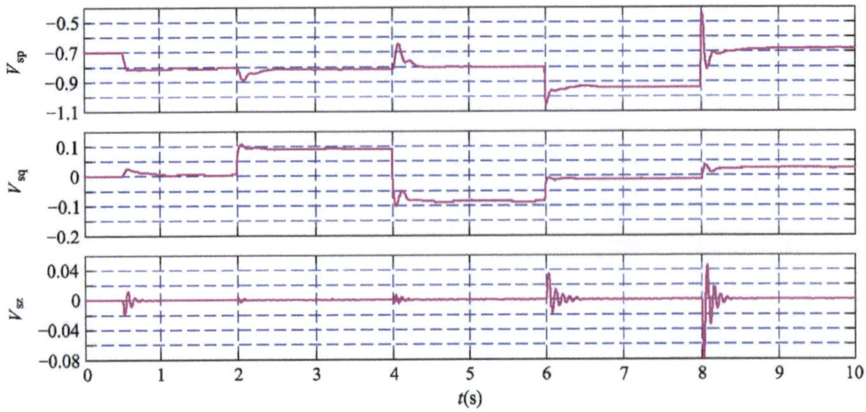

Fig. 5.33 Voltage reference values (V_{sp}, V_{sq}, and V_{sz}) of controller at single-phase side

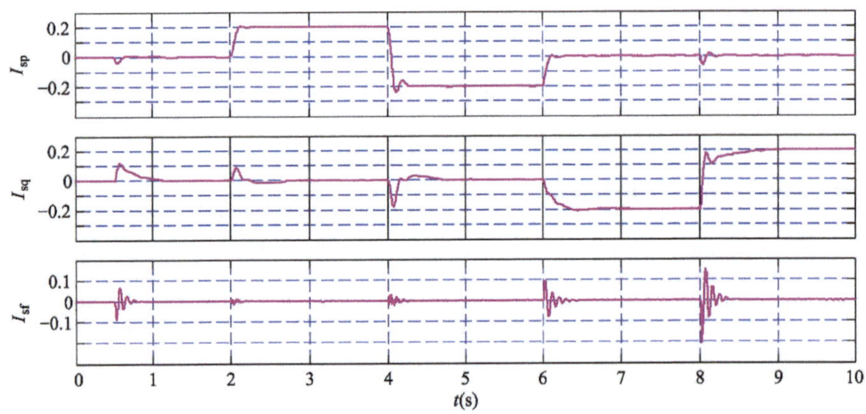

Fig. 5.34 Simulation waveforms of single-phase active current (I_{sp}), reactive current (I_{sq}), and DC bias current (I_{sf})

remains consistently stable near 0, indicating that there is no DC bias present in the single-phase current.

5.4.4 Experimental Verification and Analysis of Three-Phase-to-Single Phase System

Based on Fig. 5.3, an experimental platform was established for three-phase-to-single-phase converters, featuring the diode-clamped five-level circuit configuration. With the experimental circuit topology shown in Fig. 5.35, this platform allows for various experimental verifications, including three-phase rectifiers, single-phase rectifiers, and grid-connected three-phase-to-single-phase systems.

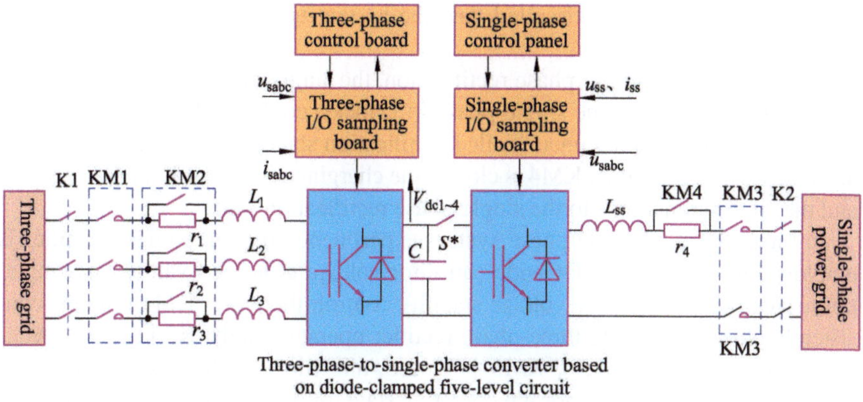

Fig. 5.35 Topology of experimental system for power source substation

Table 5.3 Experimental parameters for grid connection of three-phase-to-single-phase system

Parameter	Value
Three-phase power line voltage	110 V
RMS of single-phase power supply (voltage source substation)	105 V
Primary-side voltage/secondary-side voltage of isolation transformer on single-phase side	105 V/105 V
Total DC-side voltage reference	220 V
Three-phase AC-side inductor	15 mH
Single-phase AC-side inductor	15 mH
Single DC-link capacitor	2.35 mF
Secondary filter circuit inductor/capacitor	1 mH/2.2 mF
Voltage-sharing inductor	1 mH
Carrier frequency on three-phase side	5 kHz
Carrier frequency on single-phase side	3 kHz
FPGA model/clock frequency	EP3C55F484C8/32MHz
IGBT model/clamping diode model	IKW20N-60T/ MUR30120

The main parameters of the experimental system are shown in Table 5.3.

To prevent excessive charging current during startup, a charging resistor is utilized to pre-charge the DC-side capacitor in both three-phase and single-phase rectification experiments. In the figure, r1–r3 represent the charging resistors of the three-phase rectifier, while r4 indicates the charging resistor of the single-phase rectifier. K1 and K2 denote the main power air switches on the three-phase and single-phase sides, respectively, and KM1, KM2, and KM3 represent the contactor switches. During operation of the three-phase rectifier, K1 is closed first. After the phase-locked loop (PLL) locks and the phase sequence are determined, KM1 is closed. The circuit is in the uncontrolled rectification stage, with switching signals still blocked; the circuit charges the DC capacitor through the body diode of the switching tube. Once the DC capacitor voltage stabilizes, KM2 is closed, the charging resistor is short-circuited, and the system transitions to the three-phase rectification state.

Similarly, during single-phase rectification, the single-phase main power switch K2 is closed first. After the single-phase PLL locks, contactor KM3 is closed, and the circuit operates in the uncontrolled rectification stage. Once the capacitor voltage reaches a stable value, KM4 is closed, the charging resistor r4 is short-circuited, and the circuit transitions to the single-phase rectification state.

In three-phase-to-single-phase system experiments, the three-phase rectifier is employed to stabilize the DC-side capacitor voltage at a fixed value. The single-phase side does not require voltage stability control; therefore, KM4 remains consistently closed. Once the three-phase rectifier operates stably, K2 is closed. After the single-phase PLL locks, KM3 is closed, allowing the three-phase-to-single-phase converter to enter its normal operating state following a brief delay. In experiments focusing solely on three-phase PWM rectification, both steady-state and dynamic tests of the three-phase five-level diode-clamped rectifier circuit can be conducted by opening switches S* and K2, which, respectively, connect the three-phase rectifier and the single-phase inverter.

5.4.4.1 Steady-State Experiment and Analysis of Three-Phase Rectifier

Figure 5.36 displays the experimental waveforms of the three-phase rectifier during the unloaded steady state, with the DC voltage (V_{dc}) stabilizing at around 220 V, as shown in Figure (a). The port line voltage in phases ab exhibits a typical seven-level stepwise waveform. In this state, the rectifier operates without any load. Due to the no-load losses of the converter, the AC current amplitude is slightly greater than 0 A. In Figure (b), Ch1–Ch3 illustrate that the waveforms of the three-phase line voltage ports all present seven-level stepwise patterns, with a phase difference of 120°.

Figure 5.37 displays the experimental waveforms of the three-phase rectifier during the loaded steady state. In Figure (a), the DC voltage remains stable at 220 V after the load is connected. The ab-phase port line voltage (v_{ab}) on the AC side maintains a seven-level configuration, while the amplitude of the a-phase current approaches 5 A. The third channel indicates the line voltage (v_{sab}), which exhibits a phase difference of 30° with respect to the a-phase current. Figure (b) shows that the currents in phases a, b, and c are equal after the load is connected, each with an amplitude of approximately 5 A.

5.4.4.2 Dynamic Experiment and Analysis of Three-Phase Rectifier

Figure 5.38 displays the dynamic waveforms of the three-phase rectifier during the process of load connection and removal. After the load is either connected or disconnected, the total DC voltage returns to the reference level within 1 s. The ab-phase port line voltage (v_{ab}) also stabilizes within a short time. The a-phase AC

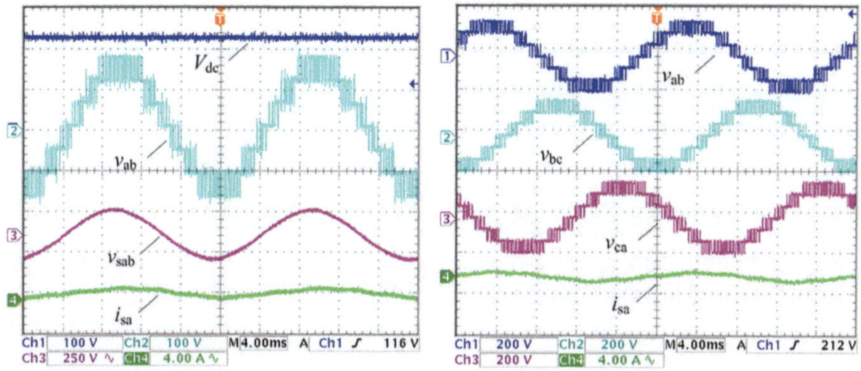

(a) Waveforms in Unloaded State Ch1: total DC voltage (V_{dc}), Ch2: ab-phase port line voltage (v_{ab}), Ch3: ab-phase AC line voltage (v_{sab}), Ch4: a-phase AC current (i_{sa})

(b) Ch1~Ch3: ab-phase, bc-phase, ca-phase port line voltages (v_{ab}, v_{bc}, v_{ca}), Ch4: a-phase AC current (i_{sa})

Fig. 5.36 Experimental waveforms of three-phase rectifier in unloaded steady state. (**a**) Waveforms in unloaded state: Ch1: total DC voltage (V_{dc}), Ch2: ab-phase port line voltage (v_{ab}), Ch3: ab-phase AC line voltage (v_{sab}), Ch4: a-phase AC current (i_{sa}). (**b**) Ch1–Ch3: ab-phase, bc-phase, ca-phase port line voltages (v_{ab}, v_{bc}, v_{ca}), Ch4: a-phase AC current (i_{sa})

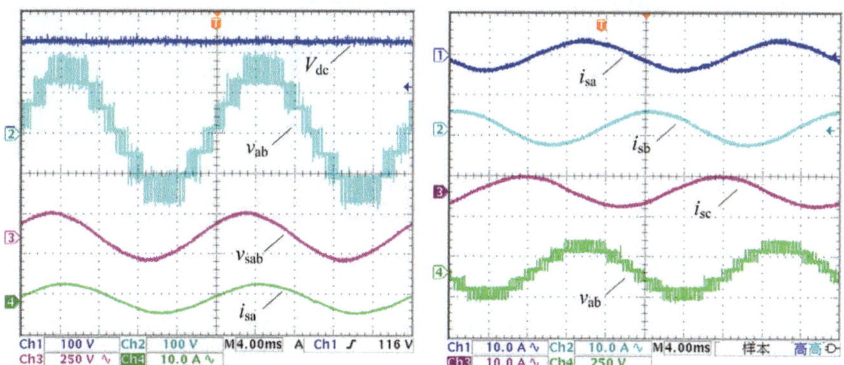

(a) Waveforms in Loaded State Ch1: total DC voltage (V_{dc}), Ch2: ab-phase AC port voltage (v_{ab}), Ch3: ab-phase AC line voltage (v_{sab}), Ch4: a-phase AC current (i_{sa})

(b) Ch1~Ch3: three-phase AC currents (i_{sa}, i_{sb}, i_{sc}), Ch4: ab-phase AC port voltage (v_{ab})

Fig. 5.37 Experimental waveforms of three-phase rectifier in loaded steady state. (**a**) Waveforms in loaded state: Ch1: total DC voltage (Vdc), Ch2: ab-phase AC port voltage (vab), Ch3: ab-phase AC line voltage (vsab), Ch4: a-phase AC current (isa). (**b**) Ch1–Ch3: three-phase AC currents (isa, isb, isc), Ch4: ab-phase AC port voltage (vab)

current (i_{sa}) measures approximately 1 A at no load, stabilizing at about 7.4 A after the load is connected.

Figure 5.39 (a) displays the dynamic waveforms of the four sub-capacitor voltages when the load is connected. The four sub-capacitor voltages stabilize at

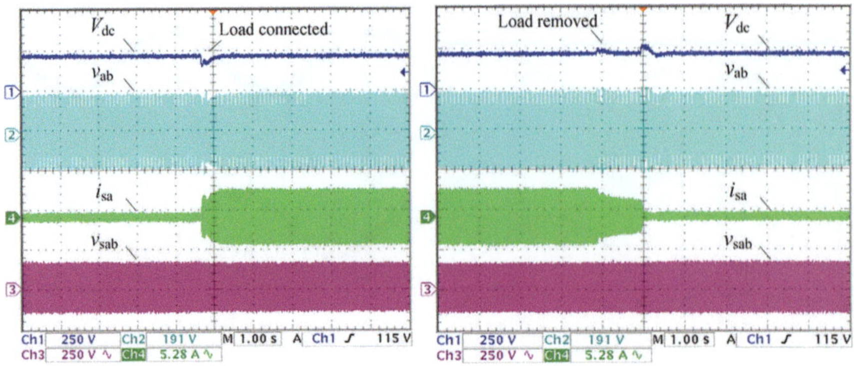

(a) Dynamic Waveform During Load Connection

(b) Dynamic Waveform During Load Removal

Ch1: total DC voltage (V_{dc}), Ch2: ab-phase AC port voltage (v_{ab}), Ch3: ab-phase AC line voltage (v_{sab}), Ch4: a-phase AC current (i_{sa})

Fig. 5.38 Dynamic waveforms of three-phase rectifier. (**a**) Dynamic waveform during load connection. (**b**) Dynamic waveform during load removal

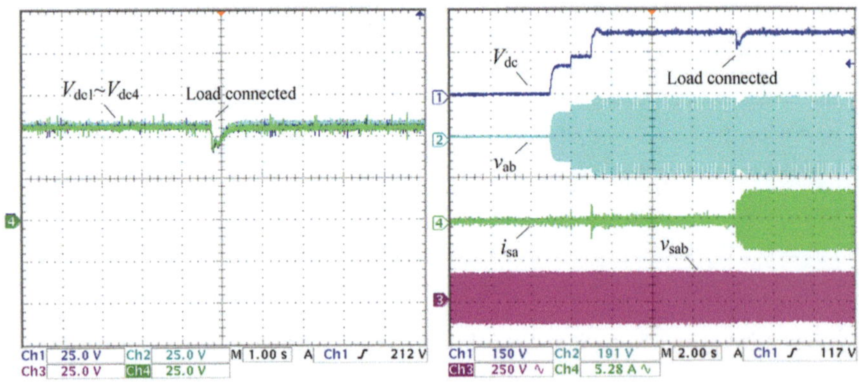

(a) Dynamic Waveforms of Four DC Sub-capacitor Voltages in Loaded State

(b) Dynamic Waveforms During Startup and Load Connection

Fig. 5.39 Dynamic waveforms of four sub-capacitor voltages. (**a**) Dynamic waveforms of four DC sub-capacitor voltages in loaded state. (**b**) Dynamic waveforms during startup and load connection

approximately 55 V in the unloaded state. Once the load is connected, the voltages return to stability after about 0.3 s. Figure 5.39b illustrates the dynamic waveforms during the startup of the three-phase rectifier and the load connection. The startup process is divided into three stages. In stage 1, the system charges through the charging resistor, with the total DC voltage stabilizing at around 90 V. In stage 2, the charging resistor is removed, and the voltage stabilizes at 130 V. Stage 3 marks the PWM rectification process, during which no load is connected, and the DC

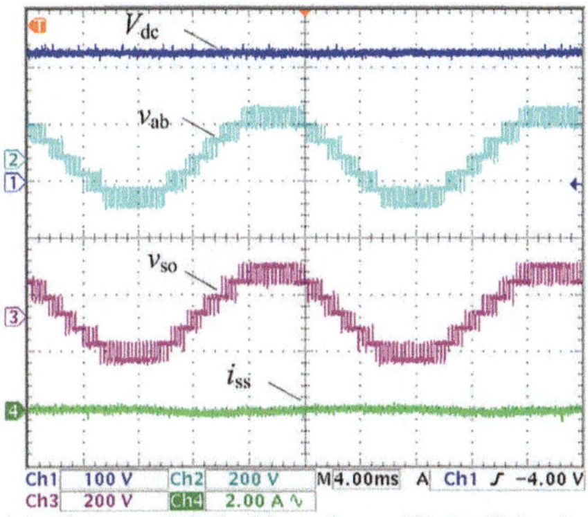

(a) Ch1: total DC-side voltage (V_{dc}), Ch2: ab-phase port voltage on the three-phase side (v_{ab}), Ch3: single-phase AC port voltage (v_{so}), Ch4: single-phase AC current (i_{ss})

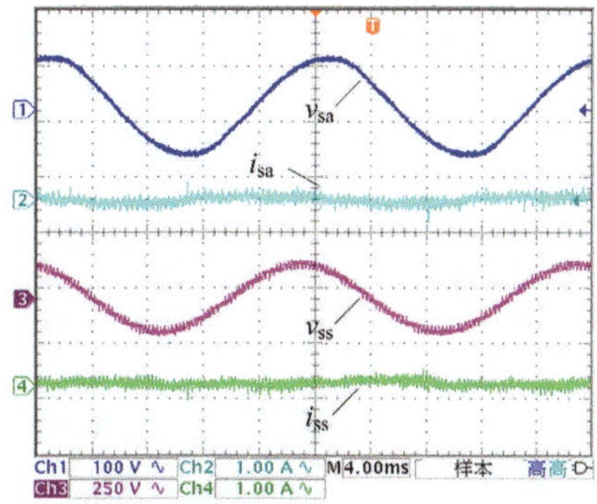

(b) Ch1: a-phase voltage (v_{sa}), Ch2: a-phase current (i_{sa}), Ch3: single-phase AC voltage (v_{ss}), Ch4: single-phase AC current (i_{ss})

Fig. 5.40 Experimental waveforms of three-phase-to-single-phase converter in unloaded state

voltage stabilizes at 220 V after a brief adjustment period. Following the completion of the startup process, the AC current increases significantly after the load is connected.

5.4.4.3 Steady-State Experiment and Analysis of Single-Phase Rectifier

Figure 5.40 displays the experimental waveforms of the three-phase-to-single-phase converter during the unloaded state, with the DC-side voltage stabilizing at 220 V. The ab-phase AC port voltages on the three-phase side exhibit seven-level stepwise waveforms, and the AC port voltage on the single-phase side (vso) also shows a seven-level stepwise configuration. As illustrated in Figure (b), the amplitudes of the a-phase voltage on the three-phase side and the AC voltage on the single-phase side are 100 V and 150 V, respectively. Meanwhile, the AC current on both the three-phase side and single-phase side is close to 0 A in the unloaded state.

5.4.4.4 Dynamic Experiment and Analysis of Single-Phase Rectifier

Figure 5.41 displays the dynamic waveforms during the connection and removal of a resistive load. Following the connection of a resistive load, the DC voltage returns to 220 V after 1 second, exhibiting an overshoot of about 10% during the adjustment process, while both the a-phase current and the single-phase output current quickly stabilize. When the load is connected, the amplitude of the single-phase port voltage increases; conversely, when the load is removed, the amplitude of the single-phase port voltage decreases. In both scenarios, the voltage reaches a stable value in a short time.

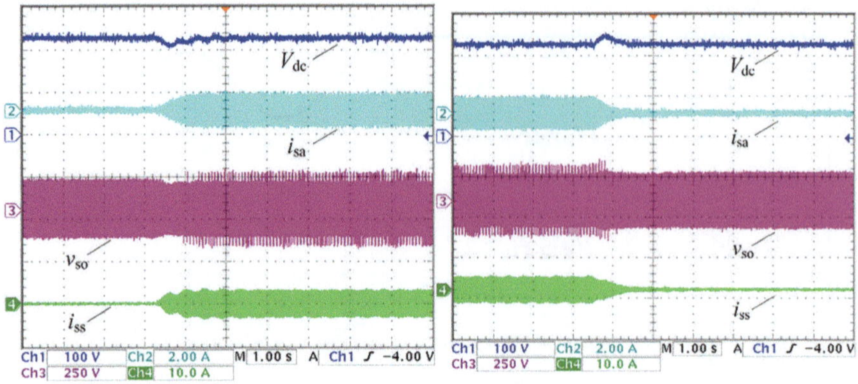

(a) Connection of Active Power (b) Removal of Active Power

Ch1: total DC-side voltage (V_{dc}), Ch2: a-phase current (i_{sa}), Ch3: single-phase AC port voltage (v_{so}), Ch4: single-phase AC current (i_{ss})

Fig. 5.41 Dynamic waveforms during connection and removal of resistive load. (**a**) Connection of active power. (**b**) Removal of active powerCh1: total DC-side voltage (Vdc), Ch2: a-phase current (isa), Ch3: single-phase AC port voltage (vso), Ch4: single-phase AC current (iss)

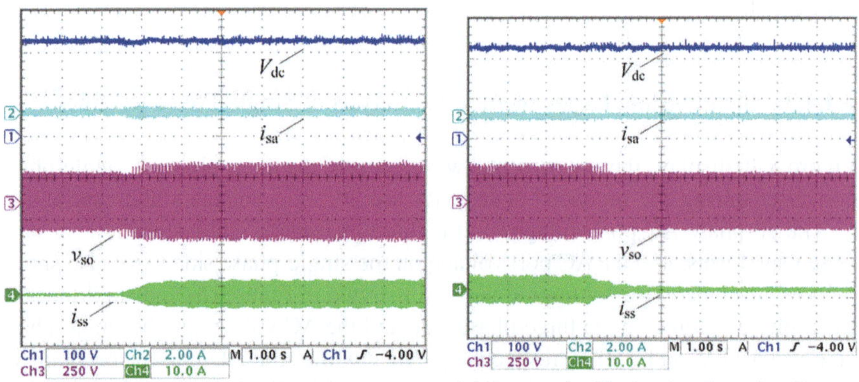

(a) Connection of Inductive Reactive Power (b) Removal of Inductive Reactive Power

Ch1: total DC-side voltage (V_{dc}), Ch2: a-phase current (i_{sa}), Ch3: single-phase AC port voltage (v_{so}), Ch4: single-phase AC current (i_{ss})

Fig. 5.42 Dynamic waveforms during connection and removal of inductive load on single-phase side. (**a**) Connection of inductive reactive power. (**b**) Removal of inductive reactive power

5.4.4.5 Experiment and Analysis of Three-Phase-to-Single-Phase System

Ch1: total DC-side voltage (Vdc), Ch2: a-phase current (isa), Ch3: single-phase AC port voltage (vso), Ch4: single-phase AC current (iss)

Figure 5.42 displays the waveforms during the connection and removal of an inductive load on the single-phase side. Without reactive power exchange with the DC capacitor, the DC voltage remains relatively stable despite changes in inductive reactive power. Similarly, the a-phase current on the three-phase side shows minimal

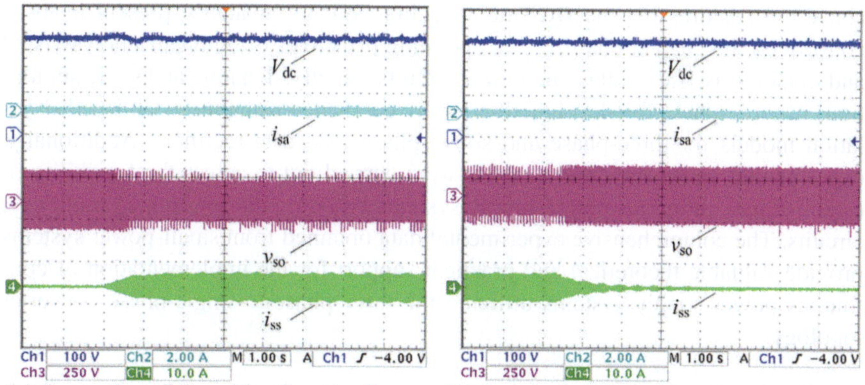

(a) Connection of Capacitive Reactive Power (b) Removal of Capacitive Reactive Power

Ch1: total DC-side voltage (V_{dc}), Ch2: a-phase current (i_{sa}), Ch3: single-phase AC port voltage (v_{so}), Ch4: single-phase AC current (i_{ss})

Fig. 5.43 Dynamic waveforms during connection of capacitive reactive load on single-phase side. (**a**) Connection of capacitive reactive power. (**b**) Removal of capacitive reactive powerCh1: total DC-side voltage (V_{dc}), Ch2: a-phase current (i_{sa}), Ch3: single-phase AC port voltage (v_{so}), Ch4: single-phase AC current (i_{ss})

variation. The single-phase port voltage reaches a new stable value at approximately $t = 0.7$ s, accompanied by increases in both amplitude and modulation degree. The reactive current on the single-phase side stabilizes after a brief adjustment period. When the inductive reactive output is removed, the single-phase port voltage returns to its unloaded state stable value after a short adjustment, resulting in a decrease in voltage amplitude and a reduction of the single-phase output current to 0.

Figure 5.43 displays the dynamic waveforms during the connection and removal of a capacitive reactive load on the single-phase side. Similar to the experiment involving inductive reactive power, the DC-side voltage shows minimal fluctuation when the capacitive reactive power is connected and removed. Since the reactive power does not need to be transmitted or provided by the three-phase-side converter, the three-phase-side current remains relatively stable despite the changes in reactive power on the single-phase side. When the capacitive reactive power is connected, the amplitude of the single-phase port voltage decreases (indicating a reduction in modulation degree). Conversely, when the capacitive reactive power is removed, the amplitude of the single-phase port voltage increases, and the single-phase output current stabilizes at 0 after a brief adjustment period.

5.5 Summary

This chapter focuses on the operating principles and control modes of diode-clamped five-level converters. It begins with an introduction to advanced co-phase traction power supply systems based on three-phase-to-single-phase converters, followed by the establishment of the circuit model for diode-clamped five-level converters. Addressing the DC-side capacitor voltage unbalance present in these converters, the chapter analyzes the operating principles of both capacitor-assisted and inductor-assisted voltage equalizer circuits. Finally, dynamic and static simulations of three-phase five-level rectifiers were conducted using the established simulation models for three-phase and single-phase five-level rectifiers. Additionally, experimental verification and analysis were carried out to assess the feasibility of establishing continuous power supply systems based on diode-clamped multilevel circuits. The comprehensive experimental data obtained from small-power systems provide valuable theoretical and practical support for the implementation of continuous power supply systems utilizing the three-phase-to-single-phase converter topology.

References

Dixon J W, Boon-Teck O. Indirect current control of a unity power factor sinusoidal current boost type three-phase rectifier[J]. IEEE Transactions on Industrial Electronics, 1988, 35(4): 508-515.

He X, Chen A. Theory and application technology of multilevel converters[M]. Beijing: China Machine Press, 2006.

Jie F. Research on three-phase PWM rectifier and its control[D]. Hangzhou: Zhejiang University, 2006.

Li J, Jiang J, Su Q. A space vector pulse width modulation for five-level nested neutral point piloted converter[J]. IEEE Transactions on Power Electronics, 2016, PP(99): 1.

Liu Z, Wang Y, Tan G, et al. A novel SVPWM algorithm for five-level active neutral-point-clamped converter[J]. IEEE Transactions on Power Electronics, 2016, 31(5): 3859-3866.

Pelicer J C, Seixas F J M D, Lourenço A C D, et al. Novel isolated multi-pulse rectifiers with low current distortion using three-phase half-controlled boost converters[M]. In: 2015 IEEE 13th Brazilian Power Electronics Conference and 1st Southern Power Electronics Conference (COBEP/SPEC). 2015: 1-5.

Shu Z, Tang J, Guo Y, et al. An efficient SVPWM algorithm with low computational overhead for three-phase inverters[J]. IEEE Transactions on Power Electronics, 2007, 22(5): 1797-1805.

Shu Z, Ding N, Chen J, et al. Multilevel SVPWM with DC-link capacitor voltage balancing control for diode-clamped multilevel converter based STATCOM[J]. IEEE Transactions on Industrial Electronics, 2013a, 60(5): 1884-1896.

Shu Z, He X, Wang Z, et al. Voltage balancing approaches for diode-clamped multilevel converters using auxiliary capacitor-based circuits[J]. IEEE Transactions on Power Electronics, 2013b, 28(5): 2111-2124.

Shu Z, Zhu H, He X, et al. One-inductor-based auxiliary circuit for DC-link capacitor voltage equalisation of diode-clamped multilevel converter[J]. IET Power Electronics, 2013c, 6(7): 1339-1349.

Zheng Z. Research on control strategies and application of three-phase voltage-source PWM rectifier[D]. Dalian: Dalian University of Technology, 2011.

Chapter 6
Modular Multilevel Three-Phase-to-Single-Phase Converter

6.1 Modular Multilevel Three-Phase-to-Single-Phase Converter for Continuous Power Supply

Modular multilevel converters (MMCs) represent one of the multilevel converter structures currently under extensive research and application (Shu et al. 2013). With advantages such as compact structure and easy extensibility, these converters have been widely studied and implemented in various fields, including high-voltage, high-power, large-capacity reactive power compensation (Ambrozic et al. 2003) and DC power transmission (Shu et al. 2011; Xing 2016). These systems are typically based on a three-phase-to-three-phase configuration (Liu 2007). However, traction networks, which generally operate in a single-phase configuration, require a single-phase bridge arm output from MMCs. Given the current lack of studies on three-phase-to-single-phase converters based on MMCs, further research is needed, particularly regarding applications in advanced co-phase traction power supply scenarios.

Figure 6.1 illustrates a three-phase-to-single-phase MMC designed for advanced co-phase traction power supply. This MMC circuit employs a three-phase-to-single-phase AC-DC-AC conversion topology, featuring structurally identical submodules that are cascaded to enhance extensibility. This cascade structure allows for an increase in the voltage level at the converter interfaces and provides the necessary conditions for transformer-free direct connection. The incorporation of a common DC bus enables decoupled control of power between the three-phase and single-phase systems. Additionally, by adjusting the number of cascaded submodules that are switched on or bypassed for each phase, the voltage level at the AC port can be altered, achieving controllability over both the amplitude and phase of the AC port voltage. This design meets the characteristic requirements of high-voltage and high-power advanced co-phase traction power supply systems for grid connection and energy transmission. This chapter primarily discusses the operating principles and control strategies of three-phase-to-single-phase MMCs for advanced co-phase traction power supply

Z. Shu, *Power Electronic Converters in Advanced Co-Phase Traction Power Supply System*, https://doi.org/10.1007/978-981-95-3276-6_6

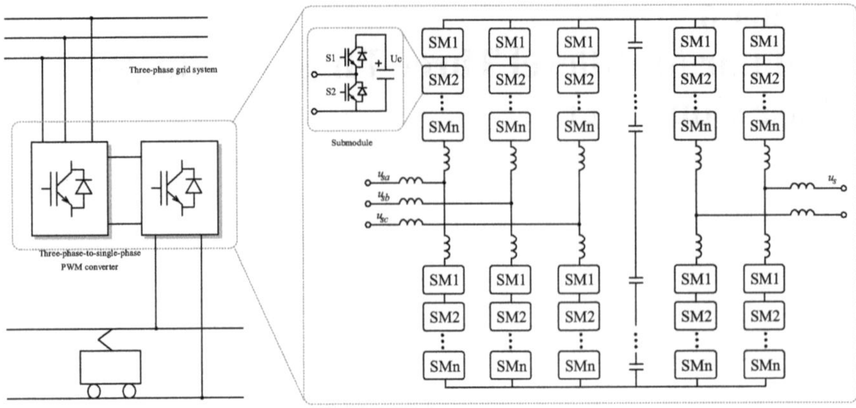

Fig. 6.1 Continuous power supply system of three-phase-to-single-phase PWM converter based on MMC topology

6.2 Operating Principles of MMC Submodule and Single Bridge Arm

Each submodule of the three-phase-to-single-phase MMC shown in Fig. 6.1 consists of two switching half-bridges and a parallel circuit with a DC capacitor. Relying on different switch combinations, the submodule can output two voltage levels: UC and 0. The upper and lower bridge arms are cascaded with *n* submodules each. At any given time, the number of submodules operating for each bridge arm is *n*, and the sum of the output voltages of the operating modules equals the DC voltage. Additionally, the MMC requires the submodules of the upper and lower bridge arms to operate in a complementary manner (Liu 2008). When 0 to *n* modules are switched on for the upper bridge arm, *n* to 0 modules will be activated for the lower bridge arm. Consequently, the AC port can output from 0 to nUC levels (Zhou et al. 2006).

Three-phase-to-single-phase MMCs for advanced co-phase traction power supply systems must enable direct connection between the three-phase electric power system and the single-phase traction network, necessitating a large number of submodules. To simplify analysis and experiments in this chapter, a configuration of four cascaded submodules per phase bridge arm is selected. Additionally, to mitigate the impact of secondary equal-frequency voltage pulsations on the DC side caused by the single-phase converter, an LC secondary filter circuit is employed to filter out the pulsation effects on the DC side.

6.2.1 Operating Principles of MMC Submodule and Single Bridge Arm

6.2.1.1 Operating Principles of Submodule

As shown in Fig. 6.2, the submodule of the MMC consists of two switches (S1 and S2) and a DC capacitor (C) connected in parallel. D1 and D2 are antiparallel diodes across the switching devices.

The midpoint of the switching devices (S1, S2) and the negative terminal of capacitor C form the AC output terminals (a, b) of the submodule. Let UC denote the DC-side capacitor voltage, V_{ab} the AC output port voltage, and IAB the AC current.

As illustrated in Fig. 6.2, different switching combinations of S1 and S2 enable the AC port voltage V_{ab} to output two levels: 0 or UC. This results in three operating modes:

1. Locking mode: both S1 and S2 are turned off

   As shown in Fig. 6.3a, when AC current flows into port A and passes through diode D1 and the DC capacitor (C) to reach the negative terminal of the output port, the AC output reflects the DC capacitor voltage, which is deemed an uncontrolled rectification state. When the AC current reverses, it flows to port A through port B and diode D2. In this scenario, the current does not pass through the DC capacitor, which does not affect the DC voltage, and the output voltage is 0. Therefore, the port level can either be 1 or 0 in this mode, depending on the direction of the AC current. Due to the uncontrollable nature of the output levels, this mode is solely used for protection.
2. Enabling mode: S1 is turned on, and S2 is turned off

   As shown in Fig. 6.3b, when AC current flows into port A, reaches the positive terminal of capacitor C through diode D1, and subsequently flows to port B, the DC capacitor is charged. Conversely, when current flows from the negative

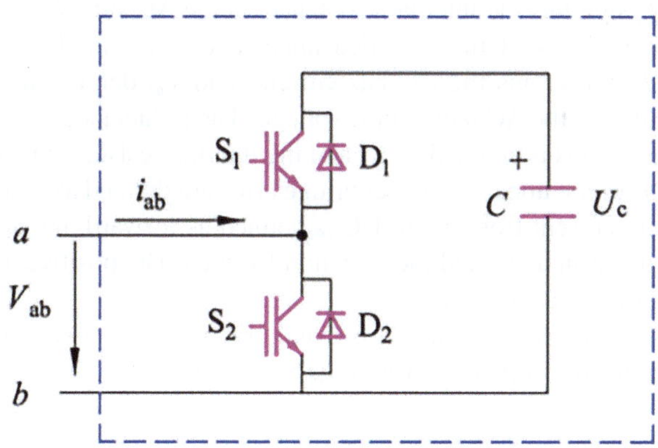

Fig. 6.2 Topology of the MMC submodule

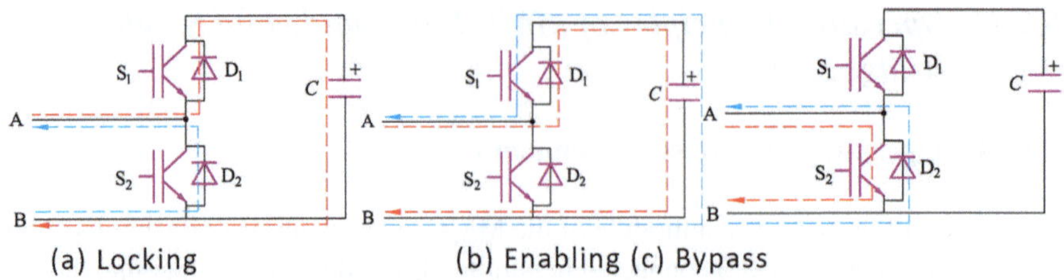

Fig. 6.3 Operating modes of MMC submodule. (**a**) Locking. (**b**) Enabling. (**c**) Bypass

terminal of the capacitor and exits through S1 at the AC port, the DC capacitor discharges. In this mode, the AC port level remains at 1 regardless of the direction of the current flow.

3. Bypass mode: S2 is turned on, and S1 is turned off

   As shown in Fig. 6.3c, when current flows into port A and exits through S2 at port B, the submodule is short-circuited, with capacitor C disconnected from the circuit. Conversely, when current flows into port B and exits at port A through diode D2, the submodule remains short-circuited, with the capacitor still bypassed, leaving the capacitor voltage unaffected. As a result, the output AC port level in this mode is 0, regardless of the direction of the current flow.

6.2.1.2 Circuit Model and Analysis of MMC Single Bridge Arm

Since each phase bridge arm circuit of the three-phase-to-single-phase PWM converter with an MMC structure is identical and operates independently, this section analyzes the mathematical model of one phase bridge arm, denoted as j ($j = a, b, c, d, e$) (Wang 2007).

The equivalent circuit model of bridge arm j is shown in Fig. 6.5, where a represents the AC output port of the single-phase half-bridge. i_{sj} and u_{sj} denote the AC output current and phase voltage, respectively. L_{sj} represents the AC output inductance, while R_{sj} signifies the internal resistance on the AC side. U_{Pj} and U_{Nj} denote the AC output voltages of the cascaded upper and lower modules, respectively. Additionally, I_{dc} represents the DC bus current, and V_{dc} denotes the DC voltage, whereas u_{vj} indicates the AC output port voltage. The inductance for both the upper and lower bridge arms is denoted as L_o, and the internal resistance is represented as R_0. The currents of the upper and lower bridge arms are defined as i_{Pj} and i_{Nj}, respectively. The input current from AC to DC is defined as forward, the circulating current of the bridge arm as i_{zj}, and the direction from the DC positive terminal to the DC negative terminal as forward.

Assuming that the AC current (i_s) is evenly distributed between the upper and lower bridge arms, the derivation from Fig. 6.4 yields

$$\begin{cases} i_{zj} = \dfrac{1}{2}\left(i_{Pj} + i_{Nj}\right) \\ i_{sj} = i_{Nj} - i_{Pj} \end{cases} \tag{6.1}$$

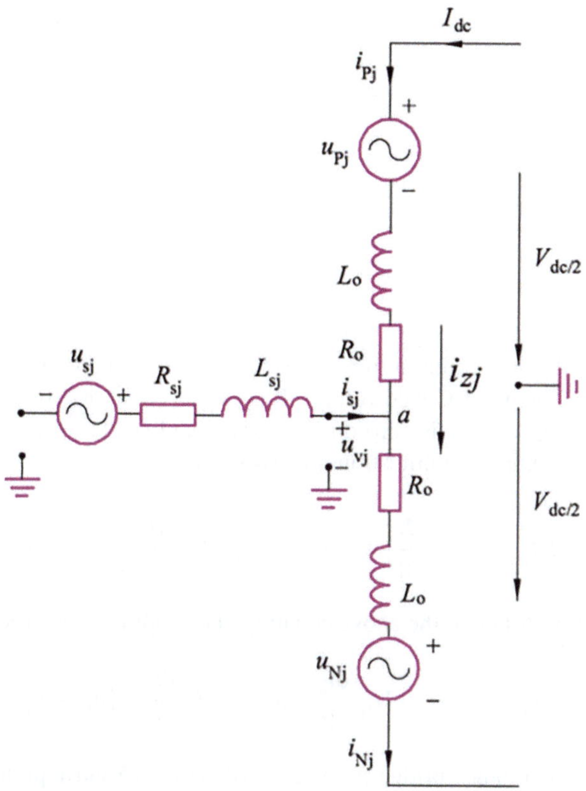

Fig. 6.4 Equivalent circuit model of MMC half-bridge

According to Kirchhoff's voltage law, the voltage circuit equations for the upper and lower bridge arms are expressed as follows:

$$u_{sj} = R_{sj} \cdot i_{sj} + L_{sj}\frac{di_{sj}}{dt} - R_o i_{Pj} - L_o\frac{di_{Pj}}{dt} - u_{Pj} + \frac{V_{dc}}{2} \tag{6.2}$$

$$u_{sj} = R_{sj} \cdot i_{sj} + L_{sj}\frac{di_{sj}}{dt} + R_o i_{Nj} + L_o\frac{di_{Nj}}{dt} + u_{Nj} - \frac{V_{dc}}{2} \tag{6.3}$$

After subtracting Eq. (6.3) from Eq. (6.2), the simplification process yields

$$V_{dz} = R_o i_{Zj} + L_o\frac{di_{zj}}{dt} = \frac{V_{dc}}{2} - \frac{u_{Pj} + u_{Nj}}{2} \tag{6.4}$$

where V_{dz} denotes the unbalanced voltage generated by the circulating current (i_{zj}) in the bridge arm inductor and its internal resistance. The upper and lower bridge arm inductances (L_o) can mitigate the impact of this voltage unbalance to some extent. Equation (6.4) indicates that the voltage unbalance is also related to the DC voltage.

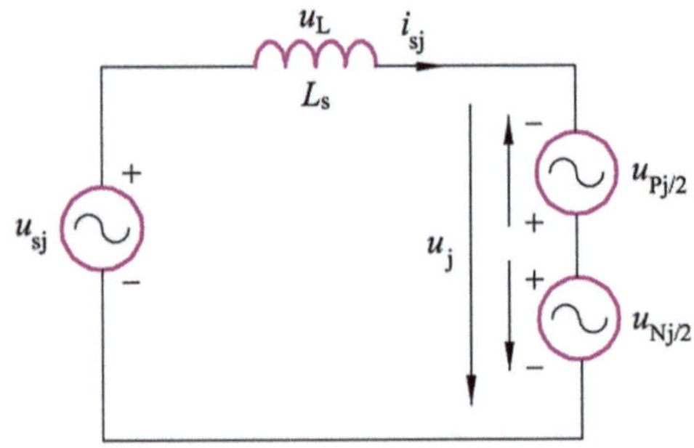

Fig. 6.5 Equivalent circuit of single-phase MMC

Therefore, the discussion in this chapter focuses on suppressing voltage unbalance by controlling the DC-side voltage. It is assumed that the voltage drop caused by the circulating current is zero during normal and stable operation. By adding Eqs. (6.2) to (6.3), the simplification process yields

$$2u_{sj} = 2R_{sj} \cdot i_{sj} + 2L_{sj}\frac{\mathrm{d}i_{sj}}{\mathrm{d}t} + R_o\left(i_{Nj} - i_{Pj}\right) + L_o\frac{\mathrm{d}\left(i_{Nj} - i_{Pj}\right)}{\mathrm{d}t} + \left(u_{Nj} - u_{Pj}\right) \tag{6.5}$$

Substituting Eq. (6.1) into the above equation, the simplification process yields

$$2u_{sj} = 2R_{sj} \cdot i_{sj} + 2L_{sj}\frac{\mathrm{d}i_{sj}}{\mathrm{d}t} + R_o \cdot i_{sj} + L_o\frac{\mathrm{d}i_{sj}}{\mathrm{d}t} + \left(u_{Nj} - u_{Pj}\right) \tag{6.6}$$

Based on the single-phase bridge arm illustrated in Fig. 6.5 and applying Kirchhoff's voltage law, the following equation is derived:

$$u_{vj} = u_{sj} - R_{sj} \cdot i_{sj} - L_{sj}\frac{\mathrm{d}i_{sj}}{\mathrm{d}t} \tag{6.7}$$

Combining Eqs. (6.6) and (6.7), the AC output port voltage of single-phase MMC can be expressed as follows:

$$u_{vj} = u_j + \frac{R_o}{2} \cdot i_{sj} + \frac{L_o}{2}\frac{\mathrm{d}i_{sj}}{\mathrm{d}t} \tag{6.8}$$

where u_j denotes the port equivalent output voltage of the single-phase converter, expressed as follows:

$$u_j = \frac{u_{Nj} - u_{Pj}}{2} \tag{6.9}$$

According to Eqs. (6.7)–(6.9), the following equation is derived:

$$u_j = u_{sj} - R_s \cdot i_{sj} - L_s\frac{di_{sj}}{dt} \tag{6.10}$$

where the equivalent AC inductance and its equivalent internal resistance are expressed as follows:

$$\begin{cases} R_{\mathrm{s}} = \left(R_{\mathrm{s}j} + \dfrac{R_{\mathrm{o}}}{2} \right) \\ L_{\mathrm{s}} = \left(L_{\mathrm{s}j} + \dfrac{L_{\mathrm{o}}}{2} \right) \end{cases} \tag{6.11}$$

Based on the above analysis, the inductor connecting the upper and lower bridge arms of the MMC, along with its internal resistor, is equivalent to a portion of the AC connecting reactor, with the internal resistance value represented as half of the original value.

The circulating current (i_{zj}) flows between the bridge arms of the phases, generating an unbalanced voltage drop (V_{dz}) between the upper and lower arms, as expressed by Eq. (6.4). Therefore, the internal circulating current (i_{zj}) can be suppressed by controlling V_{dz}; that is, it can be managed to maintain the balance of the DC-side capacitor voltage. Based on Eqs. (6.4) and (6.10), the reference voltages for the upper and lower bridge arms can be expressed as follows:

$$\begin{cases} u_{Pj} = \dfrac{V_{\mathrm{dc}}}{2} - u_{\mathrm{j}} - V_{\mathrm{dz}} \\ u_{Nj} = \dfrac{V_{\mathrm{dc}}}{2} + u_{j} - V_{\mathrm{dz}} \end{cases} \tag{6.12}$$

Based on Eq. (6.12), an appropriate control strategy can be implemented for MMCs to independently control the upper and lower bridge arms of each phase circuit. The reference voltage for the upper arm is opposite in sign to the AC port voltage, while the reference voltage for the lower arm is the same as the AC port voltage.

6.2.1.3 Analysis of MMC Four-Quadrant Operation

Figure 6.5 displays the equivalent circuit of the MMCs based on Eqs. (6.9) and (6.10), with the inductor's internal resistance (R_{s}) neglected.

In Fig. 6.5, all variables are represented as phasors, with the voltage drop across the AC inductance denoted as $\dot{u}_L$, and the equivalent AC output port voltage represented as $\dot{u}_j$. Under steady-state conditions, the phasor relationship of the steady-state AC component can be determined by considering only the sinusoidal fundamental component of the AC output while ignoring higher harmonic components, as illustrated in Fig. 6.6.

Assuming that the AC voltage ($\dot{u}_{\mathrm{s}j}$) is constant and the AC current ($\left|\dot{i}_{\mathrm{s}j}\right|$) remains unchanged in amplitude, based on Fig. 6.6, the amplitude and phase of the AC current (i_{sj}) can be managed solely by controlling the amplitude and phase of the AC equivalent output port voltage ($\dot{u}_j$) of MMCs. This approach enables the operation

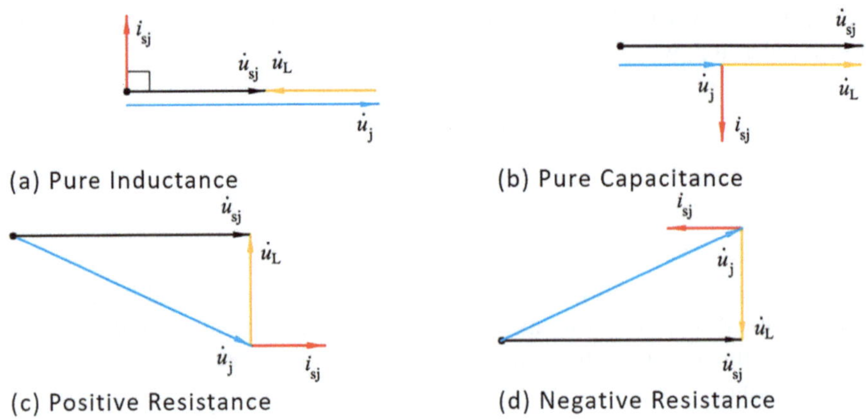

Fig. 6.6 Steady-state phasor diagram of single-phase MMC. (**a**) Pure inductance. (**b**) Pure capacitance. (**c**) Positive resistance. (**d**) Negative resistance

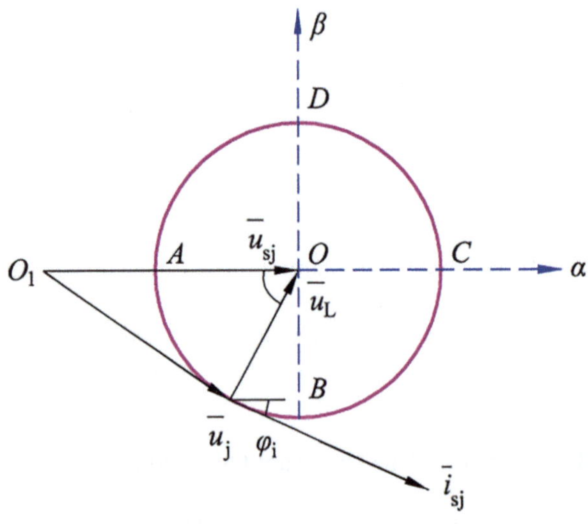

Fig. 6.7 Four-quadrant steady-state operation diagram of single-phase MMC

of single-phase converters under four conditions: pure inductance, pure capacitance, positive resistance, and negative resistance, which align with the PWM rectifier control method and the four operating conditions. When the output current maintains a consistent amplitude, the control process should ensure that the inductor voltage drops are equal in amplitude. This means controlling the motion trajectory of the port voltage vector ($\dot{u}_j$) to follow a circular path with the amplitude of the AC inductor voltage drop ($|\dot{u}_L|$) as its radius, as illustrated in Fig. 6.7.

As shown in Fig. 6.7, the four-quadrant steady-state operation diagram is divided into four segments. Point A represents the operating point in the pure inductance mode, point B signifies the operating point in the positive resistance mode, point C indicates the operating point in the pure capacitance mode, and point D denotes the operating point in the negative resistance mode. These four specific operating points correspond to the four steady-state phasor diagrams illustrated in Fig. 6.6.

When the endpoints of the equivalent AC output voltage ($\overline{u}_j$) of single-phase MMCs operate at different positions on the circumference, it is possible to achieve operating states with varying distributions of active and reactive power, thereby enabling stable operation at any position within the four quadrants (Liu 2008).

When the operating point moves in the AB section, the converter operates in a rectification state. In this state, the AC power supply transfers active power to the MMC, and as the voltage of the AC connecting reactor leads the current in phase, the power supply receives a certain amount of capacitive reactive power, indicating operation in the first quadrant.

When the operating point moves in the BC section, the MMC continues to operate in a rectification state. Unlike in the first quadrant, the inductor voltage lags behind the current in phase, and the power supply transfers inductive reactive power to the MMC, indicating operation in the second quadrant.

When the operating point moves in the CD section, the converter operates in an active inversion state. In this state, active power is transferred from the MMC to the power supply. Similarly, as the voltage of the AC connecting reactor leads the current in phase, the MMC transfers capacitive reactive power to the power supply, indicating operation in the third quadrant.

When the voltage ($\overline{u}_j$) endpoint moves in the DA section, the converter continues to operate in an active inversion state; however, inductive reactive power is transferred to the power supply, indicating operation in the fourth quadrant.

6.2.2 *Analysis of Circulating Current in Single-Phase and Three-Phase MMC*

In the MMC half-bridge equivalent circuit model shown in Fig. 6.5, the single-phase AC voltage and current are defined as $u_{sj} = U_{sm} \sin \omega t$ and $i_{sj} = I_{sm} \sin (\omega t - \varphi)$, respectively, where φ denotes the power factor angle of the loads, U_{sm} denotes the maximum AC voltage, Ism denotes the maximum AC current, and ω represents the angular frequency of the voltage and current. Assuming that the DC capacitor voltage remains constant, the single-phase circulating current can be expressed as follows:

$$i_{zj} = \frac{u_{sj} i_{sj}}{V_{dc}} = \frac{M}{4} I_{sm} \left[\cos\varphi - \cos(2\omega t - \varphi) \right] \tag{6.13}$$

where M is the modulation factor, defined as $M = 2U_{sm}/V_{dc}$. As expressed by Eq. (6.13), the circulating current of single-phase MMCs consists of two components: a DC component and an AC component at twice the fundamental angular frequency (Liu 2008), which can be expressed as follows:

$$i_{zj} = i_{zdj} + i_{zqj} \tag{6.14}$$

where

$$\begin{cases} i_{zdj} = +\dfrac{M}{4} I_{sm} \cos\varphi \\ i_{zqj} = -\dfrac{M}{4} I_{sm} \cos(2\omega t - \varphi) \end{cases} \tag{6.15}$$

According to Eq. (6.15), for three-phase MMCs, the AC component at twice the fundamental angular frequency in the three-phase circulating current can be expressed as follows:

$$\begin{cases} i_{zqa} = \dfrac{M}{4} I_{sm} \left[\cos(2\omega t - \varphi)\right] \\ i_{zqb} = \dfrac{M}{4} I_{sm} \left[\cos\left(2\omega t - \dfrac{4\pi}{3} - \varphi\right)\right] \\ i_{zqc} = \dfrac{M}{4} I_{sm} \left[c\cos\left(2\omega t - \dfrac{2\pi}{3} - \varphi\right)\right] \end{cases} \tag{6.16}$$

Based on the above equation, the reactive component of the three-phase circulating current exhibits a negative sequence distribution (Zhou et al. 2006; Wang 2007), with $\sum_{j=a,b,c} i_{zqj} = 0$, leading to the relationship $\sum_{j=a,b,c} i_{zj} = i_d$, where i_d denotes the current on the DC bus side.

For single-phase MMC systems, the AC input power (S) and the output power on the DC bus side (Pd) can be, respectively, expressed as follows:

$$P_d = V_{dc} i_d = S\cos\varphi \tag{6.17}$$

$$\sum_{j=c,\,d} i_{zdj} = i_{zdc} + i_{zdd} = \frac{U_{sm} I_{sm} \cos\varphi}{V_{dC}} = i_d \tag{6.18}$$

Assuming equal distribution of the DC-side current among the phases, the following equation can be derived:

$$i_{zdj} = \frac{1}{2}\bar{i}_d \tag{6.19}$$

Considering a nonzero sum for the superposition of reactive components in the single-phase circulating current, the following equation can be derived:

$$i_{d2} = \sum_{j=a,b} i_{Zqj} = -\frac{U_{sm} I_{sm}}{V_{dc}} \cos(2\omega t - \varphi) \neq 0 \tag{6.20}$$

Therefore, when the AC current is nonzero, the DC bus in the single-phase MMC circuit must contain a circulating component at twice the fundamental frequency, resulting in secondary fluctuations of the voltage on the DC bus side. Since the DC voltage control loop typically samples the DC voltage, the component at twice the fundamental frequency of the DC bus voltage is fed back into both the single-phase and three-phase control systems. This feedback can cause distortion in the grid-side current, affecting the power quality of the connected three-phase and single-phase systems. Consequently, it is essential to suppress these secondary ripples.

To address this issue, the ripple feedback control suppression method is discussed in Sect. 4.4. To simplify the control process, a parallel LC resonant circuit is incorporated on the intermediate DC bus side to filter out the secondary fluctuations in this chapter.

When the LC circuit resonates, the circulating current component at twice the fundamental frequency flows entirely through the LC circuit. Consequently, the secondary pulsating voltage (u_{c2}) across both ends of the filter capacitor (C) satisfies the following equation:

$$i_{\mathrm{d2}} = C\frac{\mathrm{d}u_{C2}}{\mathrm{d}t} \tag{6.21}$$

Combining Eqs. (6.20) and (6.21) yields

$$u_{C2} = -\frac{U_{sm}I_{sm}}{2C\omega V_{dc}}\sin\left(2\omega t - \varphi\right) \tag{6.22}$$

Fluctuations in the DC capacitor voltage impact the output port voltage, which in turn affects the input current, leading to a deterioration of the current waveform and an increase in circulating current. This is why it is essential to limit the fluctuation amplitude of the capacitor voltage (u_{C2}). Assuming a limit of 10% beyond its rated voltage, the following equation should be satisfied:

$$\frac{U_{\mathrm{sm}}I_{\mathrm{sm}}}{2C\omega V_{\mathrm{dc}}} \le V_{\mathrm{dc}} \cdot 10\% \tag{6.23}$$

Based on Eq. (6.23), the lower limit for the capacitance value design can be expressed as follows:

$$C \ge \frac{5U_{\mathrm{sm}}I_{\mathrm{sm}}}{\omega\, V_{\mathrm{dc}}^2} \tag{6.24}$$

6.2.3 Analysis of Capacitor Harmonics in MMC

To simplify the analysis of capacitor harmonics, the following assumptions are made:

(1) The voltage and current on the grid side are pure sine waves.
(2) All switching devices are considered ideal.
(3) The internal resistance (R_o) of the bridge arm inductors is neglected.

Since the bridge arms are independent among the phases in the MMC topology, phase j is also taken as an example to discuss the harmonic content of the capacitor voltage. Equal current (i_{Pj}) flows through all submodules of the upper bridge arm in the MMC, while equal current (i_{Nj}) flows through all submodules of the lower bridge arm. Furthermore, the capacitor voltages of both the upper and lower bridge arms exhibit the same trend, resulting in identical harmonic content. Therefore, one submodule from both the upper and lower bridge arms is chosen for analysis in this chapter. Based on Fig. 6.5, the differential equations for the capacitor voltages of the upper and lower bridge arm submodules can be expressed as follows:

$$C\frac{\mathrm{d}V_{cjP}}{\mathrm{d}t} = S_{k1}i_{Pj} \tag{6.25}$$

$$C\frac{\mathrm{d}V_{cjN}}{\mathrm{d}t} = S_{k2}i_{Nj} \tag{6.26}$$

where C represents the DC capacitance of the submodule; V_{cjP} and V_{cjN} denote the capacitor voltages of submodules corresponding to the upper and lower bridge arm, respectively; and S_{k1} and S_{k2} are the switching functions. With the dual-modulation-wave CPS-SPWM adopted as the modulation strategy, assuming that the switching frequency is much greater than the fundamental frequency and that the DC capacitor has a large capacitance with low pulsation, while ignoring switching harmonics, the switching function can be established as follows:

$$\begin{cases} S_{k1} = -M\sin\omega t \\ S_{k1} = +M\sin\omega t \end{cases} \tag{6.27}$$

Combining Eqs. (6.1), (6.13), and (6.25)–(6.27) results in the expression of fluctuating components in the capacitor voltage of the submodule, as follows:

$$\frac{\mathrm{d}U_{\mathrm{CjP}}}{\mathrm{d}t} = \frac{MI_{\mathrm{sm}}}{4C}\left[-\cos\varphi + M\sin(\omega t-\alpha) + \cos(2\omega t-\varphi) - \frac{M}{2}\sin(3\omega t-\varphi)\right] \tag{6.28}$$

$$\frac{\mathrm{d}U_{\mathrm{CjN}}}{\mathrm{d}t} = \frac{MI_{\mathrm{sm}}}{4C}\left[-\cos\varphi - M\sin(\omega t-\alpha) + \cos(2\omega t-\varphi) + \frac{M}{2}\sin(3\omega t-\varphi)\right] \tag{6.29}$$

where

$$\begin{cases} \alpha = \arctan(\tan\varphi) \\ M = 2U_{\mathrm{sm}}/V_{\mathrm{dc}} \end{cases}$$

Integrating Eqs. (6.28) and (6.29) yields the capacitor voltages of the submodules corresponding to the upper and lower bridge arms, expressed as follows:

$$\begin{aligned} U_{CjP} = U_C - \frac{M^2 I_{sm}}{4\omega C}\cos(\omega t - \alpha) + \frac{MI_{sm}}{8\omega C}\sin(2\omega t - \varphi) \\ + \frac{M^2 I_{sm}}{24\omega C}\cos(3\omega t - \varphi) \end{aligned} \tag{6.30}$$

$$\begin{aligned} U_{CjN} = U_C + \frac{M^2 I_{sm}}{4\omega C}\cos(\omega t - \alpha) + \frac{MI_{sm}}{8\omega C}\sin(2\omega t - \varphi_i) \\ - \frac{M^2 I_{sm}}{24\omega C}\cos(3\omega t - \varphi) \end{aligned} \tag{6.31}$$

where UC denotes the average DC capacitor voltage.

As indicated by Eqs. (6.30) and (6.31), the capacitor voltages of the upper and lower bridge arms of the MMC submodule, in addition to their DC components, contain AC components at different frequencies, including the fundamental component (50 Hz), second harmonics (100 Hz), and a few third harmonics (150 Hz).

6.2.4 Simulation Verification

A simulation model was established using MATLAB/Simulink for low-voltage and low-power three-phase PWM rectifiers featuring a four-module MMC structure. The port voltage-current relationship and the harmonic content of the submodule capacitor voltage were analyzed, focusing specifically on phase A. The simulation parameters are detailed in Table 6.1:

Figure 6.8 displays the relationship between the bridge arm port voltages of phase A and the phase voltage of the AC output port after the three-phase MMC rectifier is connected to a load. As shown in Fig. 6.8a, b, the port voltage (u_{Pa}) of the upper bridge arm and the port voltage (u_{Na}) of the lower bridge arm both exhibit three levels, with opposite positive and negative directions, consistent with Eq. (6.12) from the aforementioned MMC mathematical model. Figure 6.8c illustrates the close alignment between the port voltage of phase a, based on Eq. (6.9), and the port output voltage (u_a) in the simulation results presented in Fig. 6.8d.

Figure 6.9 displays the waveforms of the bridge arm current and the phase current of phase A after a load is connected. The waveforms for the upper bridge arm current (i_{Pa}) in Fig. 6.9a and the lower bridge arm current (i_{Na}) in Fig. 6.9b show a phase discrepancy. However, the a-phase AC current waveform derived from Eq. (6.1), as shown in Fig. 6.9c, aligns closely with the a-phase current (i_{sa}) waveform from the simulation system shown in Fig. 6.9d.

Figures 6.10 and 6.11 present the spectrum analysis of the voltage waveforms (u_{ca1} to u_{ca4}) for the four sub-capacitors in phase A and the voltage of capacitor C1

Table 6.1 Simulation parameters of three-phase PWM rectifier

Parameter	Value
RMS line voltage of three-phase grid	400 V
DC-side voltage reference	700 V
Three-phase AC input inductance	3 mH
Connection inductance of three-phase bridge arms	0.5 mF
DC-side capacitance	3 mF
Load	10 Ω
Triangular carrier frequency	3 kHz
Simulation step size	1 μs

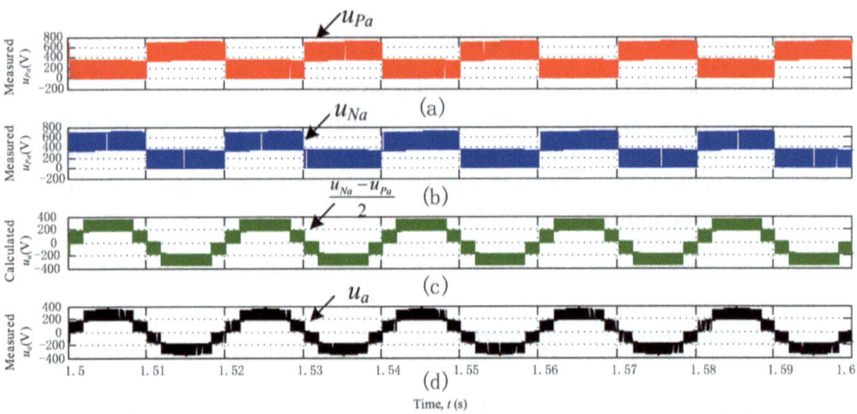

Fig. 6.8 Upper and lower port voltages of phase A and output phase voltage waveforms: (**a**) upper arm port voltage, u_{Pa}; (**b**) lower arm port voltage, u_{Na}; (**c**) calculated AC phase voltage; (**d**) measured AC phase voltage

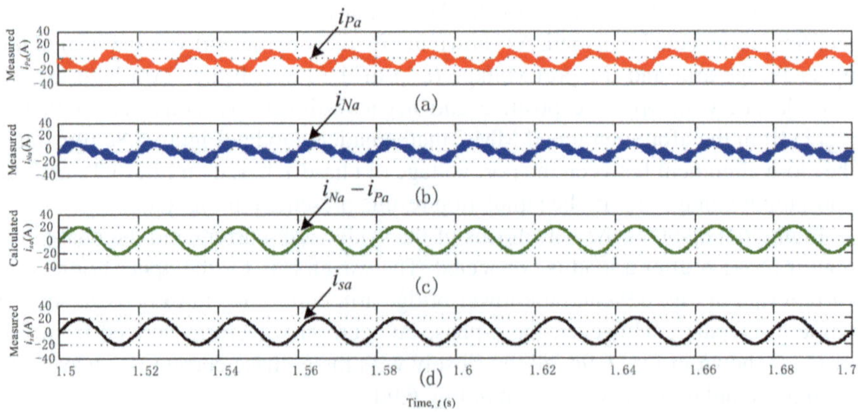

Fig. 6.9 Bridge arm current and phase current of phase A: (**a**) upper bridge arm, i_{Pa}; (**b**) lower bridge arm, i_{Na}; (**c**) calculated value; (**d**) simulation value, ISA

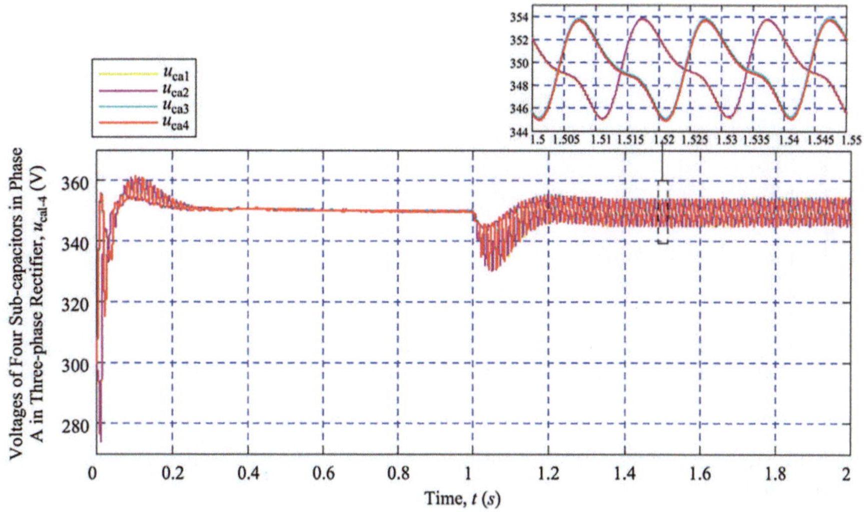

Fig. 6.10 Voltage waveforms of four sub-capacitors in phase A

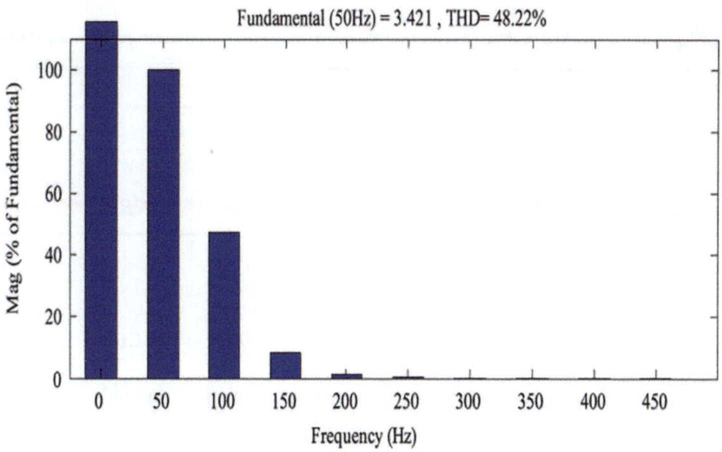

Fig. 6.11 Spectrum analysis of voltage of capacitor C (u_{Ca1}) in phase A during on-load operation

(u_{ca1}), respectively. The voltage of each of the four sub-capacitors is half of the total DC bus voltage, with voltage fluctuations of approximately 8 V. In the no-load interval from t = 0 to t = 1 s, there is no active power transfer, and the sub-capacitor voltage stabilizes without fluctuations. During the on-load operation from t = 1 to t = 2 s, the three-phase rectifier transfers active power from the three-phase grid to the load side, resulting in significant fluctuations in the sub-capacitor voltage. The two capacitors in the upper bridge arm exhibit consistent fluctuations, as do the two in the lower bridge arm, with a fluctuation amplitude of about 5% of the DC value. Figure 6.11 displays the spectrum analysis of the capacitor C1

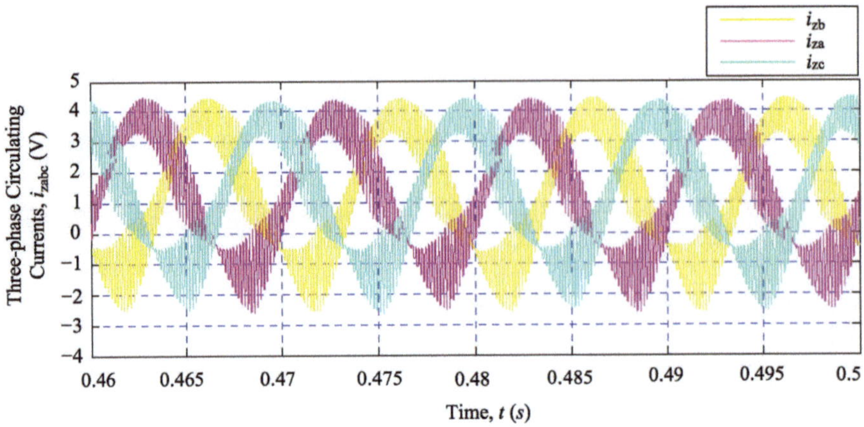

Fig. 6.12 Waveforms of three-phase circulating currents (i_{zabc}). (**a**) Sum of three-phase circulating currents; (**b**) DC-side current

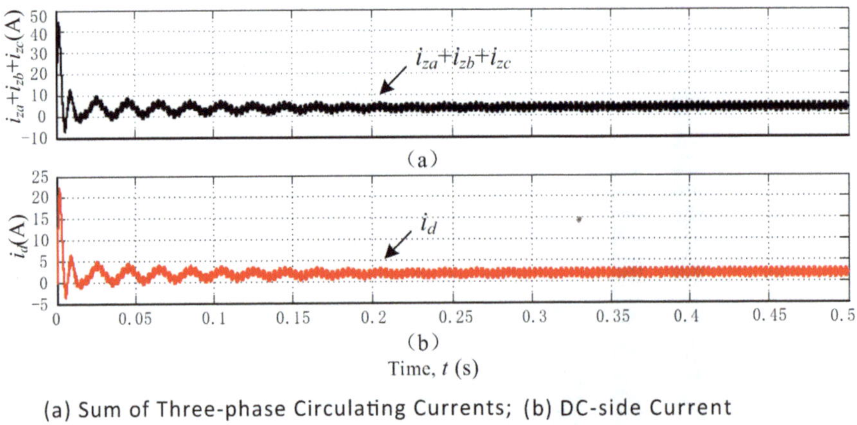

Fig. 6.13 Waveforms of sum of three-phase circulating currents (i_{zabc}) and DC-side current (i_d)

voltage during the on-load period. In addition to the DC components, there are significant harmonics at twice the fundamental frequency and a few harmonics at three times the fundamental frequency, which align with the theoretical analysis results provided by Eqs. (6.30) and (6.31) in Sect. 6.2.3.

The waveforms of the three-phase circulating current (i_{zabc}) and the DC bus current (i_d) are presented in Figs. 6.12 and 6.13, respectively. As illustrated in Fig. 6.12, the three-phase circulating current exhibits pulsations at twice the fundamental frequency and follows a negative sequence distribution, consistent with Eq. (6.16). As indicated in Fig. 6.13, the sum of the three-phase circulating currents is close to zero and aligns with the DC current (i_d).

To further analyze the characteristics of circulating currents at twice the fundamental frequency in single-phase MMCs, simulations were conducted using

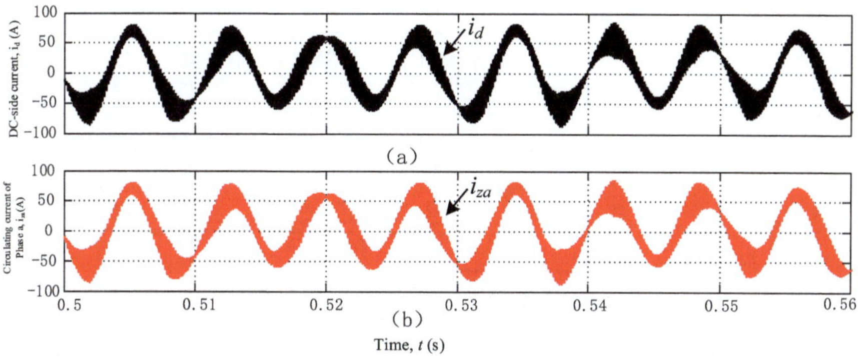

(a) DC -side Current; (b) Circulating Current of Phase A (i_{za})

Fig. 6.14 Waveforms of circulating current and DC-side current in single-phase MMC circuit. (**a**) DC-side current; (**b**) circulating current of phase A (i_{za})

MATLAB/Simulink for single-phase four-module cascaded inverter circuits. Figure 6.14 illustrates the circulating current waveform of the single-phase MMC inverter circuit, with Fig. 6.14a displaying the DC bus current (i_d) and Fig. 6.14b showing the a-phase circulating current (i_{za}) waveform, calculated using Eq. (6.1), along with the detected upper and lower bridge arm currents (i_{Pa} and i_{Na}). Both the bridge arm circulating current and the DC current exhibit secondary pulsations of equal magnitude, indicating that the circulating current flows through both the AC bridge arm and the DC loop. This result demonstrates that the single-phase MMC bridge arm experiences power pulsation exchanges at twice the fundamental frequency between the AC side and the DC side.

6.3 Principles and Control of Three-Phase-to-Single-Phase Converter Based on MMC Structure

6.3.1 Operating Principles and Control Strategies of MMC on Three-Phase Side

The initial analysis focuses on three-phase PWM converters with an MMC structure that consists of four cascaded submodules (SM1–SM4), as shown in Fig. 6.15. While the characteristics of power control are similar to those of three-phase PWM rectifiers, the control modules—such as current control for the upper and lower bridge arms, circulating current control, modulation, and voltage equalization—exhibit distinct characteristics.

In Fig. 6.15, u_{sa}, u_{sb}, and u_{sc} represent the three-phase AC voltages, while i_{sa}, i_{sb}, and i_{sc} indicate the three-phase AC currents. The AC port voltages of the MMC are denoted as u_a, u_b, and u_c. V_{dc} represents the total DC bus voltage. L_{sa}, L_{sb}, and L_{sc} indicate the AC connecting reactors on the three-phase side, while R_{sa}, R_{sb}, and R_{sc}

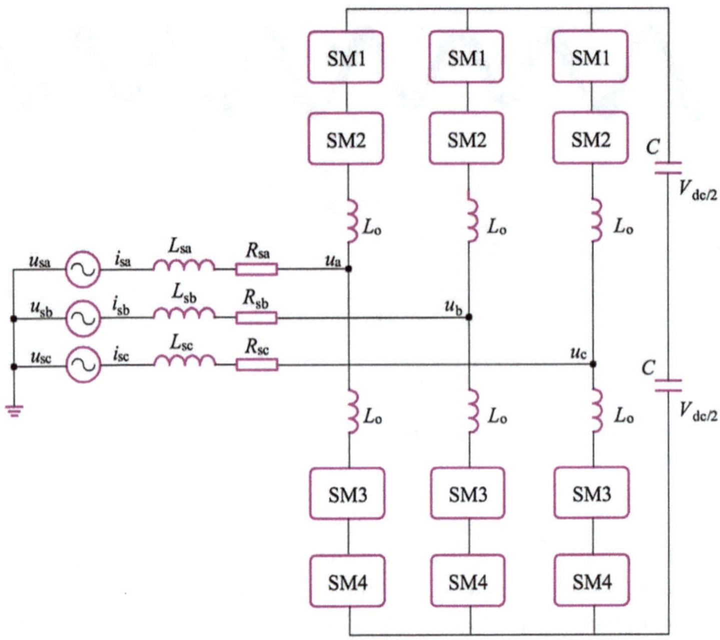

Fig. 6.15 Topology of three-phase PWM rectifier

are the internal resistances of these AC connecting reactors. Lo denotes the reactor in the upper and lower bridge arms of each phase, with the internal resistance (R_o) of the bridge arm inductor neglected. *C* represents the DC bus capacitor, and SM1–SM4 are the submodules in the cascaded structure of the upper and lower bridge arms for each phase.

Based on Eq. (6.12), the port voltages (u_{Pj} and u_{Nj}) of the upper and lower bridge arms of each phase can be controlled independently, and the AC port voltage (*uj*, where *j* = *a*, *b*, *c*) of each phase can be calculated using Eq. (6.9). Therefore, rewriting Eq. (6.12) results in the reference voltage equation for the upper and lower arms:

$$\begin{cases} u_{Pj} = \dfrac{V_{dc}}{2} + R_{sj} \cdot i_{sj} + L_{sj} \dfrac{di_{sj}}{dt} - L_o \dfrac{di_{Pi}}{dt} - u_{sj} \\ u_{Nj} = \dfrac{V_{dc}}{2} - R_{sj} \cdot i_{sj} - L_{sj} \dfrac{di_{sj}}{dt} - L_o \dfrac{di_{Ni}}{dt} + u_{sj} \end{cases} \tag{6.32}$$

According to Eq. (6.32), the equation for calculating current values for upper and lower bridge arms can be derived:

$$\begin{cases} i_{Pj} = i_{zj} - \dfrac{i_{sj}}{2} \\ i_{Nj} = i_{zj} + \dfrac{i_{sj}}{2} \end{cases} \tag{6.33}$$

By substituting Eq. (6.33) into Eq. (6.32), the simplification process yields

$$\begin{cases} u_{Pj} = \dfrac{V_{dc}}{2} + R_{sj} \cdot i_{sj} + \left(L_{sj} + \dfrac{L_o}{2}\right)\dfrac{di_{sj}}{dt} - L_o\dfrac{di_{zj}}{dt} - u_{sj} \\ u_{Nj} = \dfrac{V_{dc}}{2} - R_{sj} \cdot i_{sj} - \left(L_{sj} + \dfrac{L_o}{2}\right)\dfrac{di_{sj}}{dt} - L_o\dfrac{di_{zj}}{dt} + u_{sj} \end{cases} \tag{6.34}$$

With the typical design featuring $L_{sj} >> L_o$, the result of $L_{sj} + L_o/2$ can be approximated as L_{sj}. Assuming that the voltage drop generated by the circulating current (i_{zj}) on the bridge arm inductor during stable operation is zero, Eq. (6.34) can be simplified into the following:

$$\begin{cases} u_{Pj} = \dfrac{V_{dc}}{2} + R_{sj} \cdot i_{sj} + L_{sj}\dfrac{di_{sj}}{dt} - u_{sj} \\ u_{Nj} = \dfrac{V_{dc}}{2} - R_{sj} \cdot i_{sj} - L_{sj}\dfrac{di_{sj}}{dt} + u_{sj} \end{cases} \tag{6.35}$$

Based on Eq. (6.35), the AC modulation signal for each bridge arm can be derived by controlling the voltage at the upper and lower bridge arm ports. Following the discussion in Chap. 3, the expression for the AC port voltage of the upper arm in the *dq* coordinate system can be established as follows:

$$\begin{bmatrix} u_{Pd} \\ u_{Pq} \end{bmatrix} = \begin{bmatrix} L_{sabc}\dfrac{di_d}{dt} - \omega L_{sabc} i_q + R_{sabc} i_d - \sqrt{\dfrac{3}{2}} \cdot U_{max} \\ L_{sabc}\dfrac{di_q}{dt} + \omega L_{sabc} i_d + R_{sabc} i_q - 0 \end{bmatrix} \tag{6.36}$$

where u_{Pd} and u_{Pq} represent the active and reactive components of the upper bridge arm port voltage (U_{Pabc}) after d_q transformation, respectively. The relationship between the port voltage and the AC-side voltage of the lower bridge arm can be established as follows:

$$\begin{cases} u_{Na} = \dfrac{V_{dc}}{2} - R_{sa} \cdot i_{sa} - L_{sa}\dfrac{di_{sa}}{dt} + u_{sa} \\ u_{Nb} = \dfrac{V_{dc}}{2} - R_{sb} \cdot i_{sb} - L_{sb}\dfrac{di_{sb}}{dt} + u_{sb} \\ u_{Nc} = \dfrac{V_{dc}}{2} - R_{sc} \cdot i_{sc} - L_{sc}\dfrac{di_{sc}}{dt} + u_{sc} \end{cases} \tag{6.37}$$

Similarly, the expression for the AC port voltage of the lower bridge arm in the d_q coordinate system can be established as follows:

$$\begin{bmatrix} u_{\rm Nd} \\ u_{\rm Nq} \end{bmatrix} = \begin{bmatrix} -L_{\rm sabc}\dfrac{{\rm d}i_{\rm d}}{{\rm d}t} + \omega L_{\rm sabc} i_{\rm q} - R_{\rm sabc} i_{\rm d} + \sqrt{\dfrac{3}{2}} \cdot U_{\max} \\ -L_{\rm sabc}\dfrac{{\rm d}i_{\rm q}}{{\rm d}t} - \omega L_{\rm sabc} i_{\rm d} - R_{\rm sabc} i_{\rm q} + 0 \end{bmatrix} \tag{6.38}$$

where $u_{\rm Nd}$ and $u_{\rm Nq}$ represent the active and reactive components of the three-phase lower bridge arm port voltage ($U_{\rm Nabc}$) after $d_{\rm q}$ transformation.

Comparing Eqs. (6.47) and (6.49) indicates that the port voltages of the upper and lower arms are equal in the $\underline{d}_{\rm q}$ coordinate system but opposite in direction. By using Eq. (6.49) for the lower bridge arm port voltage in the $d_{\rm q}$ coordinate system as the control variable, the AC reference voltage for the lower bridge arm can be derived by inversely transforming the control result from the $d_{\rm q}$ coordinate system to the abc coordinate system. The AC reference voltage for the upper bridge arm can be obtained by negating the value of the lower bridge arm voltage.

6.3.2 *Capacitor Voltage Control of MMC Submodule*

MMC bridge arm currents act to charge or discharge the submodule capacitors enabled in each bridge arm. Unbalance exists in the capacitor voltages among the submodules due to variability in the number of enabled submodules and the moments at which they are switched on, which affects the DC bus voltage and system power control (Xie 2006; Martins et al. 1995). To address this issue, literature (Yadong et al. 2003) suggests superimposing the voltage stability control output with the converter regulator output before generating the modulation component for the MMC submodule. Building on this method, this chapter introduces independent carrier phase-shifted modulation (CPS-SPWM) for the MMCs on both the three-phase and single-phase sides. This approach divides the DC-side capacitor voltage control of the submodules into two parts: voltage stability control and voltage equalization control. Consequently, this method facilitates the even distribution of voltage among the submodule capacitors while ensuring the stability of the DC bus voltage. Figure 6.16 presents the block diagram of the capacitor voltage control strategies for the *i*th submodule ($i = 1, 2, 3, 4$) in phase j ($j = a, b, c, d, e$).

1. Voltage stability control

   Figure 6.16a illustrates the voltage stability control algorithm, consisting of voltage outer loop control and current inner loop control (Dixon and Boon-Teck 1988). The voltage outer loop control ensures that the average capacitor voltage ($\overline{V}_{\rm dc}$) of the four submodules in each phase bridge arm remains aligned with the reference value ($V_{\rm dref}$), which can be expressed as follows:

$$\overline{V}_{\rm dc} = \frac{1}{4}\sum_{i=1}^{4} U_{Ci} \tag{6.39}$$

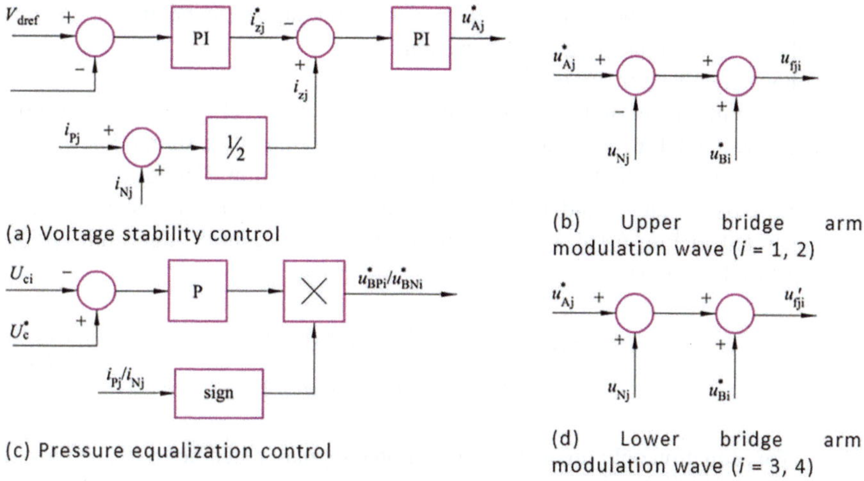

Fig. 6.16 Block diagram of capacitor voltage control strategy for MMC submodule. (**a**) Voltage stability control. (**b**) Upper bridge arm modulation wave ($i = 1, 2$). (**c**) Pressure equalization control. (**d**) Lower bridge arm modulation wave ($i = 3, 4$)

The current inner loop is used to calculate the circulating current (i_{zj}) based on the upper and lower bridge arm currents (i_{Pj} and i_{Nj}) and using Eq. (6.1) and to control i_{zj} to be equal to the reference value (i^*zj). This reference value is generated by the voltage outer loop control, by processing errors between the actual average value and the reference value through a proportional-integral regulator (PI):

$$i_{zj}^{*} = K_p\left(V_{\mathrm{dcf}} - \overline{V}_{\mathrm{dc}}\right) + K_{\mathrm{i}}\int\left(V_{\mathrm{dcf}} - \overline{V}_{\mathrm{dc}}\right)\mathrm{d}t \tag{6.40}$$

The final voltage stability control output, denoted as u^*Aj, is calculated using the following equation:

$$u_{\mathrm{A}j}^{*} = K_p\left(i_{zj} - i_{zj}^{*}\right) + K_{\mathrm{i}}\int\left(i_{zj} - i_{zj}^{*}\right)\mathrm{d}t \tag{6.41}$$

As indicated by Eq. (6.41), when the average capacitor voltage of the submodules is less than the reference value, the circulating current reference generated by a PI regulator increases. According to Eq. (6.42), the circulating current value rises with the increase in the reference value. When the circulating current is greater than zero, the submodule capacitor is charged, resulting in an increase in the total capacitor voltage of the bridge arm. Conversely, when the average capacitor voltage of the submodules exceeds the reference value, the circulating current reference generated by the PI regulator decreases. As shown in Eq. (6.42), the circulating current value decreases with the reduction in the reference value. When the current is less than zero, the submodule capacitor discharges, leading to a decrease in the total capacitor voltage of the bridge arm. Through this control loop, the average voltage of the submodules can be effectively controlled without impacting the AC current.

2. Voltage equalization control

   Since the upper and lower bridge arm currents of each phase differ in both amplitude and direction, the charging and discharging of the submodule capacitors can be controlled according to the current direction. Voltage equalization control based on i_{Pj} and i_{Nj} for the capacitor voltage of each submodule is illustrated in Fig. 6.16b (Martins et al. 1995). The control algorithm can be expressed as follows:

$$\begin{cases} u_{BPi}^{*} = K_{\mathrm{p}}\left(U_{\mathrm{C}}^{*} - U_{\mathrm{C}i}\right) \cdot sign\left(i_{Pj}\right) \\ u_{BNi}^{*} = K_{\mathrm{p}}\left(U_{\mathrm{C}}^{*} - U_{\mathrm{C}i}\right) \cdot sign\left(i_{Nj}\right) \end{cases} \tag{6.42}$$

   This equation only involves a proportional regulator, denoted as *P*, and a sign function expressed as sign (*).

   An output of 1 corresponds to a positive bridge arm current output, while an output of −1 corresponds to a negative current. When the submodule voltage is lower than the reference value, it is necessary to either charge the capacitor or reduce the discharging time. If the bridge arm current is greater than zero, the regulator output is positive, which reduces the discharging time. Conversely, if the bridge arm current is less than zero, the regulator output is negative, which increases the charging time.

   Similarly, when the submodule voltage value is higher than the reference value, it is necessary to either discharge the submodule or reduce the charging time. If the bridge arm current is greater than zero, the regulator output is negative, which reduces the charging time. Conversely, if the bridge arm current is less than zero, the regulator output is positive, which increases the discharge time.

3. Synthesis of modulation waves for upper and lower bridge arms

   Figure 6.16c illustrates how modulation waves are generated for the upper bridge arm submodule. According to Eq. (6.12), the reference voltages of the upper and lower bridge arms are in antiphase. Therefore, the three-phase active and reactive decoupling control can be used to generate the AC reference voltage for the three-phase lower bridge arm (u_{Nj}), which is then negated to yield the reference voltage for the upper bridge arm. Additionally, the AC reference voltage of the upper bridge arm is generated by superimposing the output components from the voltage stability and equalization control. The modulation signal is ultimately generated as $u_{fji} = -u_{Nj} + u_{\mathrm{A}j}^{*} + u_{\mathrm{B}i}^{*}$ $(i = 1, 2)$. Figure 6.16d illustrates how modulation waves are generated for the lower bridge arm submodule, following a principle similar to that of the upper bridge arm, expressed as $u_{fji}^{'} = u_{Nj} + u_{\mathrm{A}j}^{*} + u_{\mathrm{B}i}^{*}$ $(i = 3, 4)$.

6.3.3 Modulation Strategies for Three-Phase-to-Single-Phase MMCs

High switching frequency modulation is typically employed for MMCs, which can be categorized into two main types: space vector pulse width modulation (SVPWM) and multi-carrier sinusoidal pulse width modulation (SPWM). The latter is further divided into two types: carrier disposition SPWM (CD-SPWM) and carrier phase-shifted SPWM (CPS-SPWM) (Yadong et al. 2003). Considering the large number of MMC modules, conventional SPWM is more suitable than the SVPWM method, which generates more vectors and complicates implementation.

CD-SPWM technology involves the side-by-side arrangement of N triangular carriers that are identical in amplitude and frequency. By comparing these triangular carriers with sinusoidal modulation waves, a drive signal is generated for each switching tube. Based on the different phases of the N triangular carriers, CD-SPWM can be categorized into three types: phase disposition SPWM (PD-SPWM), phase opposition disposition SPWM (POD-SPWM), and alternate phase opposition disposition SPWM (APOD-SPWM).

CD-SPWM is widely used in diode-clamped multilevel converters and H-bridge cascaded converters. However, some triangular carriers may not intersect with the sinusoidal waves at low modulation degrees, leading to failures in generating the correct multilevel output (Dixon and Boon-Teck 1988). Developed by extending two-level SPWM, CPS-SPWM involves phase shifts of multiple triangular carriers, which are identical in amplitude and frequency, by a specified angle. These shifted carriers are then compared with sinusoidal modulated waves to generate the switching pulse signals. The phase shift differences depend on the number of levels in the multilevel output and the number of cascaded modules. CPS-SPWM enables a high equivalent switching frequency output while maintaining a low carrier switching frequency, significantly reducing the harmonic content of the output waveform and making it suitable for high-power cascaded multilevel applications (Hu 2013; Guerrero et al. 2005).

Since the reference voltage (u_{Pj}) of the upper bridge arm in MMCs is antiphase with the equivalent AC output voltage (u_j), while the reference voltage (u_{Nj}) of the lower bridge arm is co-phase with the equivalent AC output voltage (u_j), the upper and lower bridge arms can be modulated independently. In other words, the dual-modulation-wave CPS-SPWM strategy (Fig. 6.17) can be employed, as the modulation waves for the upper and lower bridge arms differ by 180° from each other (Mao et al. 2014).

With antiphase modulation waves between the upper and lower bridge arms, the triangular carriers corresponding to the two submodules of the upper bridge arm differ by 180° from each other. Similarly, the two triangular carriers of the lower bridge arm also differ by 180°. The triangular carriers in the submodules corresponding to the lower bridge arm and the upper bridge arm—specifically, the first module and the third module, as well as the second module and the fourth

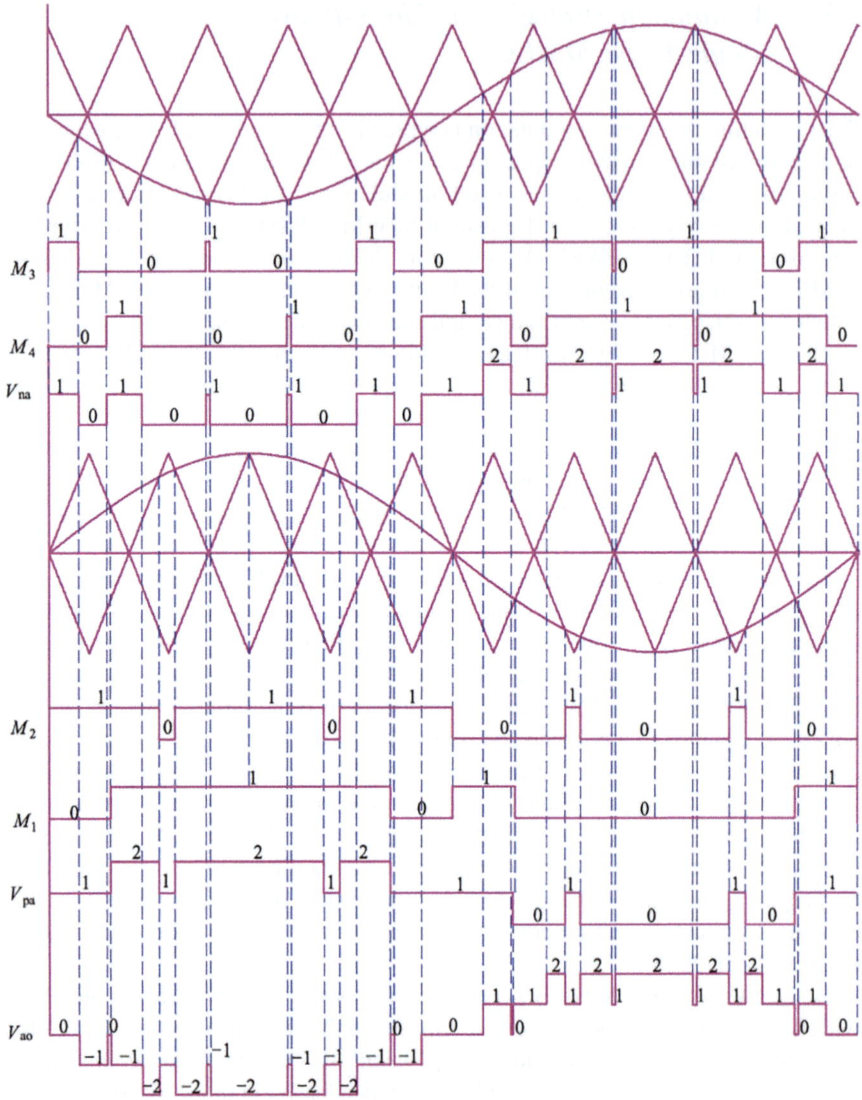

Fig. 6.17 Dual-modulation-wave CPS-SPWM strategy

module—differ by 90°. This results in triangular carrier phases for the four submodules of each phase being at 0°, 180°, 90°, and 270°.

According to Eq. (6.9), the equivalent port output (u_a) for each phase can be obtained by subtracting the AC voltage output from the lower bridge arm from that of the upper bridge arm, resulting in five distinct levels at the port of each phase.

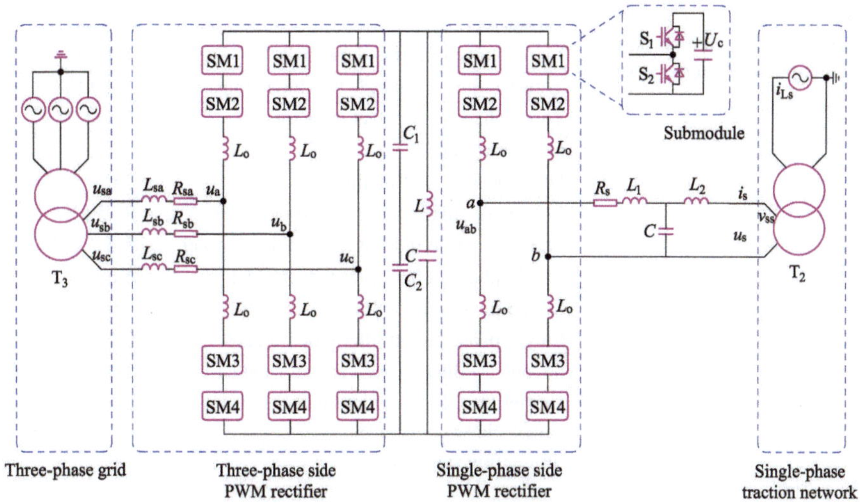

Fig. 6.18 Topology of simulation system for three-phase-to-single-phase converter based on MMC structure

6.4 System Simulation Modeling and Analysis

6.4.1 Topology and Parameters for System Simulation

To verify the feasibility of three-phase-to-single-phase converters based on the MMC structure for continuous co-phase power supply and to examine the interconnection control characteristics of the converters, a simulation system was established, as shown in Fig. 6.18. The input is connected to the three-phase grid, while the output is linked to the single-phase traction network through step-down and step-up transformers. On the DC-side link, an LC filter circuit is employed to filter out secondary ripple generated by the single-phase converter, and an LCL filter circuit is utilized to reduce high harmonics of the switching frequency on the single-phase output side.

A control system for the three-phase and single-phase converters was established in accordance with Sect. 6.2, utilizing three-phase active and reactive decoupling control on the three-phase side and single-phase active and reactive decoupling control on the single-phase side. The simulation parameters are presented in Table 6.2.

6.4.2 Simulation Analysis of Three-Phase PWM Rectifier

Figure 6.19 displays the waveform of the voltage (V_{dc}) on the DC bus side of a three-phase rectifier. The initial 0.2-s interval represents the rectifier startup stage, during which the DC voltage continuously rises and stabilizes at the reference value

Table 6.2 Simulation parameters of three-phase-to-single-phase converter system

Parameter	Value
Line voltage of three-phase grid	110 kV
Transformation ratio of step-down transformer	110 kV/400 V
Three-phase AC contact inductor	5 mH
DC bus voltage	700 V
DC capacitance of module	3 mF
Voltage of single-phase traction network	27.5 kV
Transformation ratio of step-up transformer	500 V/27.5 kV
Single-phase output filter inductor, L1	5 mH
Single-phase output filter capacitor, C	0.05 mF
Single-phase output filter inductor, L2	0.1 mH
Secondary filter circuit capacitor/inductor	3 mF/0.84 mH
Carrier frequency of three-phase and single-phase converters	3 kHz
Simulation step size	1 μs

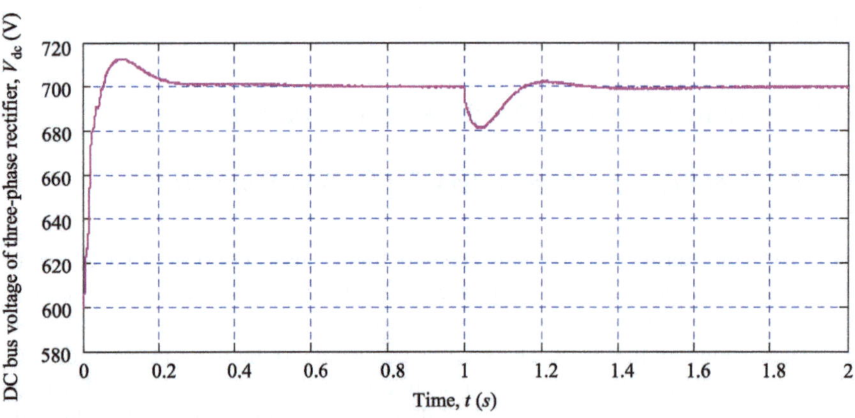

Fig. 6.19 DC bus voltage of three-phase rectifier (Vdc)

of 700 V after approximately 0.2 s. The voltage overshoot during this startup phase is about 10%. From $t = 0.2$–1 s, the rectifier operates in a no-load state, maintaining the DC voltage at 700 V. Following a load connection at $t = 1$ s, the DC voltage drops by about 20% but recovers and stabilizes at the reference value of 700 V after 0.2 s.

Figure 6.20 displays the waveform of the AC input current (isabc) of a three-phase rectifier. The interval from $t = 0.2$–1 s represents the no-load operation stage, during which the current amplitude approaches zero. Following the connection of the load at $t = 1$ s, the three-phase current stabilizes after a 0.2-s adjustment, resulting in a current amplitude of 20 A. The detail of the three-phase current waveform during the interval from $t = 1.5$–1.55 s shows a phase difference of 120°. Figure 6.21 displays the waveforms of the a-phase AC current (i_{sa}) and phase voltage (u_{sa}) after

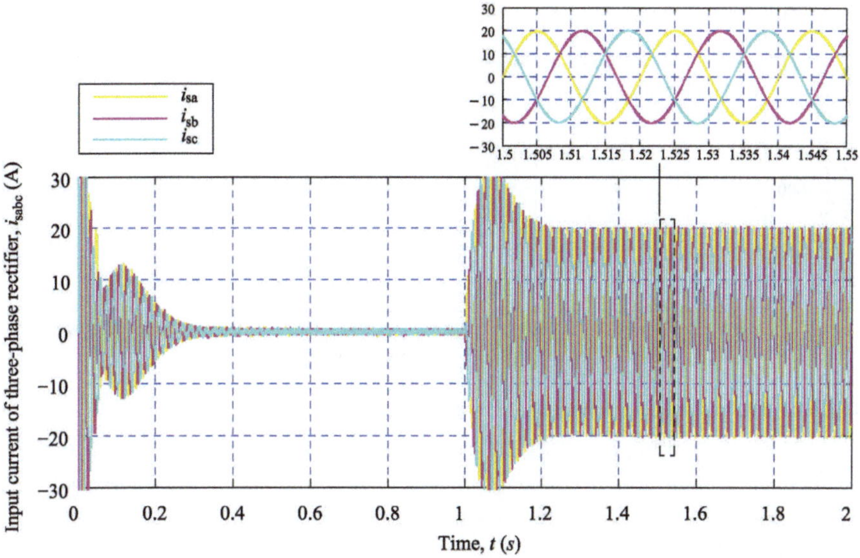

Fig. 6.20 AC input current of three-phase rectifier (i_{sabc})

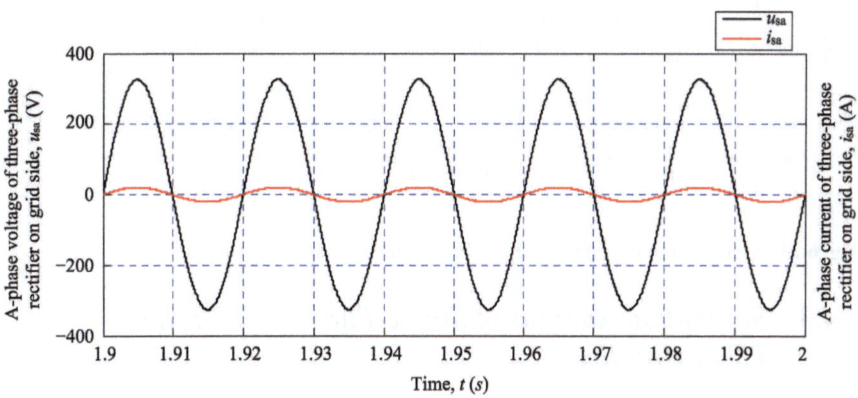

Fig. 6.21 A-phase current (i_{sa}) and phase voltage (u_{sa}) of three-phase rectifier

the load is connected. These waveforms are closely aligned in phase, indicating that the rectifier on the three-phase side operates at a unity power factor.

Figures 6.22 and 6.23 display the waveforms of the DC-side voltage (V_{dc}), AC current (i_{sa}), and AC phase voltage (u_{sa}), when the DC input of the three-phase converter is configured as a voltage source, with a load resistance of 10 Ω and the converter operating in an active inverter state. As shown in Fig. 6.23, the AC input current (i_{sa}) and AC voltage (u_{sa}) are opposite in phase. In this state, the converter operates at a power factor of -1 in active inverter mode.

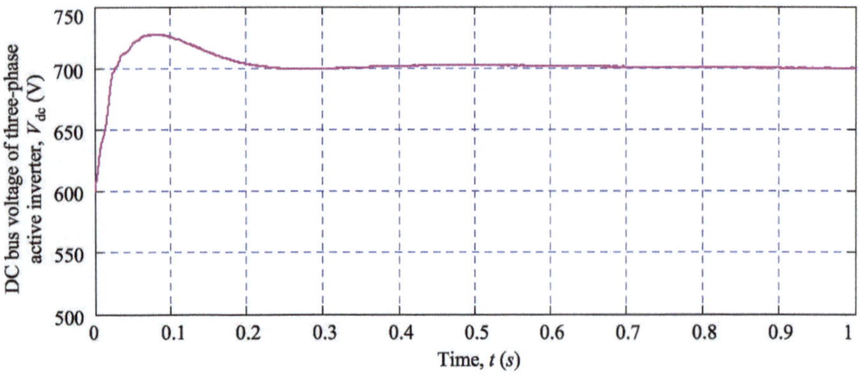

Fig. 6.22 DC bus voltage of three-phase inverter (V_{dc})

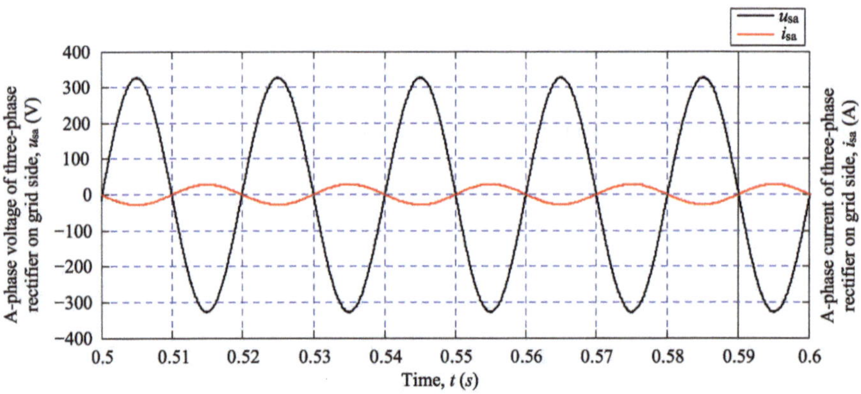

Fig. 6.23 A-phase current (i_{sa}) and voltage (u_{sa}) of three-phase active inverter

6.4.3 Simulation Analysis of Single-Phase Full-Bridge PWM Rectifier

Figure 6.24 illustrates the waveform of the voltage (V_{dc2}) on the DC bus side of a single-phase MMC rectifier. The first 0.1-s interval represents the startup stage of the single-phase rectifier, during which the DC voltage rises and stabilizes at the reference value of 700 V at $t = 0.1$ s. The DC voltage remains stable at 700 V during the subsequent no-load state from $t = 0.1$ to $t = 1$ s. Following the connection of a 50 Ω resistor at $t = 1$ s, the DC voltage returns to a stable value after approximately 0.2 s of adjustment. However, in the absence of the LC filter circuit, the waveform detail during the interval from $t = 1.5$ to 1.56 s reveals that the DC voltage contains a pulsating component at twice the angular frequency, with fluctuations of about 40 V. This observation aligns with the analysis of the circulating current component at twice the fundamental frequency of the single-phase converter discussed in Sect. 6.2.

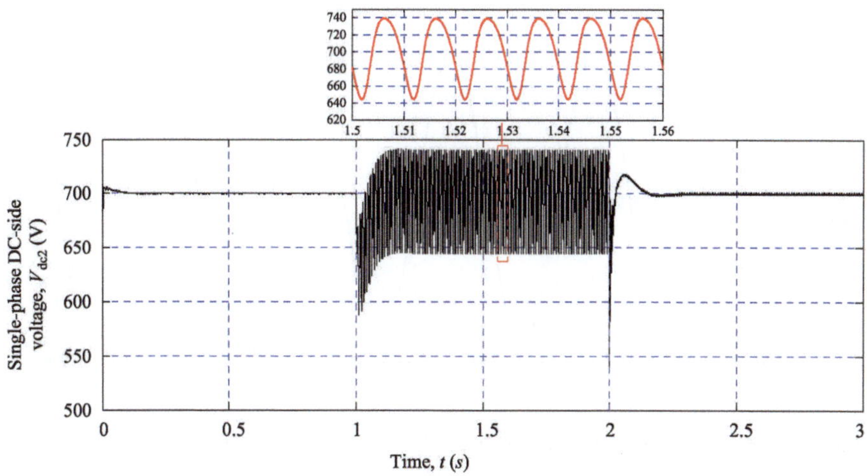

Fig. 6.24 DC voltage of single-phase PWM rectifier (V_{dc2})

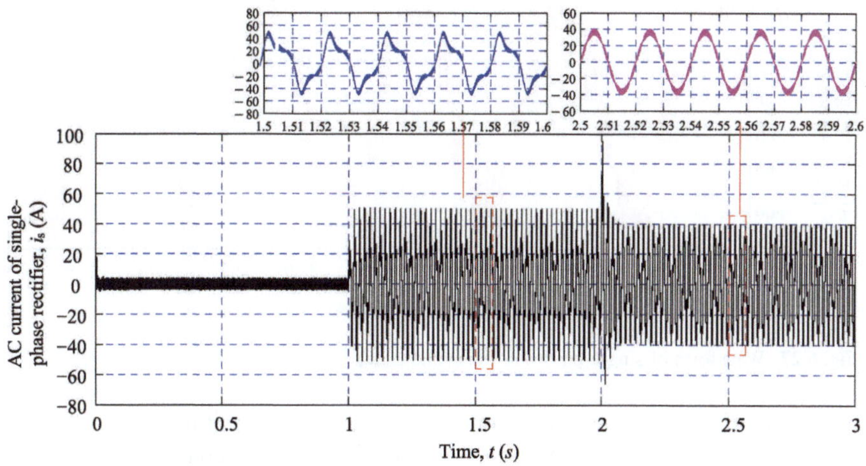

Fig. 6.25 AC current of single-phase PWM rectifier (i_s)

After the LC filter circuit is connected at t = 2 s, the secondary pulsations in the voltage are almost completely suppressed and eliminated.

Figure 6.25 displays the waveform of the AC current (i_s) in a single-phase rectifier. The interval within the first 0.1 s represents the startup stage. During the no-load operation period from t = 0.1 to 1 s, the current approaches zero, indicating the stable operation of the rectifier. After the load is connected at t = 1 s, the single-phase current quickly stabilizes following a brief adjustment period. In the absence of the LC filter circuit, the waveform detail reveals significant distortion in the AC current (i_s), indicating that secondary ripple pulsations and those at other orders

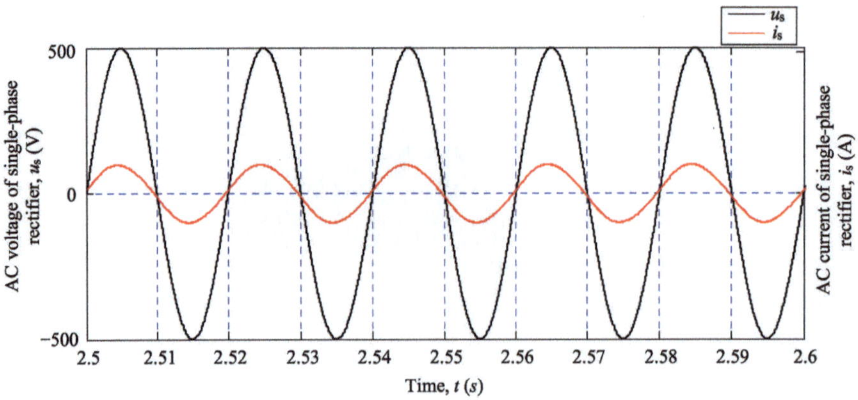

Fig. 6.26 AC voltage (u_s) and current (i_s) of single-phase PWM rectifier

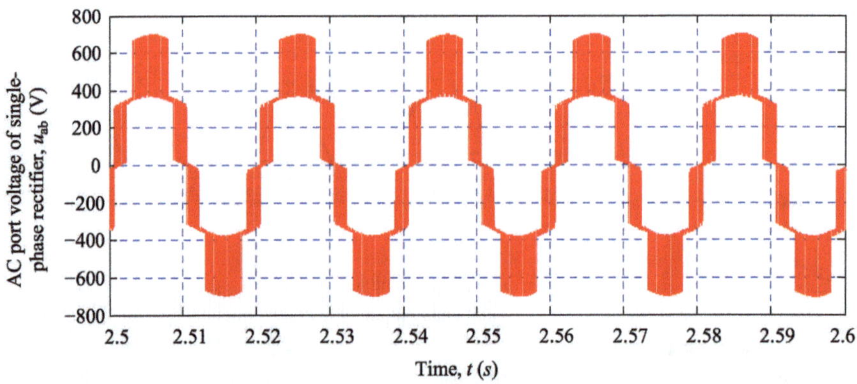

Fig. 6.27 Waveform of single-phase AC port voltage, u_{ab}

adversely affect the power quality on the AC side. The current detail also shows a noticeable reduction in distortion once the LC filter circuit is connected at $t = 2$ s.

Figure 6.25 displays the waveforms of the AC current (i_s) and AC voltage (u_s) of a single-phase rectifier after the load is connected. The current and voltage are aligned in phase, indicating that the co-phase grid side operates at a unity power factor, and active power flows from the single-phase AC side to the DC side. Figure 6.26 indicates AC voltage (u_s) and current (i_s) of single-phase PWM rectifier.

Figure 6.27 displays the waveform of the AC output port voltage after the single-phase rectifier is connected at $t = 2$ s, showing the AC port voltage at five levels. Unlike ordinary rectifier circuits, the bridge arms of MMC circuits are configured with connecting inductors, which provide a filtering effect and contribute to fluctuations in the port levels. In this state, the converter on the single-phase side operates in active inverter mode.

Figures 6.28 and 6.29 display the waveforms of the AC current (i_s) and AC voltage (us) in an active inverter state. The DC bus voltage remains stable at the

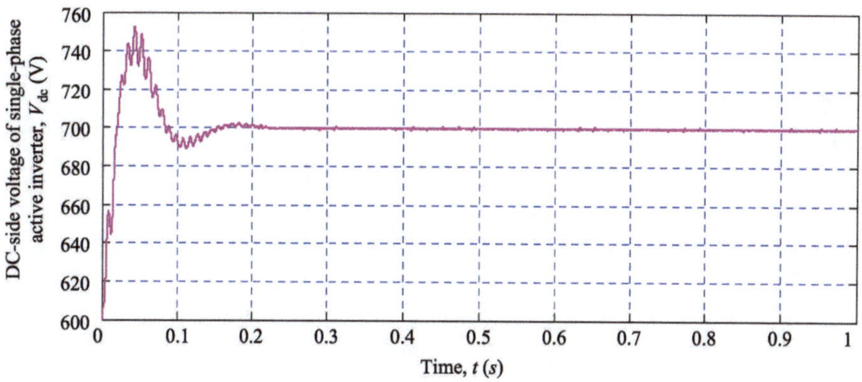

Fig. 6.28 Waveform of DC voltage in single-phase active inverter

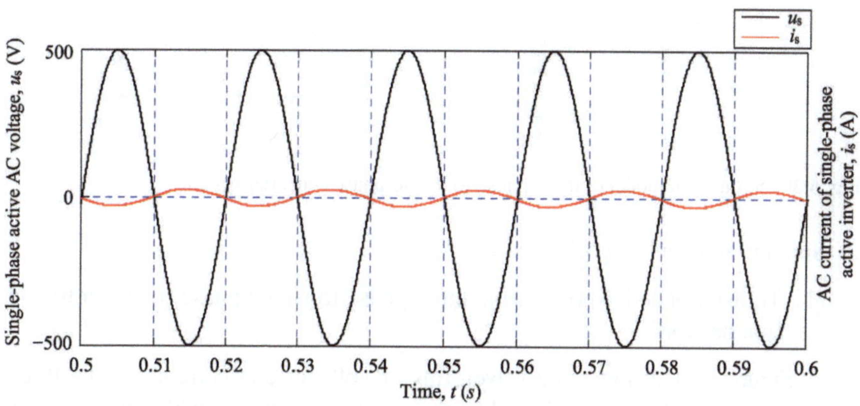

Fig. 6.29 AC voltage (u_s) and AC current (i_s) of single-phase active inverter

reference value of 700 V, and the AC current (i_s) and AC voltage (u_s) are opposite in phase, indicating a power factor of −1, with active power flowing from the DC side to the single-phase AC side.

6.4.4 Joint Simulation Analysis of Three-Phase-to-Single-Phase Converter System

1. Simulation Model

   Sections 6.2 and 6.3 independently verify the effectiveness and accuracy of the control strategies for converters on both the three-phase and single-phase sides. This section progresses with the joint simulation of three-phase-to-single-phase converters. The simulation system, based on MATLAB/Simulink, is illustrated in Fig. 6.30 (Zheng 2009).

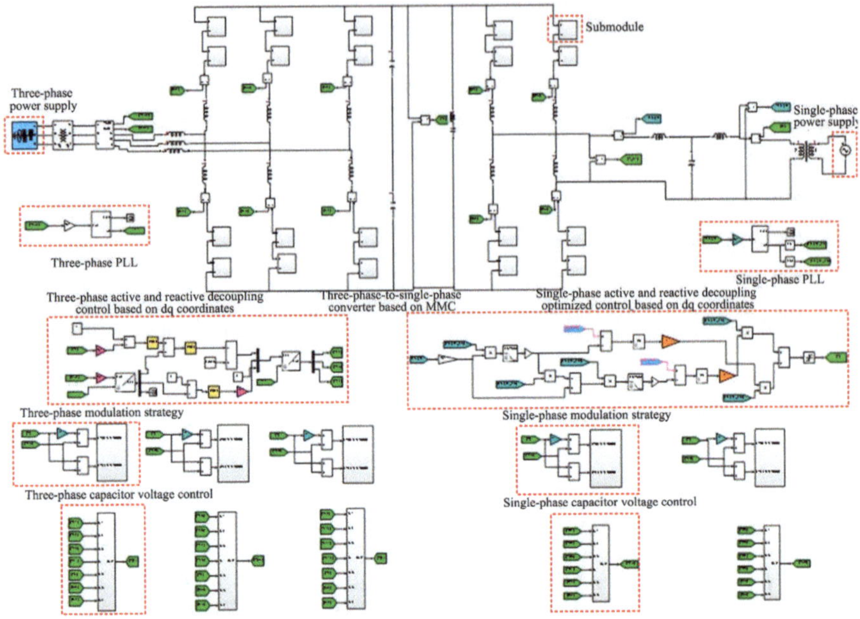

Fig. 6.30 Simulation model of three-phase-to-single-phase converter system

2. Simulation Results and Analysis

(1) Transfer of active power from three-phase-to-single-phase converter to traction network

Figure 6.31 displays the waveforms of voltages and currents on the three-phase and single-phase sides when the regulator operates stably under no-load conditions. As shown in Fig. 6.31a, the DC voltage stabilizes at the reference value of 700 V after 0.3 s of adjustment. Figure 6.31b, d indicates that the A-phase current on the three-phase side and the output current on the single-phase side are both approximately zero, signifying that there is no significant active loss on either side under no-load conditions. Figure 6.31c illustrates the AC output port voltage (u_{ab}) on the single-phase side, displaying five levels.

Figures 6.32 and 6.33 display the voltage and current waveforms during the transfer of active power from the system to the traction network. As shown in Fig. 6.32a, the DC voltage stabilizes at 700 V after a brief adjustment period. With active power flowing on both the three-phase and single-phase sides, the A-phase AC current (i_{sa}) and the AC output current on the single-phase side (i_s) are both sinusoidal, as illustrated in Fig. 6.33b, d. The AC output port voltage exhibits five levels, as shown in Fig. 6.32c. Figure 6.33a, b demonstrates that the AC voltage and current on both the three-phase and single-phase sides are aligned in phase, indicating that both sides operate at a unity power factor. Figure 6.33c displays the AC output port voltage on the three-phase side, which also shows five levels. Figure 6.33d presents the waveform of the single-phase

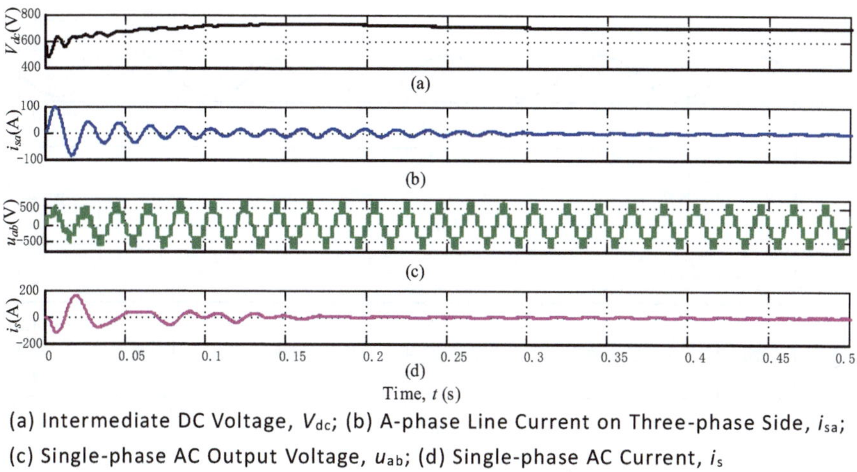

Fig. 6.31 Voltage and current waveforms of unloaded system. (**a**) Intermediate DC voltage, V_{dc}; (**b**) A-phase line current on three-phase side, i_{sa}; (**c**) single-phase AC output voltage, u_{ab}; (**d**) single-phase AC current, i_s

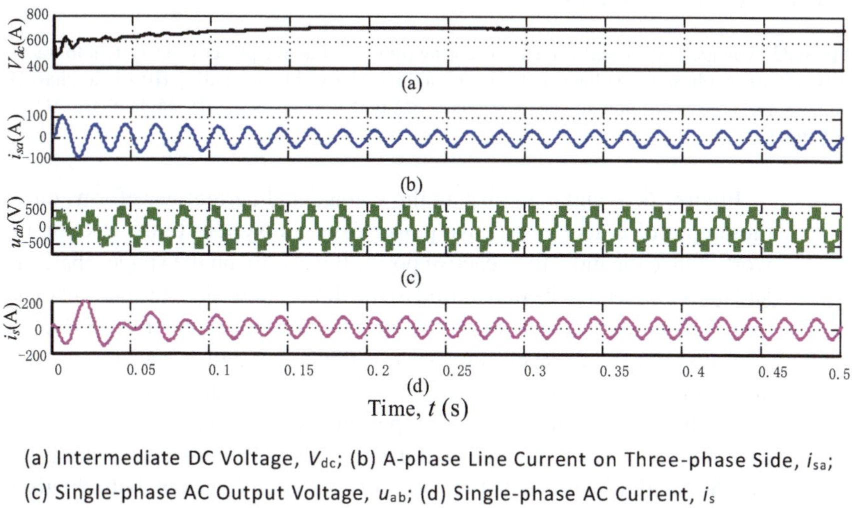

Fig. 6.32 Voltage and current waveforms during transfer of active power from system to traction network. (**a**) Intermediate DC voltage, V_{dc}; (**b**) A-phase line current on three-phase side, i_{sa}; (**c**) single-phase AC output voltage, u_{ab}; (**d**) single-phase AC current, i_s

AC output five-level voltage. Based on these simulations, it is verified that three-phase-to-single-phase converters are capable of transferring active power from the three-phase grid to the traction network as required.

(2) Output of reactive power from three-phase-to-single-phase converter to traction network

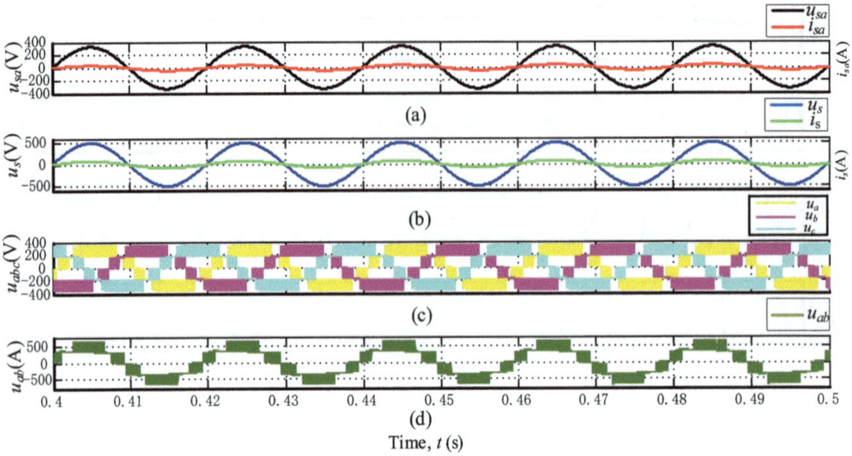

(a) A-phase AC Voltage and Current on Three-phase Side, u_{sa} and i_{sa}; (b) Single-phase AC Voltage and Current, u_s and i_s; (c) Three-phase AC Port Voltage, u_{abc}; (d) Single-phase AC Port Voltage, u_{ab}

Figure 6-33 Voltage and Current Waveforms During Transfer of Active Power from System to Traction Network

Fig. 6.33 Voltage and current waveforms during transfer of active power from system to traction network. (**a**) A-phase AC voltage and current on three-phase side, u_{sa} and i_{sa}; (**b**) single-phase AC voltage and current, u_s and i_s; (**c**) three-phase AC port voltage, u_{abc}; (**d**) single-phase AC port voltage, u_{ab}

Figures 6.34 and 6.35 display the voltage and current waveforms during the output of inductive reactive power from the system to the traction network. Since the inductive reactive power flows only on the single-phase side and does not need to be transferred to the three-phase side, the current on the three-phase side is zero, while the output current on the single-phase side displays sinusoidal waveforms. As shown in Fig. 6.34a, b, the current on the single-phase side (i_s) lags behind the single-phase supply voltage (u_{ab}) by 90°, indicating that the system is outputting inductive reactive power. During this interval, the DC voltage stabilizes at 700 V. Figure 6.34c, along with Fig. 6.35c, d, illustrates that the AC port output voltages on both the three-phase and single-phase sides exhibit five levels. Therefore, single-phase converters are deemed capable of directly delivering reactive power to the traction network without the need to provide inductive reactive power on the three-phase side, thereby avoiding any impact on the operating state of the three-phase rectifier, which maintains a unity power factor.

Similarly, the output of capacitive reactive power from the system to the traction network is illustrated in Figs. 6.36 and 6.37. As shown in Fig. 6.36b, d, the three-phase current is zero, while the single-phase side outputs a sinusoidal reactive current. In Fig. 6.37b, the single-phase current (i_s) leads the single-phase voltage (u_s) by 90°, indicating that the system is delivering capacitive reactive power to the traction network. It is important to note that

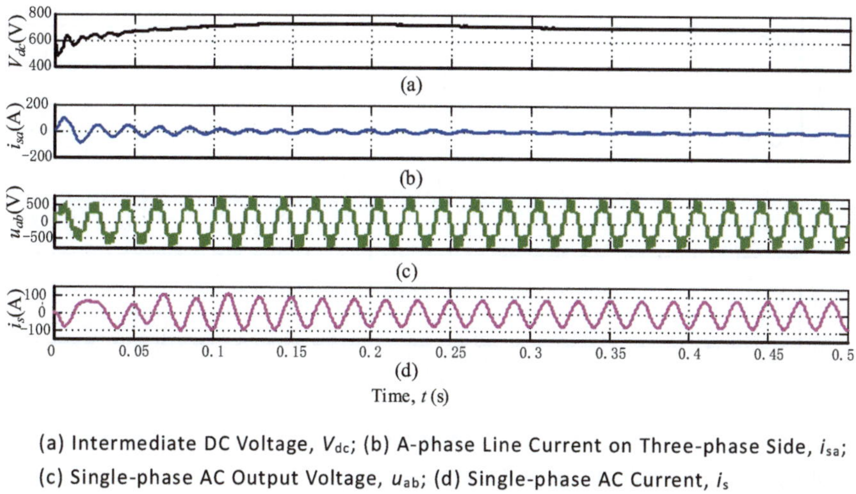

Fig. 6.34 Voltage and current waveforms during transfer of inductive reactive power from system to traction network. (**a**) Intermediate DC voltage, V_{dc}; (**b**) A-phase line current on three-phase side, i_{sa}; (**c**) single-phase AC output voltage, u_{ab}; (**d**) single-phase AC current, i_s

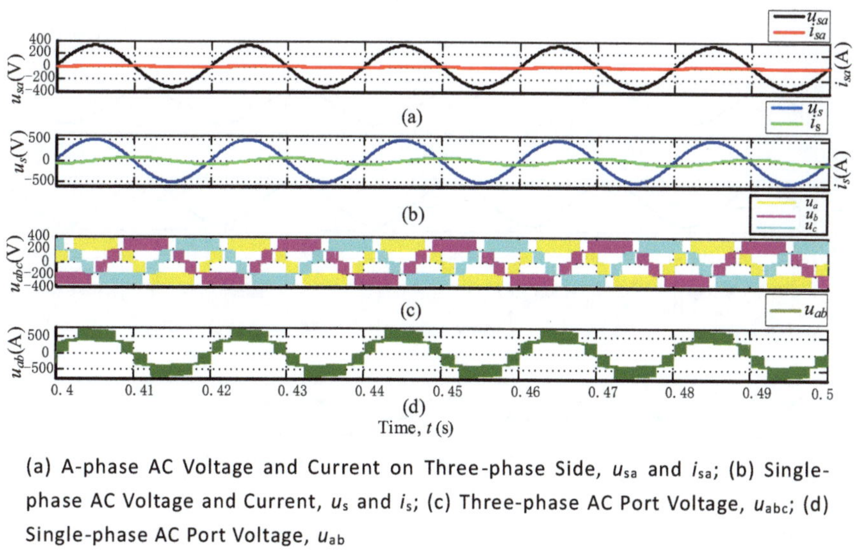

Fig. 6.35 Voltage and current waveforms during transfer of inductive reactive power from system to traction network. (**a**) A-phase AC voltage and current on three-phase side, u_{sa} and i_{sa}; (**b**) single-phase AC voltage and current, u_s and i_s; (**c**) three-phase AC port voltage, u_{abc}; (**d**) single-phase AC port voltage, u_{ab}

the single-phase AC port voltage (u_{ab}) shown in Fig. 6.37d exhibits only four levels, with one level missing. This occurs because, when the converter outputs capacitive reactive power, a low output voltage amplitude is required at the port, resulting in a low modulation degree. In this mode, the converter

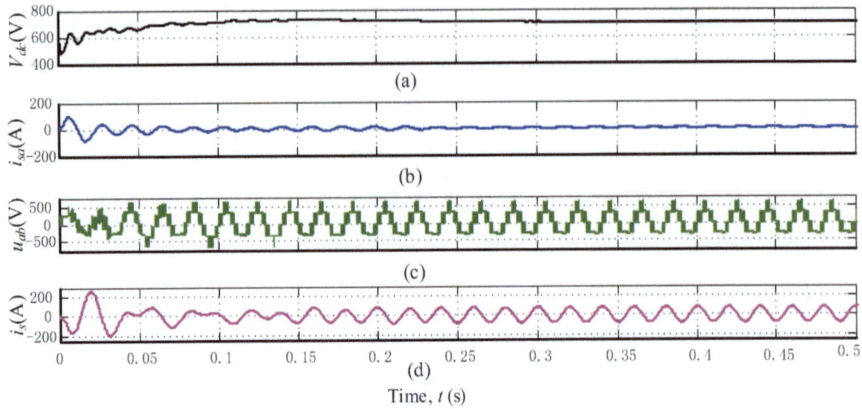

(a) Intermediate DC Voltage, V_{dc}; (b) A-phase Current, i_{sa}; (c) Single-phase Output Voltage, u_{ab}; (d) Single-phase AC Current, i_s

Fig. 6.36 Voltage and current waveforms during transfer of capacitive reactive power from system to traction network. (**a**) A-phase AC voltage and current on three-phase side, u_{sa} and i_{sa}; (**b**) single-phase AC voltage and current, u_s and i_s; (**c**) three-phase AC port voltage, u_{abc}; (**d**) single-phase AC port voltage, u_{ab}

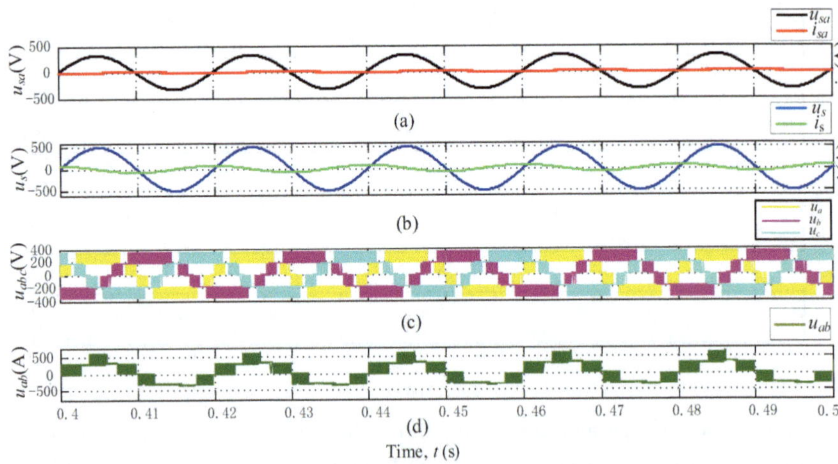

(a) A-phase Voltage and Current, u_{sa} and i_{sa}; (b) Single-phase Voltage and Current, u_s and i_s; (c) Three-phase Port Voltage, u_{abc}; (d) Single-phase Port Voltage, u_{ab}

Fig. 6.37 Voltage and current waveforms during transfer of capacitive reactive power from system to traction network. (**a**) A-phase AC voltage and current on three-phase side, u_{sa} and i_{sa}; (**b**) single-phase AC voltage and current, u_s and i_s; (**c**) three-phase AC port voltage, u_{abc}; (**d**) single-phase AC port voltage, u_{ab}

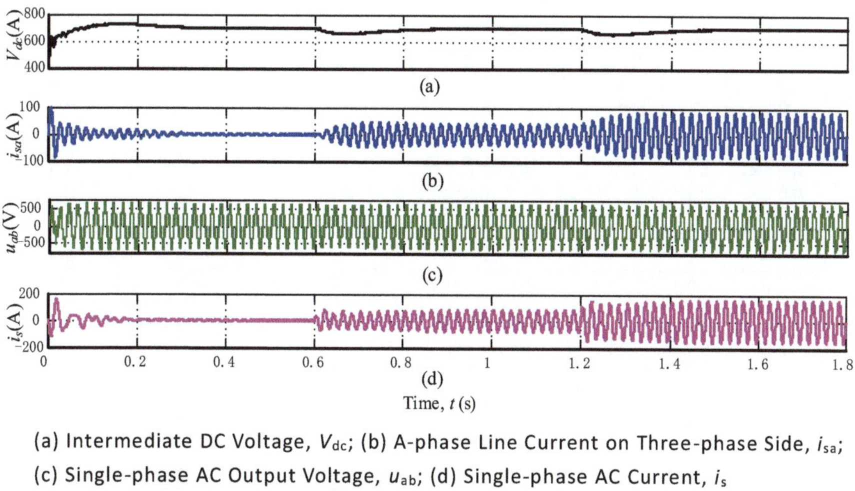

Fig. 6.38 Voltage and current waveforms during sudden active power change of system. (**a**) Intermediate DC voltage, V_{dc}; (**b**) A-phase current, i_{sa}; (**c**) single-phase output voltage, u_{ab}; (**d**) single-phase AC current, i_s

outputs reactive power, where the positive direction of current is defined as the output direction. This differs from application scenarios like STATCOMs, where the positive direction of current is defined as the input direction for reactive power compensation.

(3) Dynamic performance of converter in response to load changes

Figure 6.38 displays the voltage and current waveforms of the three-phase-to-single-phase converter transitioning from a no-load state to transferring 20% active power and then to transferring 40% active power. As shown in Figure 6.38a, in response to a sudden power change, the DC voltage stabilizes at 700 V after a brief adjustment period, with an overshoot of <10%, demonstrating a good dynamic response. Figure 6.38b, d indicates that both the three-phase and single-phase output currents are approximately zero under unloaded conditions. When the active power changes suddenly, the currents on both the three-phase and single-phase sides exhibit sinusoidal waveforms and reach a new stable value after 0.2 s of adjustment. The port voltage on the single-phase side maintains a five-level output, as shown in Fig. 6.38c.

Figure 6.39 displays the main control and voltage waveforms of the converter on both the three-phase and single-phase sides. The sudden change in active power clearly causes the regulator output values on both the three-phase and single-phase sides to change simultaneously, with both reaching new stable values within 0.2 s. This indicates the effective dynamic response and stability performance of the regulator.

Figure 6.40 displays the voltage and current waveforms as the system transitions from a no-load condition to transferring 20% inductive reactive power and then to transferring 40% inductive reactive power. Since the reactive power is output solely

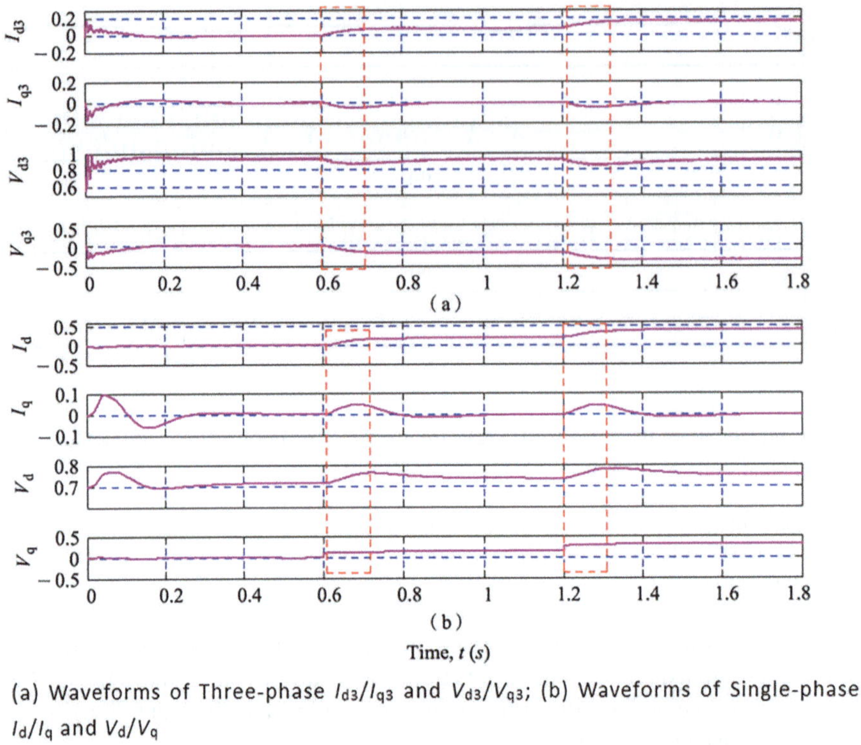

Fig. 6.39 Waveform of three-phase and single-phase regulator parameters in response to sudden active power change of system. (**a**) Waveforms of three-phase I_{d3}/I_{q3} and V_{d3}/V_{q3}; (**b**) waveforms of single-phase I_d/I_q and V_d/V_q

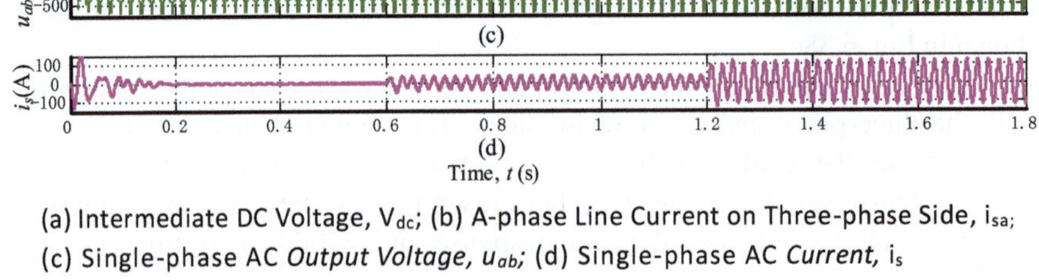

Fig. 6.40 Voltage and current waveforms during sudden inductive reactive power change of system. (**a**) Intermediate DC voltage, V_{dc}; (**b**) A-phase line current on three-phase side, i_{sa}; (**c**) single-phase AC output voltage, u_{ab}; (**d**) single-phase AC current, is

by the converter on the single-phase side, there is no impact on the current amplitude on the three-phase side, as indicated in Figure 6.40a, b. Figure 6.40d illustrates the fluctuations in the amplitude of current in the converter on the single-phase side in response to the sudden change in inductive reactive power. Figure 6.40c presents the waveform of the AC port voltage (u_{ab}) of the converter on the single-phase side, displaying five levels throughout the entire process in response to the sudden change in inductive reactive power.

Figure 6.41 displays the main control signal waveforms on the three-phase and single-phase sides when the inductive reactive power on the single-phase side suddenly changes from 20% to 40%. As indicated in Figure 6.41a, the output on the three-phase side remains almost unchanged, while the output current on the single-phase side is dynamically adjusted in response to the change in power amplitude, quickly reaching a new stable value.

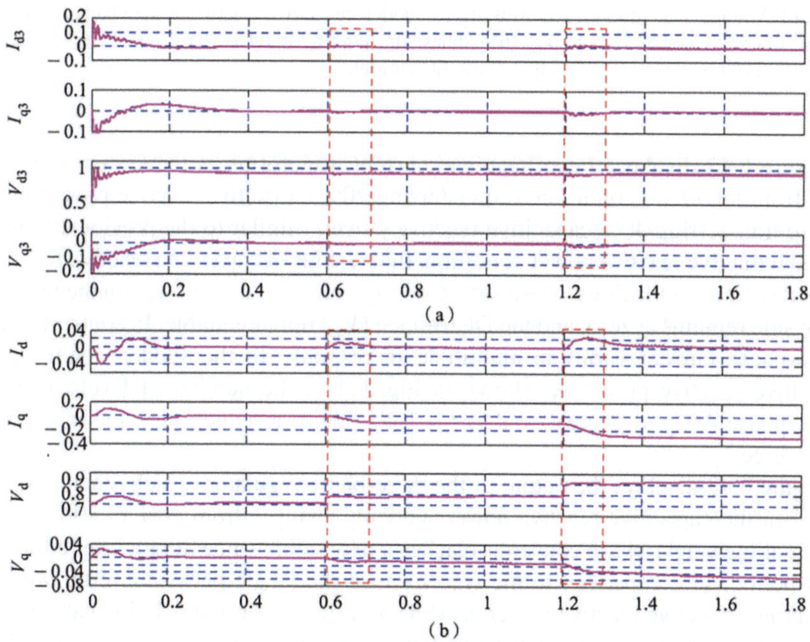

(a) Waveforms of Three-phase I_{d3}/I_{q3} and V_{d3}/V_{q3}; (b) Waveforms of Single-phase I_d/I_q and V_d/V_q

Fig. 6.41 Waveforms of three-phase and single-phase regulator parameters during sudden inductive reactive power change of system. (**a**) Waveforms of three-phase I_{d3}/I_{q3} and V_{d3}/V_{q3}; (**b**) waveforms of single-phase I_d/I_q and V_d/V_q

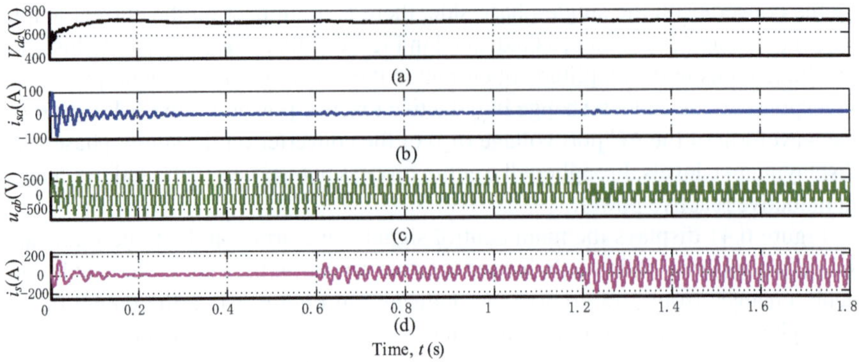

(a) Intermediate DC Voltage, V_{dc}; (b) A-phase Line Current on Three-phase Side, i_{sa}; (c) Single-phase AC *Output Voltage, uab;* (d) Single-phase AC *Current,* is

Fig. 6.42 Voltage and current waveforms during sudden capacitive reactive power change of system. (**a**) Intermediate DC voltage, V_{dc}; (**b**) A-phase line current on three-phase side, i_{sa}; (**c**) single-phase AC output voltage, u_{ab}; (**d**) single-phase AC current, is

Figure 6.42 displays the voltage and current waveforms as the system transitions from a no-load condition to transferring 20% capacitive reactive power and then to transferring 40% capacitive reactive power. Similar to the previously discussed inductive reactive power condition, the transfer of capacitive reactive power has little effect on the three-phase side. Consequently, the current (i_{sa}) on the three-phase side remains at zero, and the DC voltage (V_{dc}) remains stable. In contrast, the single-phase current varies in amplitude in response to the sudden change in capacitive reactive power. As shown in Fig. 6.42c, the number of levels at the single-phase AC port decreases after t = 1.2 s due to a reduction in the modulation degree.

Figure 6.43 displays the output of the three-phase regulator, which remains stable when the capacitive reactive power suddenly changes from 20% to 40%. The output of the single-phase converter reaches a new stable value after a brief adjustment period. In comparison to the waveform of the single-phase converter when outputting inductive reactive power, as shown in Figs. 6.41 and 6.43 indicates that the direction of change when the regulator experiences a sudden adjustment is exactly the opposite. This indicates that the system's reactive power is reversed at this time, resulting in the output of capacitive reactive power.

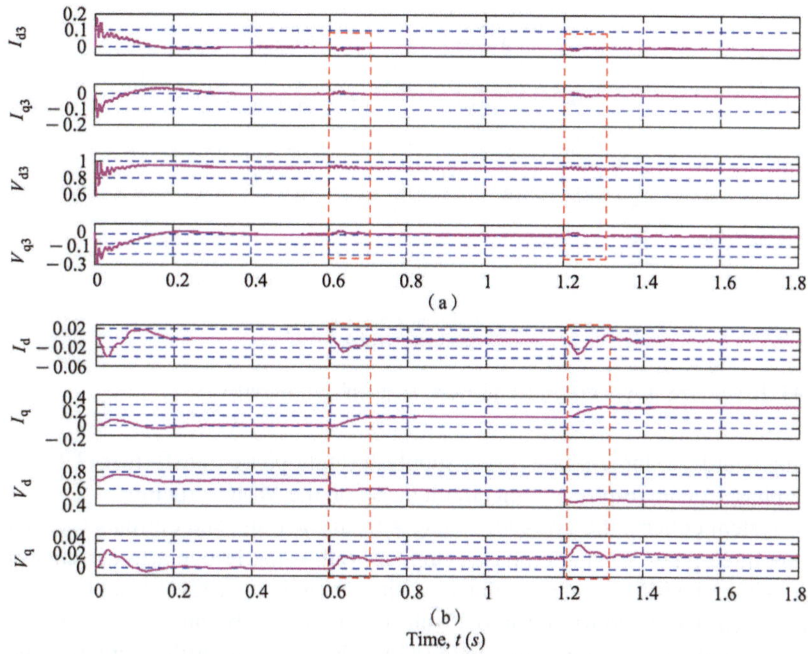

(a) Waveforms of Three-phase Id3/Iq3 and *Vd3/Vq3;* (b) Waveforms of Single-phase Id/Iq and *Vd/Vq*

Fig. 6.43 Waveforms of three-phase and single-phase regulator parameters during sudden capacitive reactive power change of system. (**a**) Waveforms of three-phase I_{d3}/I_{q3} and V_{d3}/V_{q3}; (**b**) waveforms of single-phase I_d/I_q and V_d/V_q

6.5 Experimental Verification

Based on the above simulation analysis, this section presents low-power experimental verification and analysis of three-phase-to-single-phase MMCs. It includes both dynamic and static experimental results, as well as an analysis of interconnection control in the contexts of three-phase rectifier, single-phase rectifier, and three-phase-to-single-phase converter systems.

6.5.1 Experimental Platform

A three-phase-to-single-phase low-power experimental platform featuring a four-module MMC architecture per phase was designed, as illustrated in Fig. 6.44. The system primarily consists of a three-phase AC power supply, a three-phase voltage regulator (T1), contactors (K1–K4), a charging resistor (R), FPGA control boards for both three-phase and single-phase operation, three-phase and single-phase I/O

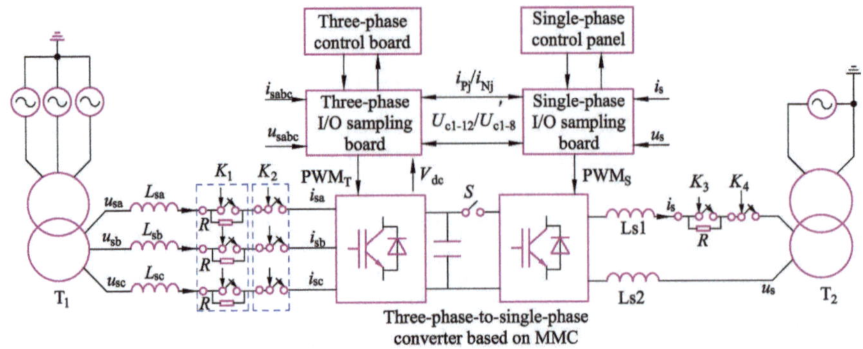

Fig. 6.44 Experimental system of three-phase-to-single-phase converter

sampling boards, a single-phase power supply, a single-phase transformer (T2), and a three-phase-to-single-phase main circuit based on the MMC structure.

The system collects the grid-side voltages (usabc and us) and currents (i_{sabc} and i_s) of the three-phase-to-single-phase setup, as well as the bridge arm currents (i_{Pj} and i_{Nj}) within the MMC and the voltages of the submodules (UC1-12 and U'C1-8). The collected data is converted into 12-bit digital signals by an analog-to-digital conversion chip and then input into the FPGA. All control algorithms are executed in the FPGA, leading to the generation of PWM signals, which are ultimately transmitted to each IGBT via optical fiber to control their turn-on and turn-off.

To prevent device damage caused by excessive initial current, the submodule capacitor is pre-charged upon the startup of the three-phase and single-phase rectifiers to ensure that the circuit operates normally. Among the four contactors in the system (K1, K2, K3, and K4), K1 functions as the short-circuit charging resistor after the three-phase rectifier is started, while K2 serves as the three-phase grid connection switch. The single-phase side is similarly configured, with K3 acting as the short-circuit charging resistor after the single-phase rectifier is started, and K4 functioning as the single-phase grid connection switch.

The photo of the low-power experimental platform is presented in Fig. 6.45, with its parameters detailed in Table 6.3.

6.5.2 Experiment of Three-Phase PWM Rectifier

Three-phase PWM converters used in continuous power supply systems typically operate in a rectification state. In this experiment, switch S, which connects the three-phase and single-phase sides as shown in Figure 6.45, is disconnected, allowing only the three-phase side of the system to operate normally.

1. Steady-state and dynamic analysis

   Figure 6.46 displays the steady-state waveforms of the DC voltage (V_{dc}), A-phase line voltage and phase current ($u_{\text{sa}}/i_{\text{sa}}$), and port voltage ($u_a$) when the

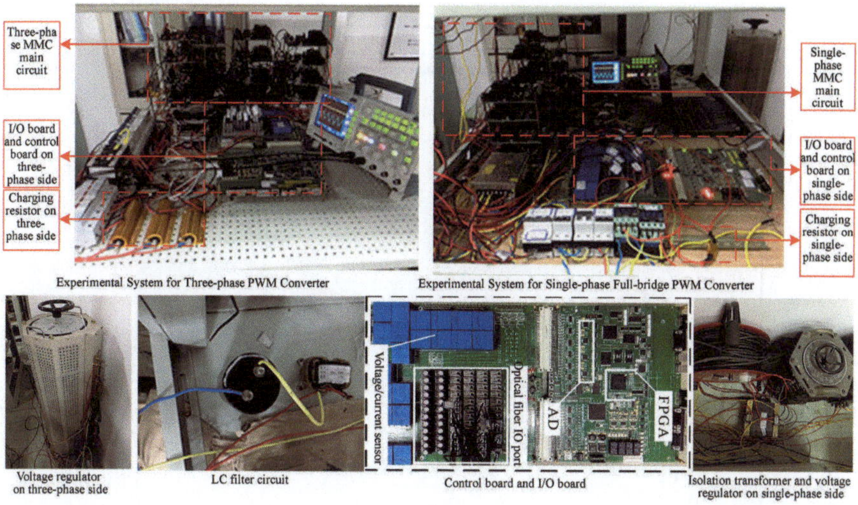

Fig. 6.45 Low-power experimental platform for three-phase-to-single-phase converter

Table 6.3 Parameters of experimental platform for three-phase-to-single-phase PWM converter

Parameter	Parameter value
Three-phase power line voltage	65 V
Line voltage on primary side/line voltage on secondary side of three-phase isolation transformer	65 V/65 V
RMS of single-phase power supply	60 V
Primary-side voltage/secondary-side voltage of single-phase isolation transformer	60 V/60 V
DC-side bus voltage reference	120 V
Three-phase input inductance	5 mH
Single-phase output inductance	15 mH
Bridge arm inductance	0.5 mH
Three-phase DC-side capacitance	470 μF
Single-phase DC-side capacitance	2.2 mF
Secondary filter circuit capacitor/inductor	2.2 mF/1 mH
Carrier frequency of converter	3 kHz
IGBT model/drive	IHW20N120R/K841L
FPGA model/clock frequency	EP3C55F484C8/32 MHz

three-phase PWM rectifier operates under both no-load and on-load conditions. As shown, regardless of the load state, the DC-side voltage (V_{dc}) stabilizes at the reference value of 120 V, and the AC port voltage (u_a) exhibits five levels. In the unloaded state, the current is recorded at 0 A, as illustrated in Fig. 6.46a; however, when a 100 Ω load is connected, the AC current becomes sinusoidal, as shown in Fig. 6.46b. The third channel in this figure captures the AB-phase line voltage, which has a 30° phase difference from the A-phase current.

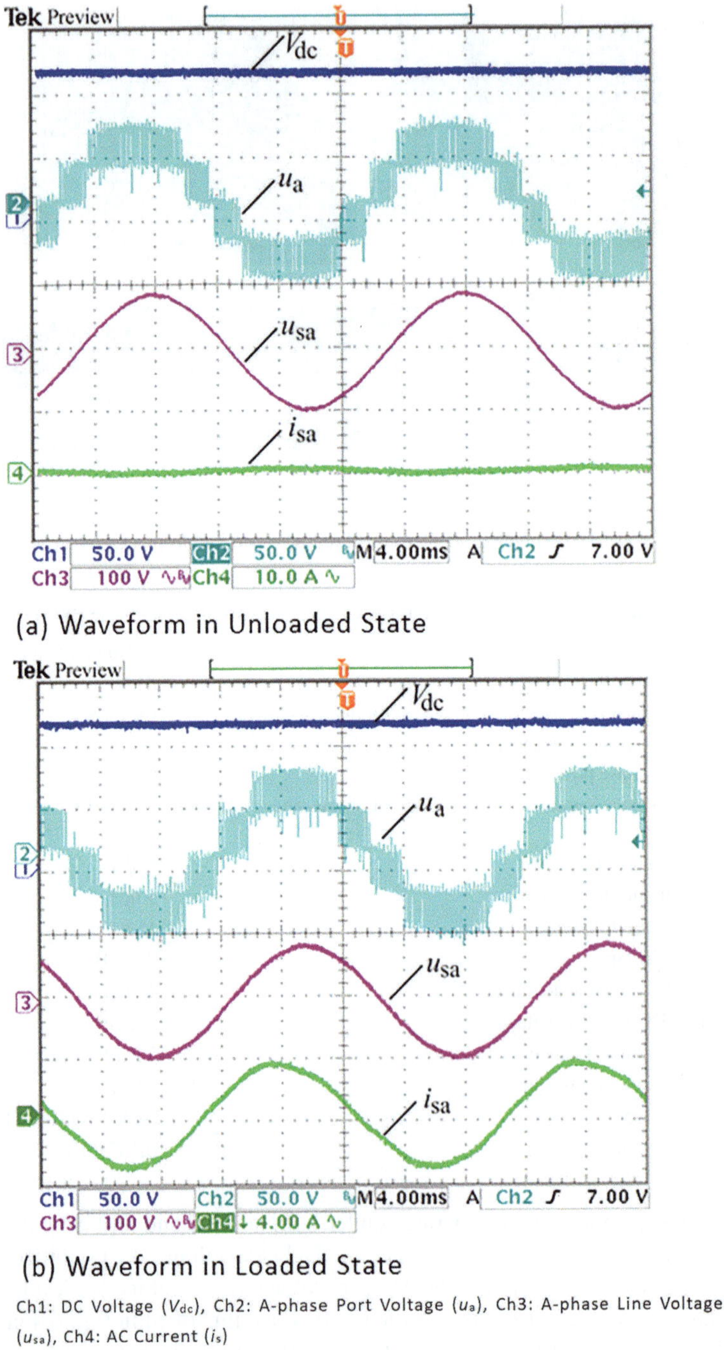

Fig. 6.46 Steady-state waveforms of three-phase rectifier. (**a**) Waveform in unloaded state. (**b**) Waveform in loaded stateCh1: DC voltage (V_{dc}), Ch2: A-phase port voltage (u_a), Ch3: A-phase line voltage (u_{sa}), Ch4: AC current (i_s)

Figure 6.47 displays the AC port voltage (u_{abc}) and AC current (i_{sabc}) when the three-phase rectifier is under load. All three-phase port AC output voltages exhibit five levels, as shown in Fig. 6.47a. The three-phase AC currents are sinusoidal and have a phase difference of 120° from each other during the transfer of active power, as illustrated in Fig. 6.47b.

Figure 6.48a displays the waveforms of the three-phase rectifier transitioning from a no-load state to an on-load state, while Fig. 6.48b shows the waveform after the load is removed. Regardless of when the load is connected or disconnected, the DC voltage (V_{dc}) quickly stabilizes at 120 V after approximately 400 ms of adjustment. The AC port voltage (u_a) also returns to a steady state following a brief adjustment period, and the AC current reaches a new stable value after adjustment.

2. Experiment and Analysis of MMC Module

   Figure 6.49 displays the waveforms of the A-phase bridge arm port voltage under no-load and on-load conditions. Regardless of the load state, both the upper arm port voltage (u_{Pa}) and the lower arm port voltage (u_{Na}) exhibit a three-level output pattern, while the A-phase port outputs a five-level voltage. This behavior aligns with the relationship of MMC port voltages discussed in Sect. 6.1.

   The current waveform of the A-phase bridge arm is shown in Fig. 6.51a. Due to the presence of circulating current in the MMC circuits, the upper and lower bridge arm currents are not perfectly sinusoidal; however, the sum of the three-phase circulating currents is zero. In this state, the output current (i_{sa}) remains sinusoidal, and the AC current exhibits a consistent trend with the lower bridge arm current (i_{Na}), aligning with the simulation verification waveforms shown in Fig. 6.9 in Sect. 6.1.

Ch1: A-phase port voltage (u_a), Ch6–Ch4: three-phase current (i_{sabc})

Figure 6.50b displays the capacitor voltage waveforms of the two submodules in the A-phase upper bridge arm, while Fig. 6.51a shows the capacitor voltage waveforms of the two submodules in the lower bridge arm. Figure 6.51b illustrates the capacitor voltages of the first and third submodules. As presented in these three figures, the capacitor voltages (Uc1–Uc4) of the four submodules are each half of the total DC-side voltage (V_{dc}), stabilizing at approximately 60 V. Although the total DC-side voltage remains stable, the capacitor voltage of each submodule fluctuates, which aligns with the analysis in Sect. 6.1. Figures 6.50b and 6.51a show that the capacitor voltages of the two submodules in the upper bridge arm and those in the lower bridge arm fluctuate in tandem. However, Fig. 6.51b indicates that the capacitor voltages of the submodules in the upper bridge arm and those in the lower bridge arm fluctuate in opposition.

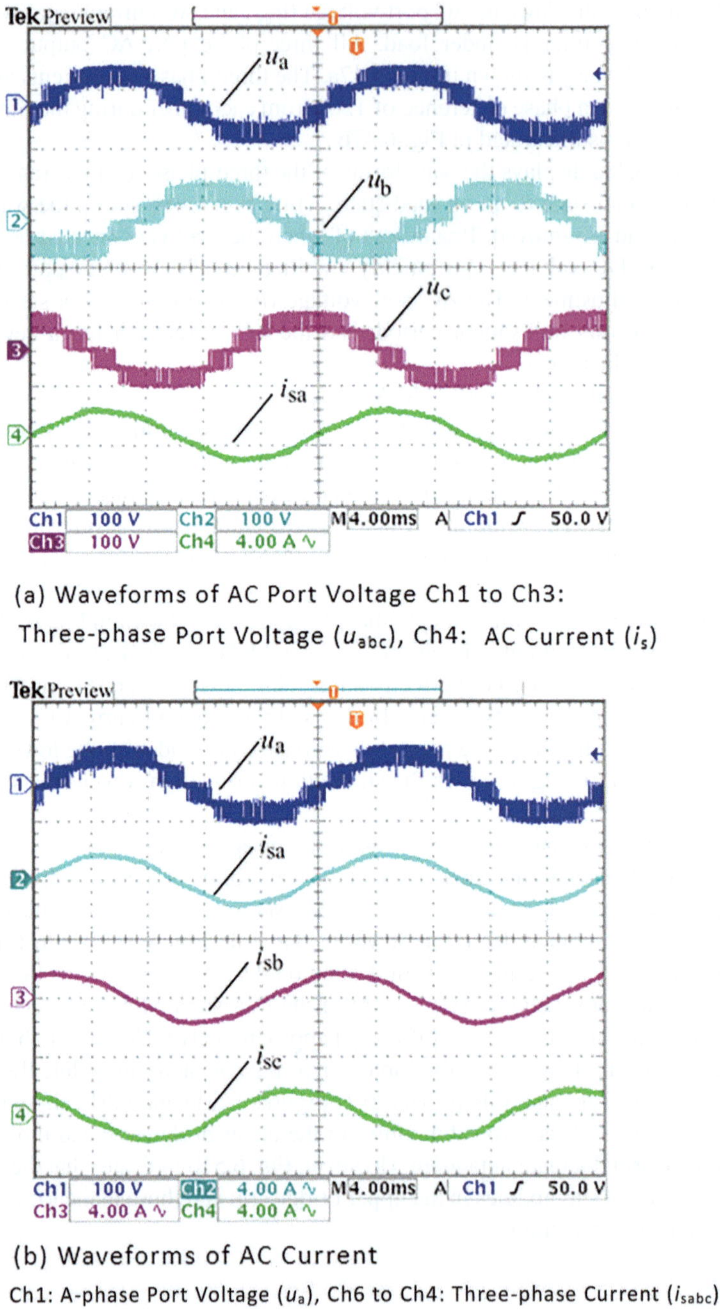

Fig. 6.47 Waveforms of three-phase AC voltage and current. (**a**) Waveforms of AC port voltage Ch1–Ch3: three-phase port voltage (u_{abc}), Ch4: AC current (i_s). (**b**) Waveforms of AC current

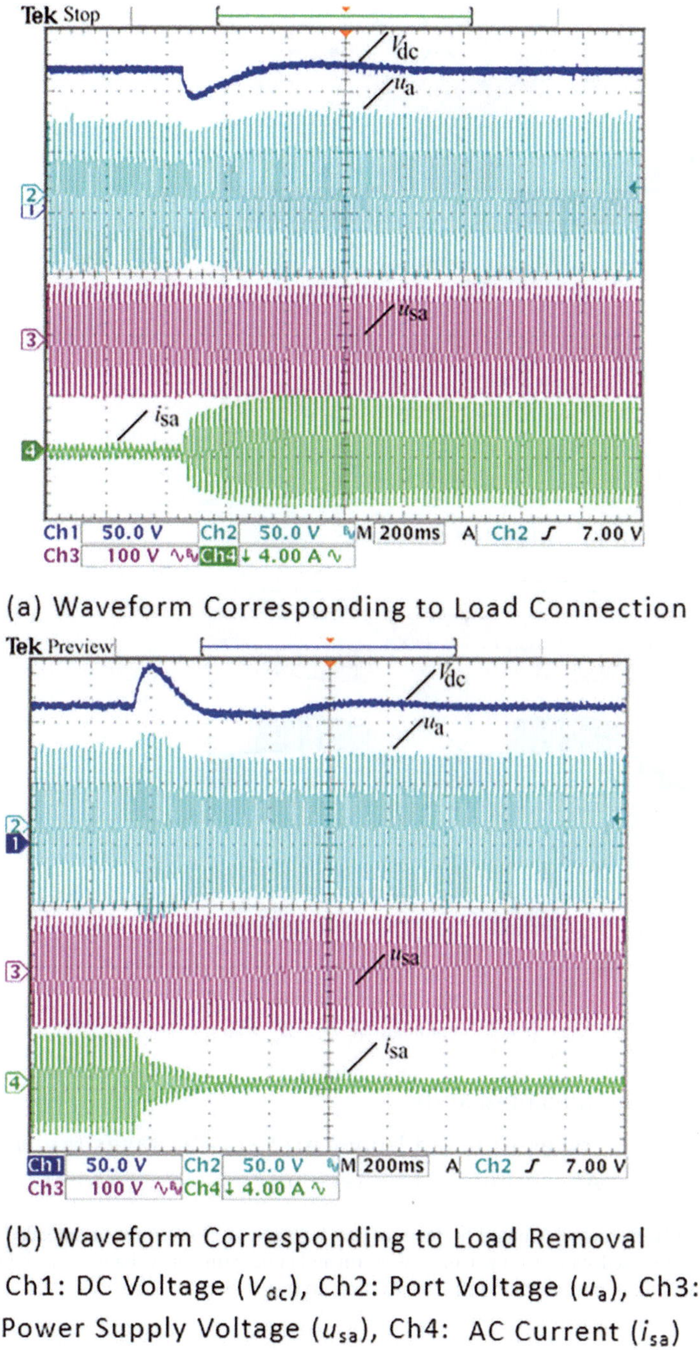

Fig. 6.48 Dynamic waveforms of three-phase rectifier. (**a**) Waveform corresponding to load connection. (**b**) Waveform corresponding to load removalCh1: DC voltage (V_{dc}), Ch2: port voltage (u_a), Ch3: power supply voltage (u_{sa}), Ch4: AC current (i_{sa})

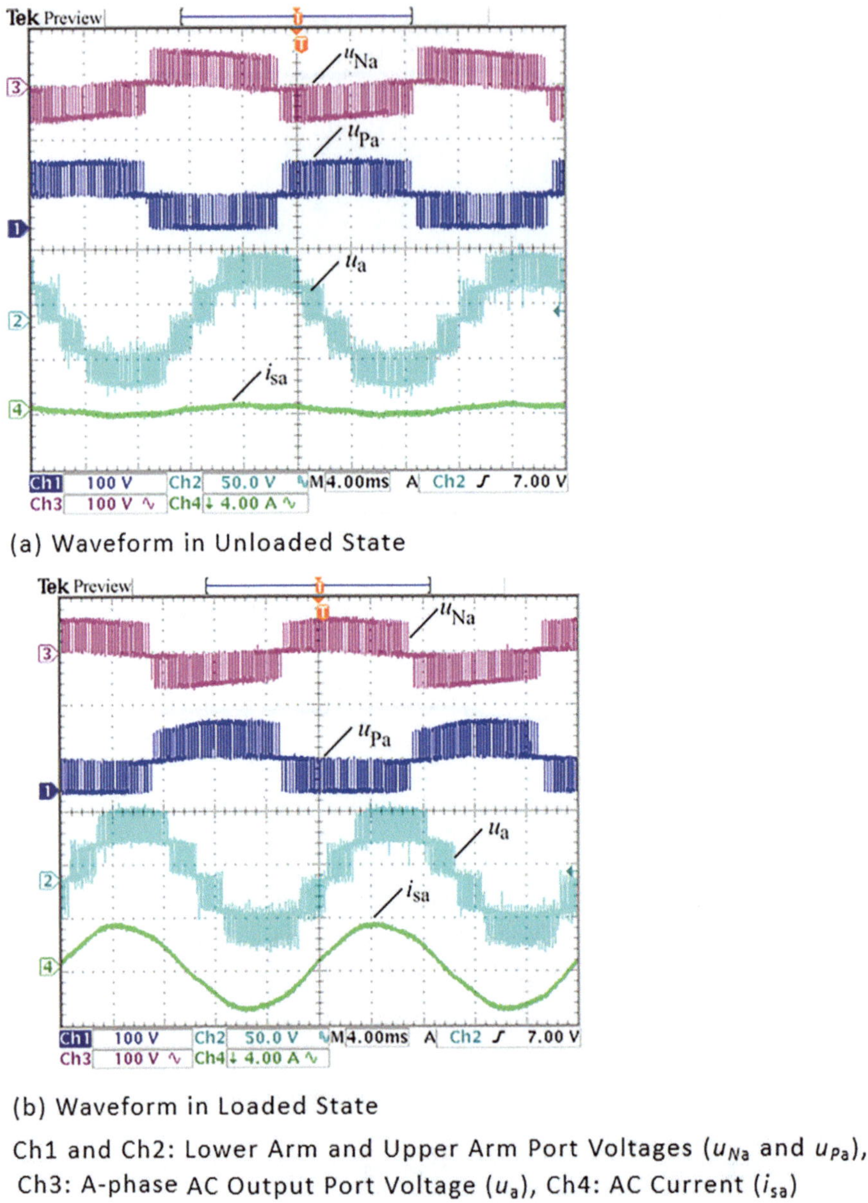

Fig. 6.49 Waveforms of A-phase bridge arm port voltages. (**a**) Waveform in unloaded state. (**b**) Waveform in loaded stateCh1 and Ch2: lower arm and upper arm port voltages (u_{Na} and u_{Pa}), Ch3: A-phase AC output port voltage (u_a), Ch4: AC current (i_{sa})

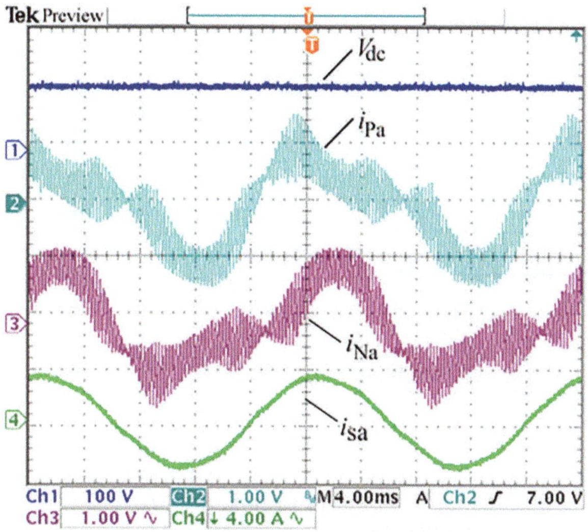

(a) Current Waveform of Bridge Arms

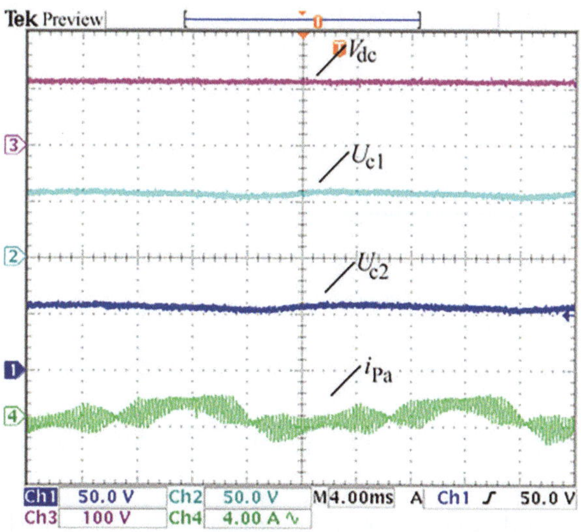

(b) Voltage Waveforms of Upper Bridge Arm Capacitors, U_{c1} and U_{c2}

Fig. 6.50 Waveforms of bridge arm currents and submodule capacitor voltages. (**a**) Current waveform of bridge arms. (**b**) Voltage waveforms of upper bridge arm capacitors, Uc1 and Uc2

6.5.3 Experiment of Single-Phase PWM Rectifier

Figure 6.52 displays the DC voltage (V_{dc2}) and AC current (i_s) on the DC side of the single-phase rectifier before and after filtering. Figure 6.52a shows the waveform when the LC filter circuit is not enabled. In this case, the voltage (V_{dc2}) on the DC bus side exhibits significant AC pulsations, and the AC current is markedly

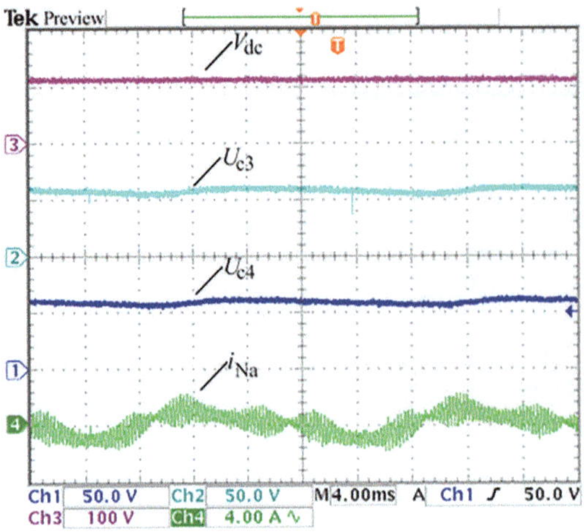

(a) Capacitor Voltages of Lower Bridge Arm Submodules, U_{c3}, U_{c4}

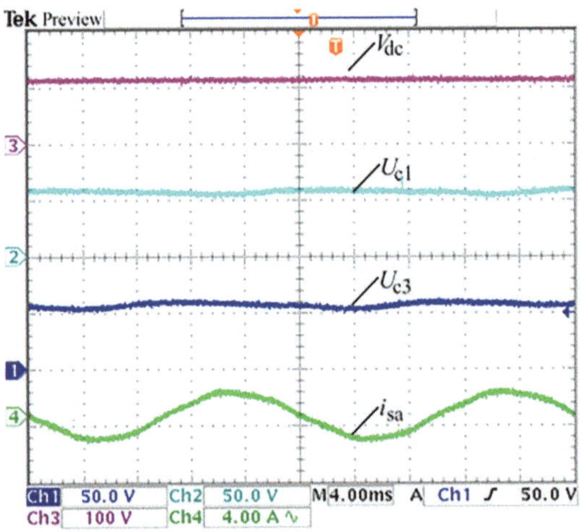

(b) Capacitor Voltages of Upper and Lower Bridge Arm Submodules, U_{c1}, U_{c3}

Fig. 6.51 Waveforms of A-phase submodule capacitor voltages. (**a**) Capacitor voltages of lower bridge arm submodules, Uc3 and Uc4. (**b**) Capacitor voltages of upper and lower bridge arm submodules, Uc1 and Uc3

distorted. In contrast, Fig. 6.52b shows the waveform after the LC filter circuit is enabled. The secondary pulsations in the DC-side voltage are nearly eliminated, resulting in a noticeably improved waveform for the AC current.

Figure 6.53 displays the steady-state waveforms of the DC voltage (V_{dc2}), AC voltage (u_s), AC current (i_s), and port voltage (u_{ab}) of the single-phase PWM rectifier under both no-load and on-load conditions. Regardless of the load state, the DC-side voltage (V_{dc}) stabilizes at 120 V, and the AC port voltage (u_{ab}) exhibits a low modulation degree. Due to the presence of circulating current (i_{zj}), fluctuations persist in the AC circuit under no-load conditions to some extent. The experimental results

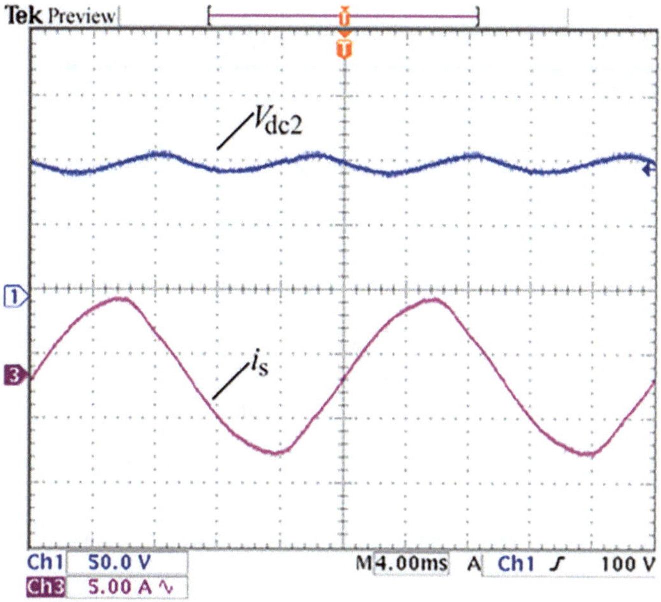

(a) Before Filtering

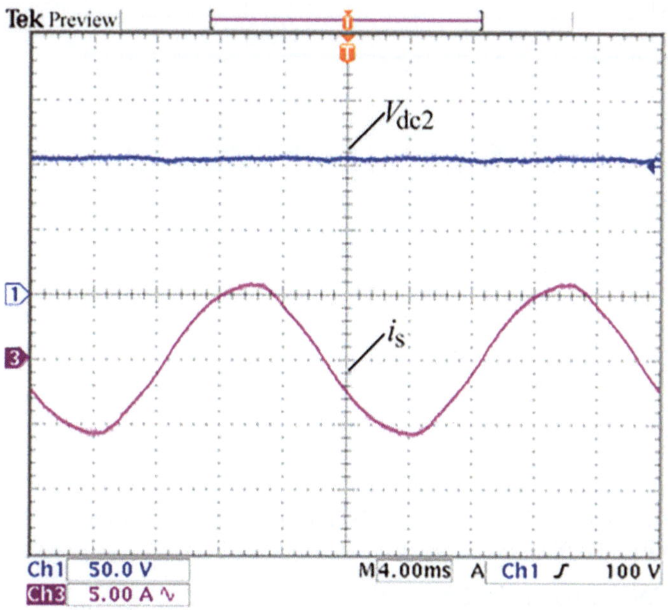

(b) After Filtering

CH1: DC-side Voltage (V_{dc2}), CH2: AC Current (i_s)

Fig. 6.52 Waveforms of DC-side voltage and AC current (i_s) of single-phase rectifier. (**a**) Before filtering. (**b**) After filteringCH1: DC-side voltage (V_{dc2}), CH2: AC current (i_s)

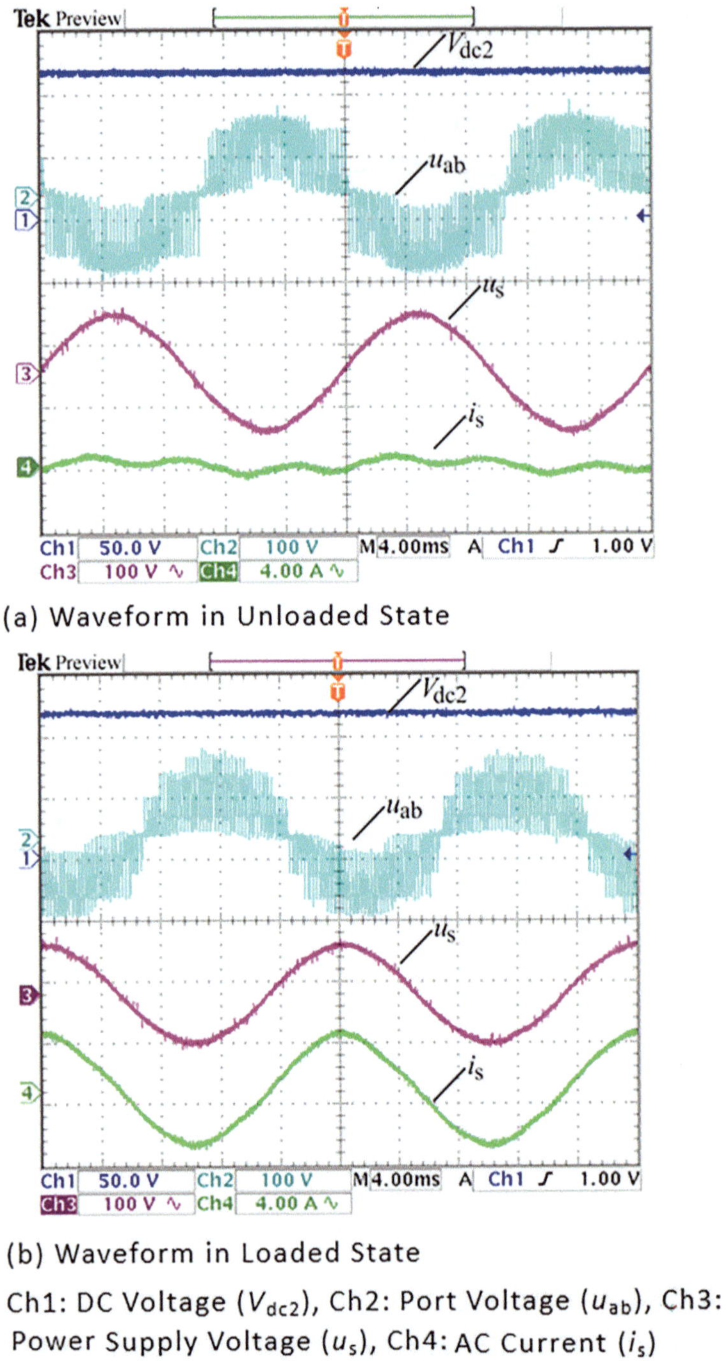

Fig. 6.53 Waveforms of single-phase rectifier transferring active power. (**a**) Waveform in unloaded state. (**b**) Waveform in loaded stateCh1: DC voltage (V_{dc2}), Ch2: port voltage (u_{ab}), Ch3: power supply voltage (u_s), Ch4: AC current (i_s)

following the connection of a 100 Ω load are shown in Fig. 6.53b. The AC current significantly increases in amplitude and maintains a sinusoidal waveform, with the AC voltage and current aligned in phase, resulting in a high power factor.

Figure 6.54 displays the steady-state waveforms of the single-phase PWM rectifier when transferring inductive and capacitive reactive power. As shown in Fig. 6.54a, when the power supply outputs inductive reactive power, the voltage leads the current by 90°, and the DC-side voltage (V_{dc2}) stabilizes at 120 V. Conversely, when capacitive reactive power is output, the current leads the voltage by 90°, as illustrated in Fig. 6.54b.

Figure 6.55a presents the experimental results when the system is switched from the no-load state to the on-load state with a resistance of 100 Ω. Figure 6.55b displays the experimental results when the load resistance of 100 Ω is switched back to the no-load state. Regardless of the load switching state, the converter is able to adjust to reach a new stable state.

6.5.4 Experiment of Three-Phase-to-Single-Phase System

The joint experiment for the three-phase-to-single-phase converter was conducted with switch S turned on to connect the three-phase and single-phase sides as shown in Fig. 6.44. The following analysis covers the steady-state performance of the system under no-load, active power transfer, and reactive power transfer conditions, as well as the dynamic performance in response to sudden load changes.

1. Steady-state analysis

   Figure 6.56 displays the steady-state performance of the three-phase-to-single-phase converter operating under no-load conditions and transferring active power to the single-phase grid. As shown in Fig. 6.56a, the current on the single-phase side is approximately 0 A when the system is unloaded. However, considering losses even in no-load conditions, small-amplitude current components are still present on the three-phase side. Figure 6.56b illustrates that when the three-phase side transfers active power to the single-phase power supply, the three-phase converter absorbs active power from the grid, with the A-phase current aligned in phase with the voltage. Simultaneously, the converter on the single-phase side outputs active power to the load, resulting in the current on the single-phase side also aligned in phase with the voltage.
2. Dynamic performance analysis

   Figure 6.57 displays the waveforms of the voltages and currents on the three-phase and single-phase sides in response to a sudden active power change on the single-phase side. As shown, the intermediate DC voltage consistently stabilizes at the reference value of 120 V. As the three-phase converter absorbs active power from the three-phase grid and transfers it to the load on the single-phase side, the input current of the three-phase converter and the output current of the single-phase converter dynamically change in amplitude in response to the active power of the load.

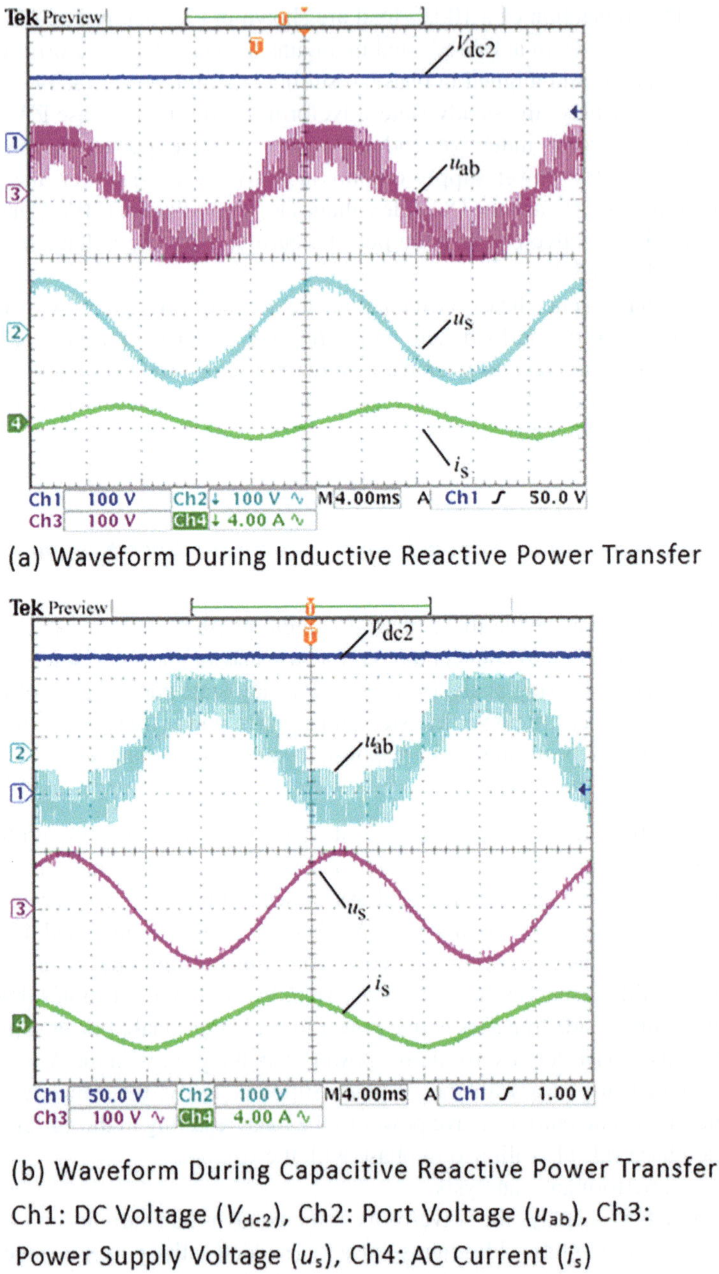

Fig. 6.54 Waveforms of single-phase rectifier transferring reactive power. (**a**) Waveform during inductive reactive power transfer. (**b**) Waveform during capacitive reactive power transferCh1: DC voltage (V_{dc2}), Ch2: port voltage (u_{ab}), Ch3: power supply voltage (u_s), Ch4: AC current (i_s)

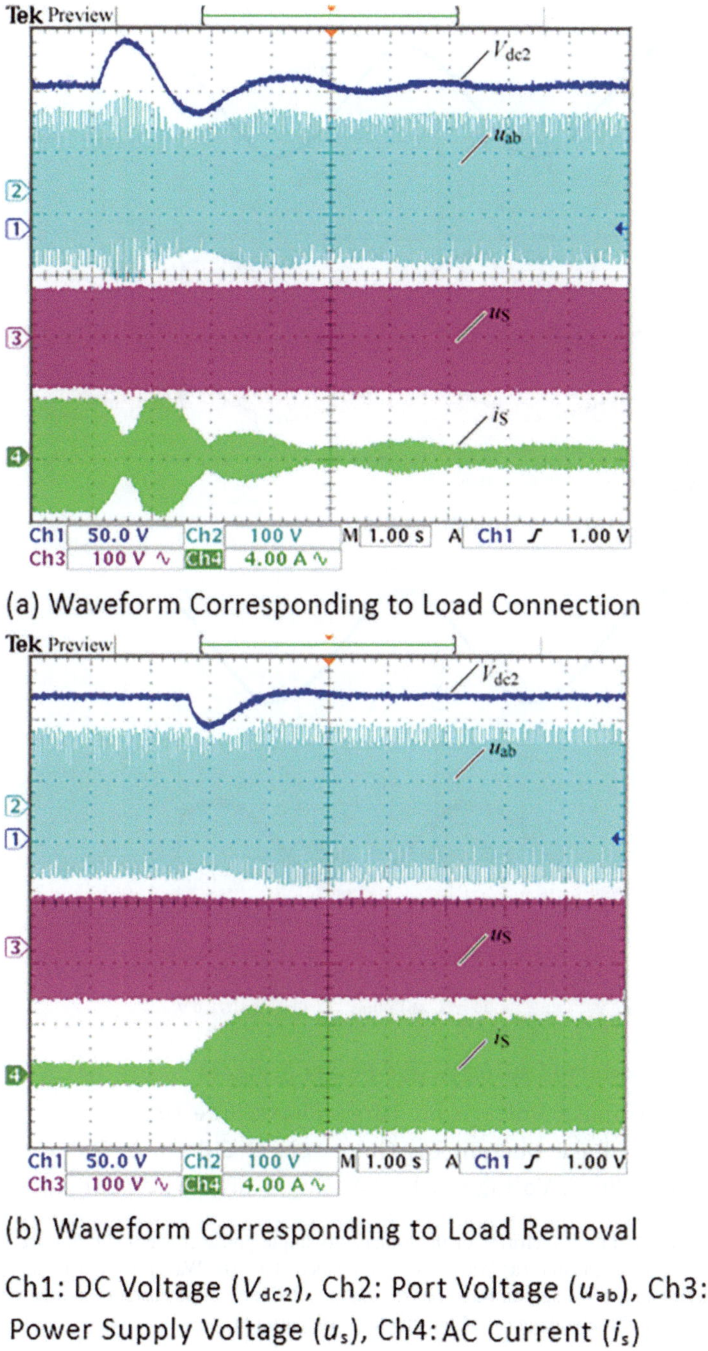

Fig. 6.55 Dynamic waveforms of single-phase rectifier. (**a**) Waveform corresponding to load connection. (**b**) Waveform corresponding to load removalCh1: DC voltage (V_{dc2}), Ch2: port voltage (u_{ab}), Ch3: power supply voltage (u_s), Ch4: AC current (i_s)

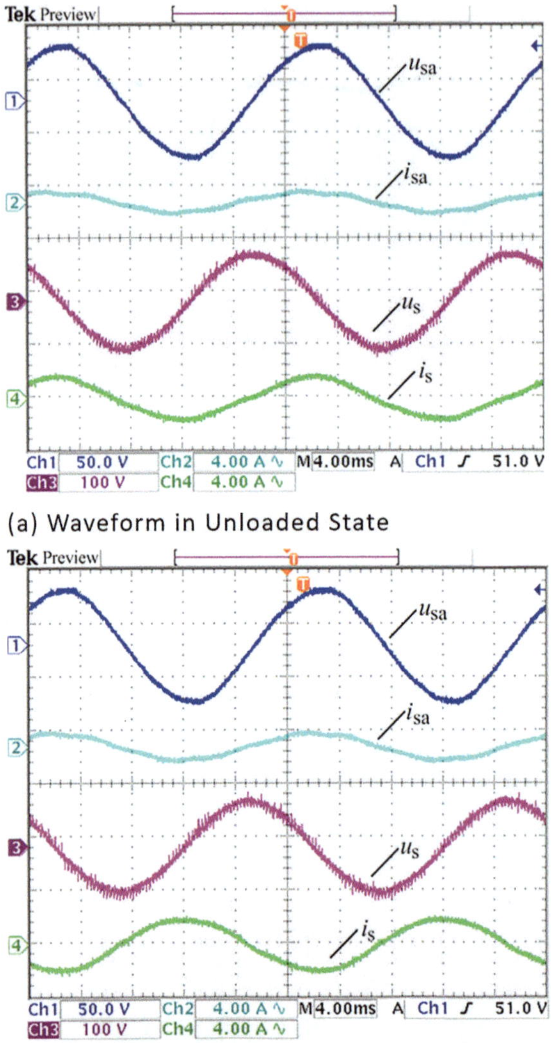

(a) Waveform in Unloaded State

(b) Waveform During Active Power Transfer

Ch1: A-phase Voltage on Three-phase Side (u_{sa}),
Ch2: A-phase Current on Three- phase Side (i_{sa}), Ch3:
Single-phase Voltage (u_s), Ch4: Single-phase Current (i_s)

Fig. 6.56 Steady-state waveforms of three-phase-to-single-phase converter in no-load state and during active power transfer. (**a**) Waveform in unloaded state. (**b**) Waveform during active power transferCh1: A-phase voltage on three-phase side (u_{sa}), Ch2: A-phase current on three-phase side (i_{sa}), Ch3: single-phase voltage (u_s), Ch4: single-phase current (i_s)

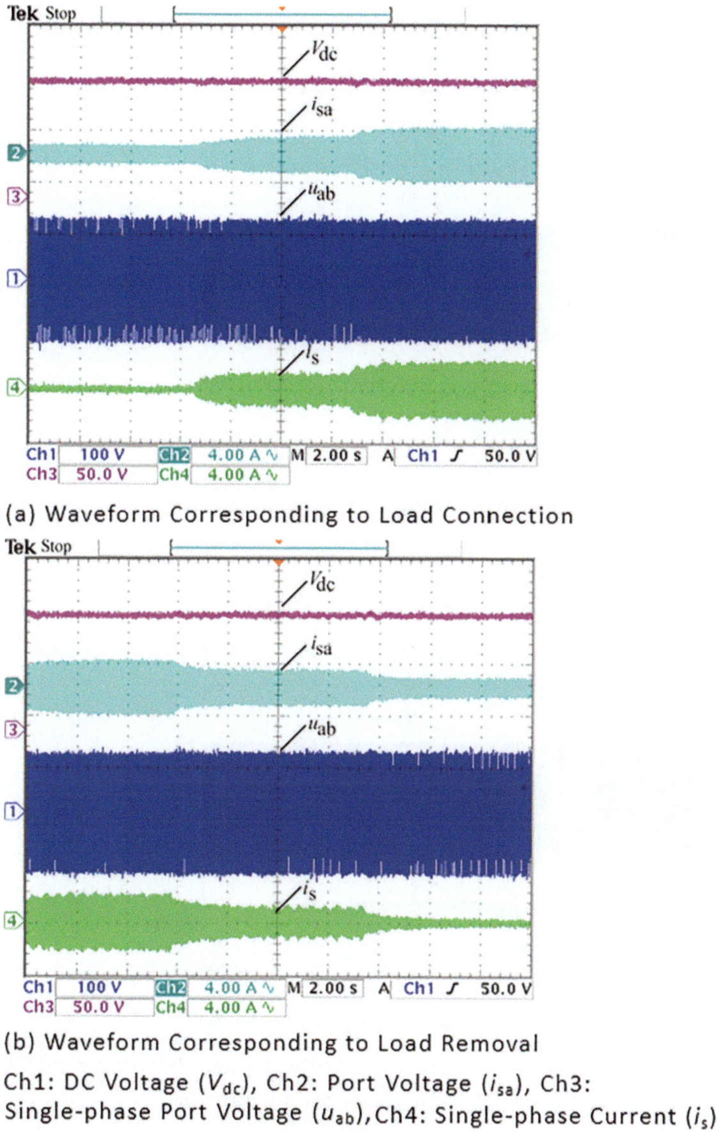

Fig. 6.57 Dynamic waveforms of three-phase-to-single-phase during active power transfer. (**a**) Waveform corresponding to load connection. (**b**) Waveform corresponding to load removalCh1: DC voltage (V_{dc}), Ch2: port voltage (i_{sa}), Ch3: single-phase port voltage (u_{ab}), Ch4: single-phase current (i_s)

Figure 6.58 displays the voltages and currents on the three-phase and single-phase sides in response to a sudden change in capacitive reactive power on the single-phase side. The intermediate DC-side voltage remains stable at the reference value of 120 V. Since the three-phase converter only needs to absorb a small amount of active power to maintain the stability of the DC voltage in this state, the

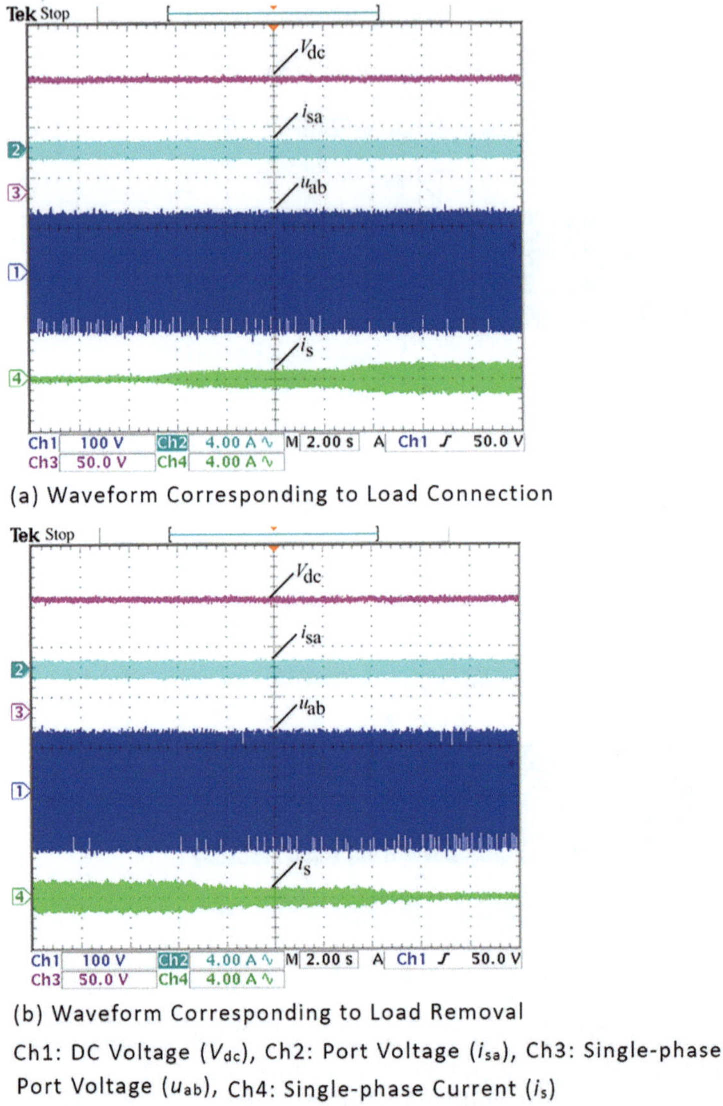

Fig. 6.58 Dynamic waveforms of three-phase-to-single-phase during capacitive reactive power transfer. (**a**) Waveform corresponding to load connection. (**b**) Waveform corresponding to load removalCh1: DC voltage (V_{dc}), Ch2: port voltage (i_{sa}), Ch3: single-phase port voltage (u_{ab}), Ch4: single-phase current (i_s)

active power only compensates for active losses such as switching losses, resulting in relatively low current amplitudes. On the single-phase side, the converter outputs reactive power, so the amplitude of the reactive current changes in accordance with the reactive power demand, demonstrating an effective dynamic response.

Figure 6.59 displays the dynamic waveforms of the voltages and currents on the three-phase and single-phase sides in response to a sudden change in the

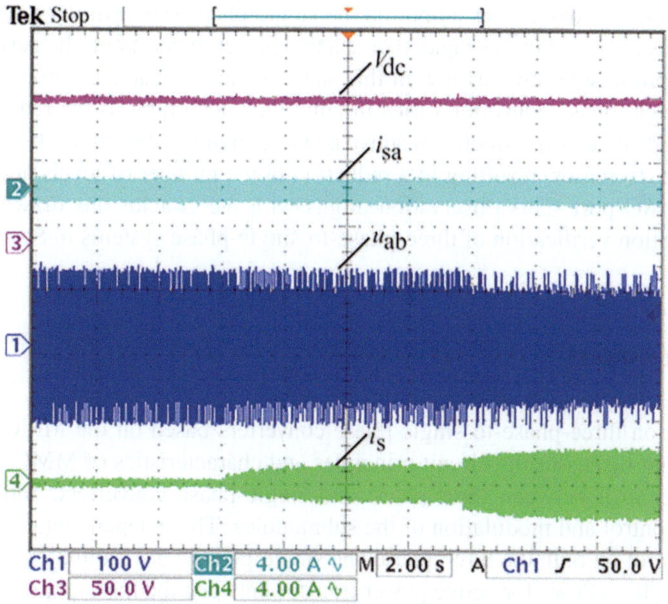

(a) Waveform Corresponding to Load Connection

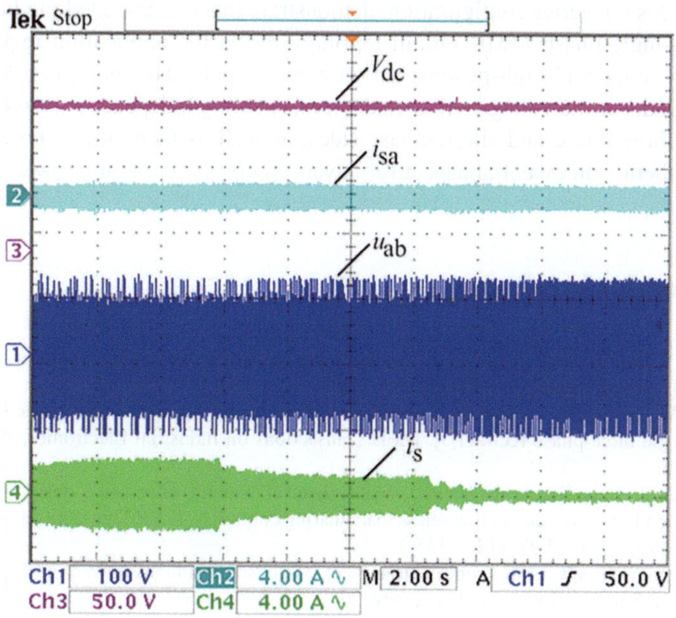

(b) Waveform Corresponding to Load Removal

Fig. 6.59 Dynamic waveforms of three-phase-to-single-phase during inductive reactive power transfer. (**a**) Waveform corresponding to load connection. (**b**) Waveform corresponding to load removalCh1: DC voltage (V_{dc}), Ch2: port voltage (i_{sa}), Ch3: single-phase port voltage (u_{ab}), Ch4: single-phase current (i_s)

inductive reactive power output on the single-phase side. Similar to the characteristics observed in the capacitive reactive power experiment, the current on the three-phase side remains low in this state, while the reactive current output by the single-phase converter varies in amplitude according to the reactive power demand. When the inductive reactive power output is substantial, the modulation degree decreases, resulting in a reduction in the number of levels at the single-phase AC port. This observation aligns with the conclusions drawn from the simulation verification of three-phase-to-single-phase systems in Sect. 6.3.

6.6 Summary

Focusing on three-phase-to-single-phase converters based on the MMC structure, this chapter discusses the circuit principles and characteristics of MMCs, interconnection control between three-phase and single-phase converters, and capacitor voltage control and modulation of the submodules. The independent power decoupling control on both the three-phase and single-phase sides facilitates connections on both sides, allows for active power transfer between the three-phase and single-phase systems, and enables reactive power output to the single-phase side.

The research results, which encompass the principles, simulations, and experiments of this converter configuration, demonstrate that the cascaded modular three-phase-to-single-phase MMC circuit topology meets the application requirements for high-voltage and high-power traction power supply. The incorporation of submodules reduces the voltage stress on devices, enabling independent connections on both the three-phase and single-phase sides, as well as facilitating power transfer between them—an essential aspect for advanced co-phase traction power supply.

References

Ambrozic V, Fiser R, Nedeljkovic D. Direct current control—a new current regulation principle[J]. IEEE Transactions on Power Electronics, 2003, 18(1): 495-503.

Dixon J W, Boon-Teck O. Indirect current control of a unity power factor sinusoidal current boost type three-phase rectifier[J]. IEEE Transactions on Industrial Electronics, 1988, 35(4): 508-515.

Guerrero J M, Luis Garcia de Vicuno, Matas J, et al. Output impedance design of parallel-connected UPS inverters with wireless load-sharing control[J]. IEEE Transactions on Industrial Electronics, 2005, 52(4): 1126.1135.

Hu J. Research on power flow control strategy of co-phase continuous power supply system[D]. Chengdu: Southwest Jiaotong University, 2013.

Liu F. Multilevel inverter technology and its application[M]. Beijing: China Machine Press, 2007.

Liu Z. Research on space vector modulation technology of multilevel inverter[D]. Changsha: Hunan University, 2008.

Mao Y, Xiong X, Chen X, et al. Research on improved dual-loop control strategy of single-phase inverter[J]. Electrical Measurement & Instrumentation, 2014(08): 69-74.

Martins A P, Carvalho A S, Araujo A S. Design and implementation of a current controller for the parallel operation of standard UPSs[M]. In: Industrial Electronics, Control, and Instrumentation, Proceedings of the 1995 IEEE IECON 21st International Conference on, vol. 1, 1995, pp. 584–589.

Shu Z, Xie S, Li Q. Single-phase back-to-back converter for active power balancing, reactive power compensation, and harmonic filtering in traction power system[J]. IEEE Transactions on Power Electronics, 2011, 26(2): 334-343.

Shu Z, He X, Wang Z, et al. Voltage balancing approaches for diode-clamped multilevel converters using auxiliary capacitor-based circuits[J]. IEEE Transactions on Power Electronics, 2013, 28(5): 2111-2124.

Wang X. Research on voltage balancing, clamping strategy and modulation technology of three-phase multilevel converters[D]. Hangzhou: Zhejiang University, 2007.

Xie M. Output control and parallel operation control of single-phase 400 Hz medium-frequency voltage source inverter[D]. Beijing: Graduate University of Chinese Academy of Sciences, 2006.

Xing L. Research on modulation strategy of NPC/H-bridge five-level inverter[D]. Xuzhou: China University of Mining and Technology, 2016.

Yadong L, Yan X, Lipei H, et al. A novel distributed control scheme for the parallel operation of digital controlled UPS[M]. In: Industrial Electronics Society, 2003 IECON '03 The 29th Annual Conference of the IEEE, 2003, vol. 1, pp. 668-672.

Zheng H. Simulation analysis and parameter study of traction auxiliary system for 350 km/h EMU[D]. Beijing: Beijing Jiaotong University, 2009.

Zhou W, Wu Z, Tang J, et al. Equivalent algorithm of SVPWM and essential relationship between SVPWM and SPWM. Proceedings of the Chinese Society for Electrical Engineering, 2006, 26(2): 133-137.

Chapter 7
Analysis and Control of Two-Phase-to-Single-Phase Converter

7.1 Two-Phase-to-Single-Phase Converter

7.1.1 Structure of Substations for Advanced Co-phase Traction Power Supply

Oettmeier et al. (2009) and Zhang (2012a) eliminated the phase break at the substation outlet in co-phase power supply transformations of existing traction substations. While this resolved power quality issues within individual substations, the phase break at the section post between substations remains, preventing mutual energy support and thus hindering the upgrade to multi-substation advanced co-phase traction power supply systems. Therefore, an improved continuous power supply system is proposed, with the structure of a single traction substation for continuous power supply shown in Fig. 7.1. A single co-phase power supply traction substation primarily consists of a traditional traction transformer, a two-phase-to-single-phase PWM converter, and a step-up transformer. The traction transformer receives electricity from the three-phase grid and converts it into two-phase voltage. Two single-phase rectifiers are connected to the two-phase input terminals, sharing a DC bus. The rectifiers then output AC power through the single-phase inverter, which is transmitted to the traction network via the step-up transformer. The two-phase-to-single-phase PWM converter operates in four quadrants, allowing bidirectional active power flow between the three-phase grid and the single-phase traction network. The reactive energy from the traction load will not be transmitted to the three-phase grid which operates at unity power factor. This significantly addresses the power quality issues caused by locomotive loads in traditional traction power supply systems. The single-phase inverter not only supplies the reactive power demanded by the locomotive load but also provides harmonic control. Its adjustable output voltage amplitude, phase, and frequency enable continuous co-phase power supply along the entire line (Zhang 2012b; Pi et al. 2008). This approach eliminates phase breaks within and between existing traction substations, reduces the

Z. Shu, *Power Electronic Converters in Advanced Co-Phase Traction Power Supply System*, https://doi.org/10.1007/978-981-95-3276-6_7

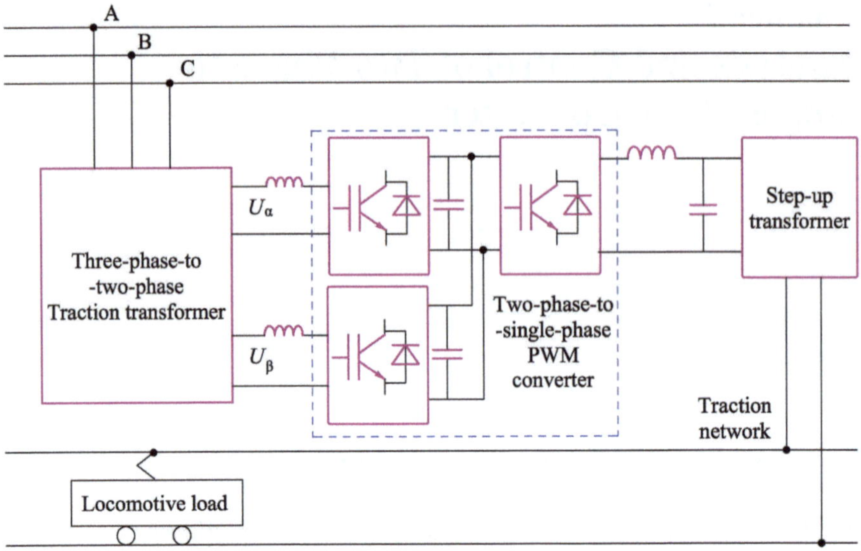

Fig. 7.1 Structure of the traction substation system based on two-phase-to-single-phase converter

requirements for the installed capacity of individual substations, and enhances the operating efficiency of the system.

7.1.2 Two-Phase-to-Single-Phase Converters for Advanced Co-phase Traction Power Supply Systems

In this chapter, all three single-phase converters in the two-phase-to-single-phase PWM converter system utilize a diode-clamped three-level PWM converter structure. Compared to two-level PWM converters, the three-level design increases system voltage and capacity, reduces switching device frequency and stress, and simplifies controller design (Liu 2010). While three-level converters offer lower voltage and capacity than five-level PWM converters, they feature a simpler structure, higher reliability, and more straightforward controller design. Both three-level and five-level PWM converters face DC-side voltage imbalance issues; however, addressing this with five-level converters requires more complex hardware and controller improvements (Wei 2010; Chen et al. 2007). In this chapter, a three-level structure is employed for the single-phase H-bridge PWM converter in the two-phase-to-single-phase PWM converter system.

Figure 7.2 illustrates the topology of the two-phase-to-single-phase PWM converters in the traction substation system. It comprises a DC-side capacitor and three single-phase H-bridge PWM converters. The three H-bridge PWM rectifiers share a common DC bus. Since the PWM rectifier supports four-quadrant operation, the two-phase-to-single-phase PWM converters can function in both traction and

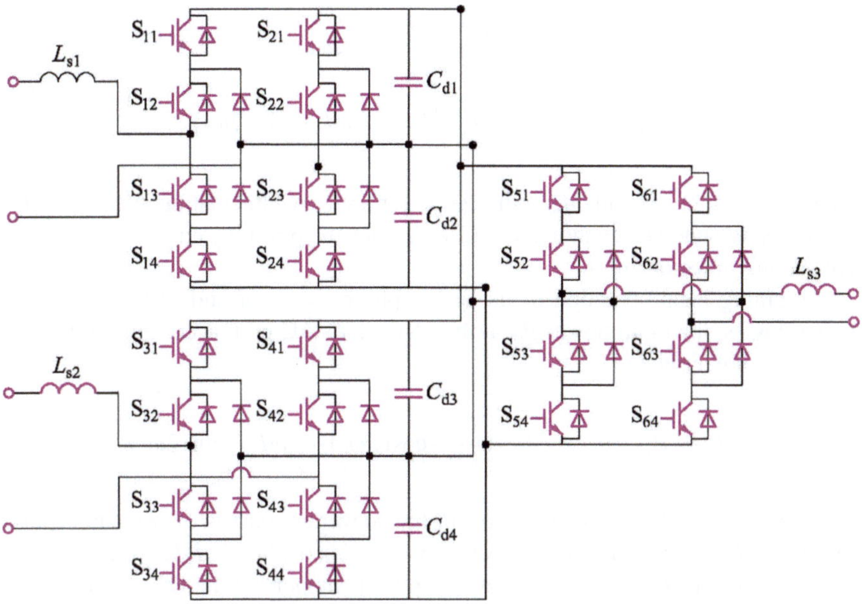

Fig. 7.2 Topology of the diode-clamped three-level two-phase-to-single-phase converter system

inversion states. When the converters function in the traction state, the system supplies active power to the locomotive load while compensating reactive power and harmonics. When the converters function in the inversion state, the system feeds the active power of the locomotive load back to the grid and simultaneously compensates reactive power and harmonics. This structure enables bidirectional energy flow between the three-phase grid and the traction network.

7.2 Analysis of Harmonic Characteristics of Two-Phase Converters

Harmonic Characteristics of Single-Phase Converters

Taking the α-phase converter as an example, the harmonic characteristics of the single-phase converter are analyzed by assuming the voltage and current on its AC side as follows:

$$\begin{cases} u_\alpha = U_s \cos \omega t \\ i_\alpha = I_s \cos \omega t \end{cases} \tag{7.1}$$

where U_s and I_s represent the amplitudes of the input voltage and current, respectively. The single-phase converter in this study supports four-quadrant operation. In the rectification state, the converter input power $P_{s\alpha}$ and the power $P_{L\alpha}$ of the connecting inductor Lα are given by

$$\begin{cases} P_{s\alpha} = u_\alpha i_\alpha = \dfrac{U_s I_s}{2} + \dfrac{U_s I_s}{2}\cos(2\omega t) \\ P_{L\alpha} = \dfrac{dE_{L\alpha}}{dt} = \dfrac{1}{2} L_s \dfrac{di_\alpha^2}{dt} = -\dfrac{1}{2}\omega L_s I_s^2 \sin(2\omega t) \end{cases} \tag{7.2}$$

where $U_sI_s\cos(2\omega t)/2$ and $\omega L_sI^2_s \sin(2\omega t)/2$ represent AC fluctuating quantity, $U_sI_s/2$ is the steady-state DC quantity, and ELα denotes the pulsating energy stored in the connecting inductor Lα.

Assuming ideal operation of the single-phase converter and neglecting device power losses, the input power P_{ab} at the converter's AC port, according to the conservation of energy, is

$$P_{ab} = P_{s\alpha} - P_{L\alpha} = \frac{U_s I_s}{2} + \frac{U_s I_s}{2}\cos(2\omega t) + \frac{1}{2}\omega L_s I_s^2 \sin(2\omega t) \tag{7.3}$$

It can be seen that the input power of the converter consists of two parts: stable DC power $U_sI_s/2$ and fluctuating reactive power $U_sI_s\cos(2\omega t)/2$, $L_sI^2_s \sin(2\omega t)/2$. The stable DC power $U_sI_s/2$ represents active power, supporting the DC-side capacitor voltage and supplying the load. The fluctuating reactive power $U_sI_s\cos(2\omega t)/2$ is a double-frequency power proportional to the input active power, while the fluctuating reactive power $\omega L_sI^2_s \sin(2\omega t)/2$ is a double-frequency power arising from the connecting inductor and is proportional to both the square of the input current I^2_s and the connecting inductor L_s.

The fluctuating reactive power causes a voltage ripple u' dcα on the converter's DC side at twice the power frequency. Therefore, the output DC voltage $u_{dc\alpha}$ can be expressed as

$$u_{dc\alpha} = U_{dc\alpha} + u'_{dc\alpha} \tag{7.4}$$

where $U_{dc\alpha}$ is the DC component of the output voltage.

The transient output power calculation formula of a single-phase converter is

$$P_{dc\alpha} = U_{dc\alpha} I_{d\alpha} + \frac{C}{2} U_{dc\alpha} \frac{du'_{dc\alpha}}{dt} \tag{7.5}$$

where I_d is the DC component of the capacitive current flowing through the DC side.

According to the law of energy conservation, the converter's output power equals its input power, expressed as

$$P_{dc} = P_{ab} \tag{7.6}$$

The steady-state and fluctuating components of the input power and output power are, respectively, equal. Combining equations (7.3), (7.5), and (7.6) yields

$$\frac{C}{2} U_{dc\alpha} \frac{du'_{dc\alpha}}{dt} = \frac{U_s I_s}{2}\cos(2\omega t) + \frac{1}{2}\omega L_s I_s^2 \sin(2\omega t) \tag{7.7}$$

The expression for the fluctuating voltage $u'_{dc\alpha}$ is then obtained as

$$u'_{dc\alpha} = \frac{1}{CU_{dc\alpha}}\left[-\frac{1}{2}L_s I_s^2 \cos(2\omega t) + \frac{U_s I_s}{2\omega}\sin(2\omega t)\right] \tag{7.8}$$

The expression for the converter's output voltage $u_{dc\alpha}$ is

$$u_{dc\alpha} = U_{dc\alpha} + u'_{dc\alpha} = U_{dc\alpha} + \frac{1}{CU_{dc\alpha}}\left[-\frac{1}{2}L_s I_s^2 \cos(2\omega t) + \frac{U_s I_s}{2\omega}\sin(2\omega t)\right] \tag{7.9}$$

It can be seen that the converter's output voltage exhibits a ripple at twice the input voltage frequency.

In practice, a double-power-frequency filter circuit is typically employed to eliminate the secondary voltage fluctuation $u'_{dc\alpha}$ in the output voltage of the single-phase converter.

7.2.1 Harmonic Characteristics of Two-Phase Converters

Since the single-phase converter is the fundamental unit of two-phase converters, secondary ripple is unavoidable. The voltage and current on the AC side of the β-phase converter are

$$\begin{cases} u_\beta = U_s \sin \omega t \\ i_\beta = I_s \sin \omega t \end{cases} \tag{7.10}$$

Like the α-phase converter, the fluctuating voltage contained in the output voltage of the β-phase converter is

$$u'_{dc\beta} = \frac{1}{CU_{dc\beta}}\left[\frac{1}{2}L_s I_s^2 \cos(2\omega t) - \frac{U_s I_s}{2\omega}\sin(2\omega t)\right] \tag{7.11}$$

where $U_{dc\beta}$ is the average DC component of the output voltage of the β-phase converter. Since the α-phase converter is connected to the DC port of the β-phase converter, $U_{dc\beta} = U_{dc\alpha}$. Thus, we have

$$u'_{dc\beta} + u'_{dc\alpha} = 0 \tag{7.12}$$

Therefore, the DC-side ripples of the two-phase converters studied cancel each other, eliminating secondary pulsation. Consequently, the secondary filter circuit can be omitted in the experiment.

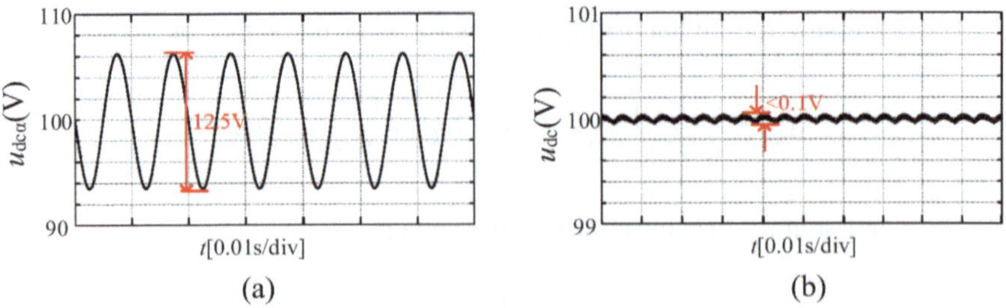

Fig. 7.3 DC-side voltage of two-phase converters

Table 7.1 Simulation parameters of two-phase converters

Experiment module	Parameter
Input phase voltage of YNvd transformer	55 V
Output phase voltage of YNvd transformer	55 V
Given DC-side voltage of two-phase converters	100 V
AC-side inductance of two-phase converters	5 mH
AC-side inductance of single-phase converter	5 mH
DC-link capacitor	1 mF

7.2.2 Verification of the Harmonic Characteristics of Two-Phase Converters

A two-phase converter model is developed on the MATLAB/Simulink platform. The parameters are listed in Table 7.1.

When the DC sides of the α- and β-phase converters are disconnected and the load of both converters is 50 Ω, the α-phase converter's output DC voltage $u_{dc\alpha}$ waveform is shown in Fig. 7.3a. This output voltage contains a 100 Hz ripple with an amplitude of approximately 12.5 V, consistent with the earlier harmonic analysis of the single-phase converter. When the DC sides of the α- and β-phase converters are connected with a combined load of 25 Ω, the secondary ripples from both converters cancel, resulting in a DC-side ripple of about 0 V. The corresponding DC-side voltage u_{dc} waveform is shown in Fig. 7.3a. Simulation results indicate that the LC secondary filter circuit can be omitted during the experiment, reducing the hardware cost associated with the LC filter circuit.

7.3 Control Strategy for Two-Phase Converters

7.3.1 Master-Slave Parallel Control Method for Two-Phase Converters

Analysis of the external characteristic droop parallel control strategy indicates that the external characteristic droop method is a straightforward, automatic current-sharing control method. By utilizing internal output impedance or external impedance, it ensures uniform load current distribution among modules without requiring inter-module communication. This self-regulating technique relies solely on the stability of individual modules. The main drawback of this control method is the trade-off between output voltage accuracy and balanced load distribution: a steeper external characteristic improves load sharing but reduces output voltage accuracy, while a smoother characteristic enhances voltage accuracy at the expense of load current distribution (Zheng 2011; Lee and Liu 2011; Zhou et al. 2010).

Due to its simplicity, this control method is widely used in low-power applications. Although the output impedance method can achieve approximate load current balance, factors such as component tolerances, aging, and changes in physical conditions over time can still lead to uneven current distribution.

For the uniform distribution of load current, the master-slave parallel control method is adopted in this study, offering high control accuracy and a simple structure. With an independent current distribution loop, this method also stabilizes the DC-side voltage and ensures load power sharing.

The master-slave parallel control method, illustrated in Fig. 7.4, consists of one master module and multiple slave modules. The master module includes two control loops: an outer voltage loop and an inner current loop, while each slave module has only an inner current loop. The master module's outer voltage loop output serves as the reference given signal for the inner current loops of the slave modules. The outer voltage loop stabilizes the system's DC-side output voltage to match the given voltage, while the inner current loop ensures load current sharing.

Figure 7.5 shows the block diagram of the two-phase H-bridge PWM rectifier system with master-slave control. The left dotted box represents the voltage loop, while the upper and lower dotted boxes correspond to current loop regulations 1 and 2, respectively. The system's DC-side reference voltage V_{dc}^* is compared with the actual DC-side voltage V_{dc1} collected by the rectifier #1, and the resulting error is processed by a PI controller in the outer voltage loop to regulate the DC voltage to the given value. Simultaneously, I_{s1}^* and I_{s2}^* generated by the voltage loop are given as the current of rectifier #1 and rectifier #2, respectively, which are then multiplied by their sinusoidal signals to generate the AC current given signals i_{s1}^* and i_{s2}^*. i_{s1}^* and i_{s2}^* are each compared with their respective sampled current signals, and the modulation signals u_{ab1} and u_{ab2} are obtained through the proportional amplifier K. The generated modulation signals are compared with the carrier to produce switching signals PWM1 and PWM2, which control the turn-on and turn-off of the switching tube to achieve uniform distribution of load current.

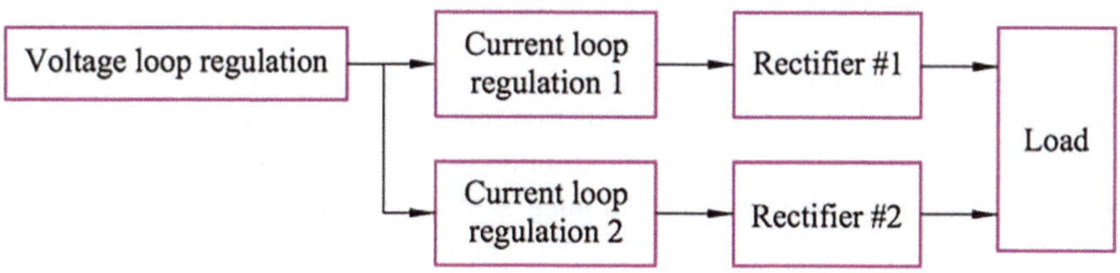

Fig. 7.4 Block diagram of the master-slave control method structure

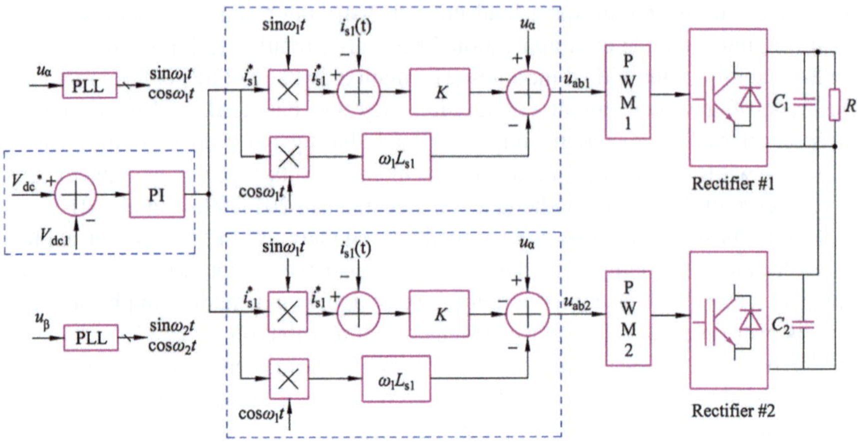

Fig. 7.5 Block diagram of the single-phase H-bridge PWM rectifier parallel system with master-slave control

In summary, for a single module, a single-phase H-bridge instantaneous current control strategy is adopted, but the two rectifiers share one outer voltage loop. Due to the phase inconsistency in the secondary voltages of the YNvd transformer, the voltages need to be multiplied by their respective sinusoidal signals as reference values for their respective currents.

7.3.2 *Control Strategy for Two-Phase Three-Level H-Bridge Converters*

The two-phase converter comprises two three-level H-bridge converters with their DC sides connected back-to-back and their AC sides connected to separate AC voltages. The topology is shown in Fig. 7.2. Consequently, the mathematical model of two-phase converters closely resembles that of single-phase converters:

$$\begin{cases} u_\alpha = L_s \dfrac{di_\alpha}{dt} + u_{ab} \\ u_\beta = L_s \dfrac{di_\beta}{dt} + u_{cd} \end{cases} \tag{7.13}$$

In this study, a two-phase DQ decoupling control strategy is employed to regulate the active power and reactive power of two-phase converters separately, enabling operation at unity power factor. More importantly, this approach treats the two independent single-phase converters as an integrated system, minimizing the impact of device errors and maximizing balance between two converters. Active power and reactive power are controlled, respectively, by PI controllers, eliminating steady-state error. Additionally, this approach reduces controller design complexity and enhances system stability compared to controlling the two converters separately.

Assume that the AC-side voltage and current of the two-phase converter are

$$\begin{cases} u_\alpha = U_s \sin\omega t \\ u_\beta = U_s \sin(\omega t + \theta) \end{cases} \tag{7.14}$$

$$\begin{cases} i_\alpha = I_s \sin\omega t \\ i_\beta = I_s \sin(\omega t + \theta) \end{cases} \tag{7.15}$$

where U_s is the amplitude of the AC-side voltage, and I_s is the amplitude of the AC-side current.

Based on a counterclockwise rotation, the coordinate system in Fig. 7.6 is established, where the α-axis and β-axis voltages represent two-phase voltages at any angle, forming a two-phase plane coordinate system. The d-axis and q-axis correspond to the system's active and reactive components, forming a two-phase rotating coordinate system. The angle between the α-axis and the β-axis is θ.

The conversion matrix between the two coordinate systems can be obtained from Fig. 7.6:

$$\begin{bmatrix} d \\ q \end{bmatrix} = \begin{bmatrix} \cos\omega t & \cos(\theta - \omega t) \\ \sin\omega t & \sin(\theta - \omega t) \end{bmatrix} \begin{bmatrix} \alpha \\ \beta \end{bmatrix} \tag{7.16}$$

$$\begin{bmatrix} \alpha \\ \beta \end{bmatrix} = \frac{1}{\cos\omega t \sin(\theta - \omega t) + \sin\omega t \cos(\theta - \omega t)} \begin{bmatrix} \sin(\theta - \omega t) & \cos(\theta - \omega t) \\ \sin\omega t & -\cos\omega t \end{bmatrix} \begin{bmatrix} d \\ q \end{bmatrix} \tag{7.17}$$

Since the coefficient does not affect the correctness of control, Eq. (7.17) can be simplified to reduce control complexity:

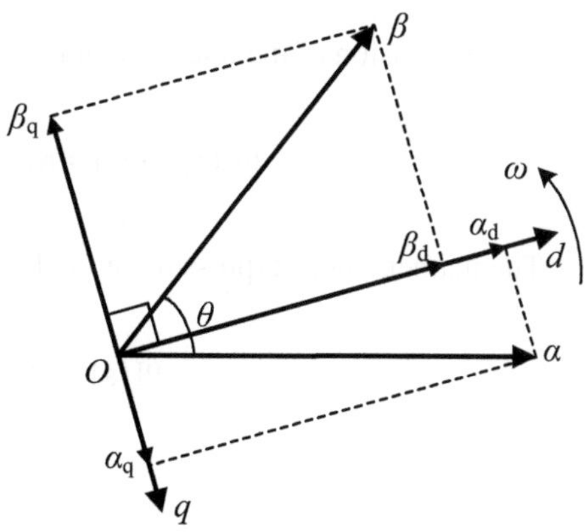

Fig. 7.6 Two-phase stationary and rotating coordinate systems at arbitrary angles

$$\begin{bmatrix} \alpha \\ \beta \end{bmatrix} = \begin{bmatrix} \sin(\theta - \omega t) & \cos(\theta - \omega t) \\ \sin \omega t & -\cos \omega t \end{bmatrix} \begin{bmatrix} d \\ q \end{bmatrix} \tag{7.18}$$

From Eqs. (7.13) and (7.16), the active and reactive components of the two-phase converter port voltage are

$$\begin{cases} V_{\mathrm{d}} = -L_{\mathrm{s}} \dfrac{\mathrm{d}i_{\mathrm{d}}}{dt} + \omega L_s i_{\mathrm{q}} + u_{\mathrm{d}} \\ V_{\mathrm{q}} = -L_{\mathrm{s}} \dfrac{\mathrm{d}i_{\mathrm{q}}}{dt} - \omega L_s i_{\mathrm{d}} + u_{\mathrm{q}} \end{cases} \tag{7.19}$$

where u_{d} and u_{q} represent the active and reactive components of the two-phase converter's AC-side voltage, while i_{d} and i_{q} denote the active and reactive components of its AC-side current. When the control system is operating at unity power factor:

$$\begin{bmatrix} u_{\mathrm{d}} \\ u_{\mathrm{q}} \end{bmatrix} = \begin{bmatrix} \sqrt{\dfrac{3}{2}} \cdot U_{\mathrm{s}} \\ 0 \end{bmatrix} \tag{7.20}$$

Equation (7.19) reveals a coupling relationship between the active and reactive components, V_{d} and V_{q}, of the port voltage, complicating the controller design. Therefore, decoupling calculation is carried out to simplify the control system.

According to the definition of mathematical differentiation:

$$\frac{\mathrm{d}i_{\mathrm{d}}}{\mathrm{d}t} = \frac{\Delta i_{\mathrm{d}}}{\Delta t} = \frac{i_{\mathrm{dref}} - i_{\mathrm{d}}}{\Delta t} \tag{7.21}$$

The expression of the proportional controller is

$$\mathrm{P}\left(i_{\mathrm{dref}} - i_{\mathrm{d}}\right) = K_{\mathrm{p}} \Delta t \frac{\mathrm{d}i_{\mathrm{d}}}{\mathrm{d}t} \tag{7.22}$$

The expression of the integral controller is

$$\mathrm{Int}\left(i_{\mathrm{dref}} - i_{\mathrm{d}}\right) = K_{\mathrm{I}} \int \Delta t \frac{di_{\mathrm{d}}}{dt} dt = K_{\mathrm{I}} \Delta t i_{\mathrm{d}} \tag{7.23}$$

The mathematical expression of the PI controller is

$$\mathrm{PI}(\mathrm{s}) = \left(\mathrm{K}_{\mathrm{p}} + \frac{K_{\mathrm{I}}}{s} \right) \Delta \varepsilon \tag{7.24}$$

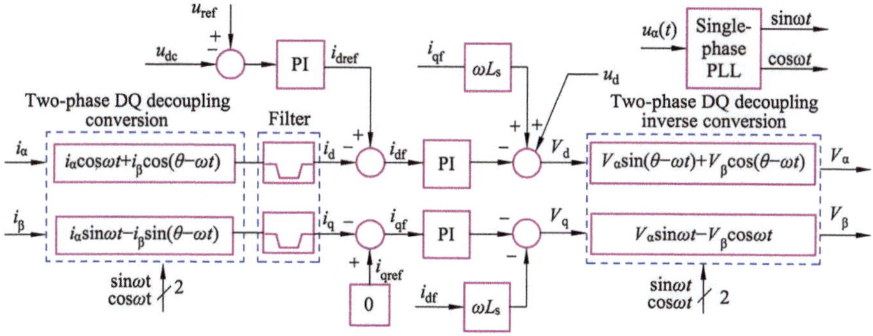

Fig. 7.7 Block diagram of two-phase DQ decoupling control

Therefore, by applying a PI controller to decouple active and reactive components, we obtain

$$\begin{cases} V_d = -\left(K_p + \dfrac{K_I}{s}\right)\left(i_{dref} - i_d\right) + \omega L_s i_q + u_d \\ V_q = -\left(K_p + \dfrac{K_I}{s}\right)\left(i_{qref} - i_q\right) - \omega L_s i_d + u_q \end{cases} \tag{7.25}$$

The block diagram of the two-phase DQ decoupling control derived from Eq. (7.25) is shown in Fig. 7.7. First, the phase information sinωt and cosωt of the two-phase converter's α-phase AC-side voltage u_α is obtained via single-phase phase-locked loop detection, which is then used to decouple the AC-side currents i_a and i_b. The active and reactive components obtained from Eq. (7.16) typically consist of a DC component superimposed with an AC component. After matrix conversion, the AC fluctuating quantity can be filtered out, yielding the active and reactive input current components, i_d and i_q. The actual DC voltage u_{dc} is then compared with the given DC voltage u_{ref}, and the resulting error is processed by the PI controller to generate the reference active current i_{dref}. This value is compared with id to obtain the active component of the current control signal, while comparing i_{dref} with 0 yields the reactive component i_{qf}. Finally, according to Eq. (7.25), the active and reactive components V_d and V_q of the voltage control signal are obtained, which are then decoupled and inversely converted to produce the system's modulation signal.

When the AC-side voltage phase difference of the two-phase converter is 90° ($\theta = 90°$), the relationship between the two-phase stationary and rotating coordinate systems is illustrated in Fig. 7.8.

The relationship between the two coordinate systems is

$$\begin{bmatrix} d \\ q \end{bmatrix} = \begin{bmatrix} \cos\omega t & \sin\omega t \\ \sin\omega t & -\cos\omega t \end{bmatrix} \begin{bmatrix} \alpha \\ \beta \end{bmatrix} \tag{7.26}$$

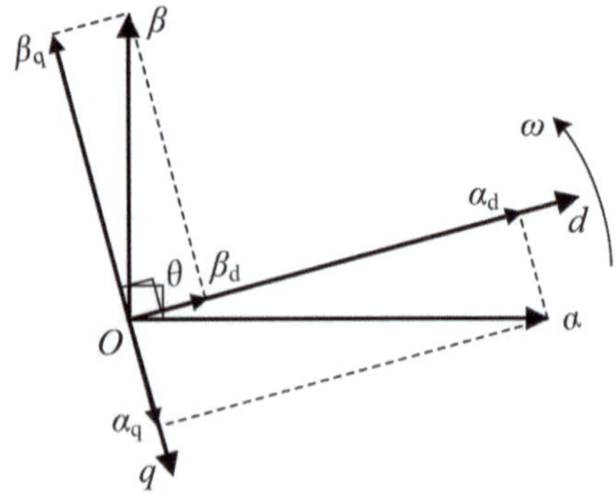

Fig. 7.8 Two-phase stationary and rotating coordinate systems with a phase difference of 90°

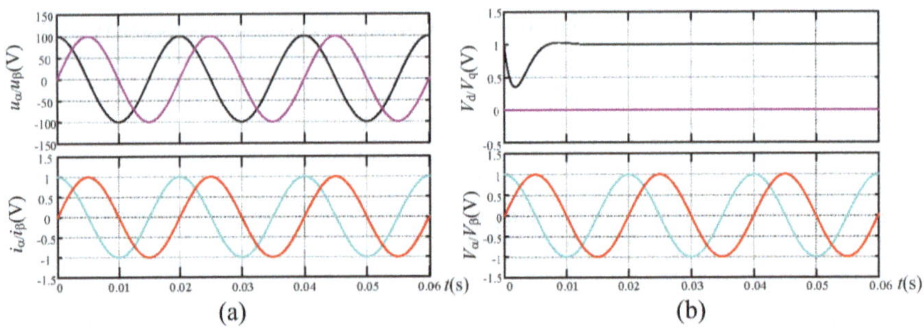

Fig. 7.9 Simulation of two-phase DQ decoupling when the input voltage and current are co-phase

$$\begin{bmatrix} \alpha \\ \beta \end{bmatrix} = \begin{bmatrix} \cos\omega t & \sin\omega t \\ \sin\omega t & -\cos\omega t \end{bmatrix} \begin{bmatrix} d \\ q \end{bmatrix} \tag{7.27}$$

Equation (7.26) shows that when the AC-side voltage phase difference of the two-phase converter is 90°, the active and reactive components contain only DC components after decoupling conversion, with no AC fluctuating quantity. Therefore, the filter block in Fig. 7.7 can be omitted to implement two-phase DQ decoupling control.

7.3.3 Simulation Verification of Two-Phase DQ Decoupling

The two-phase DQ decoupling control is validated on the MATLAB/Simulink platform by verifying the correctness of the decoupling conversion and the decoupling inverse conversion at first.

As illustrated in Fig. 7.9, when the input voltage u_α/u_β and the input current i_α/i_β of the two-phase decoupling control module are co-phase, the per-unit value V_d of active power after d_q conversion remains stable at 1, while the per-unit value V_q of

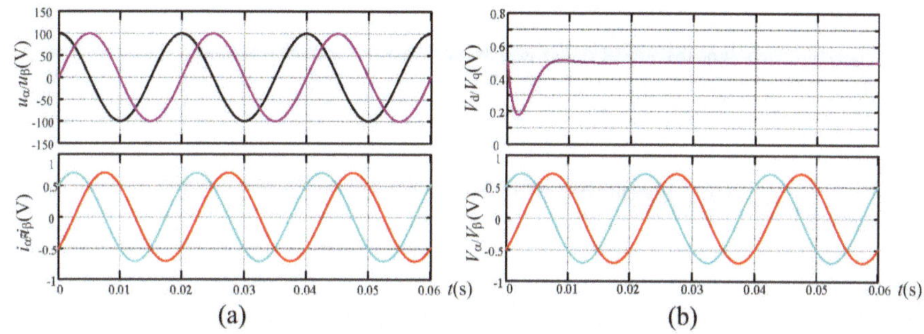

Fig. 7.10 Two-phase DQ decoupling simulation with input voltage leading input current by 45°

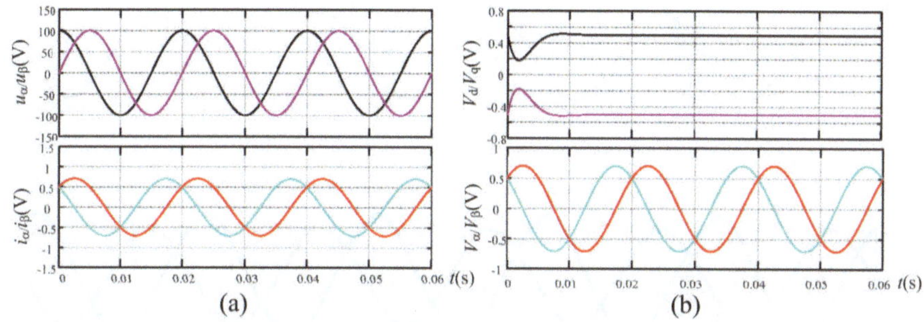

Fig. 7.11 Two-phase DQ decoupling simulation when the input voltage lags the input current by 45°

reactive power stabilizes at 0, confirming a power factor of 1 for the input signal. After d_q inverse conversion, the waveforms of the reference voltages V_a and V_b perfectly match the input current waveforms.

As shown in Fig. 7.10, when the voltage u_α/u_β of the two-phase decoupling control module leads the current *iα*/*iβ* by 45°, the per-unit values V_d and V_q of active power and reactive power both remain stable at 0.5 after d_q conversion, consistent with the phase relationship of the input signal. After d_q inverse conversion, the waveforms of the reference voltages V_a and V_b perfectly match the input current waveforms.

As illustrated in Fig. 7.11, when the voltage u_α/u_β of the decoupling module lags the current *iα*/*iβ* by 45°, the per-unit value V_d of the active power stabilizes at 0.5, while the per-unit value V_q of the reactive power stabilizes at −0.5 after d_q conversion, consistent with the input signal's phase relationship. Following the d_q inverse conversion, the waveforms of the reference voltages V_a and V_b perfectly match the input current waveforms.

The simulation results demonstrate that the active power and reactive power of input voltage and current signals can be effectively calculated through the two-phase DQ decoupling conversion and inverse conversion proposed in this study, and the reference voltage signal can be solved inversely.

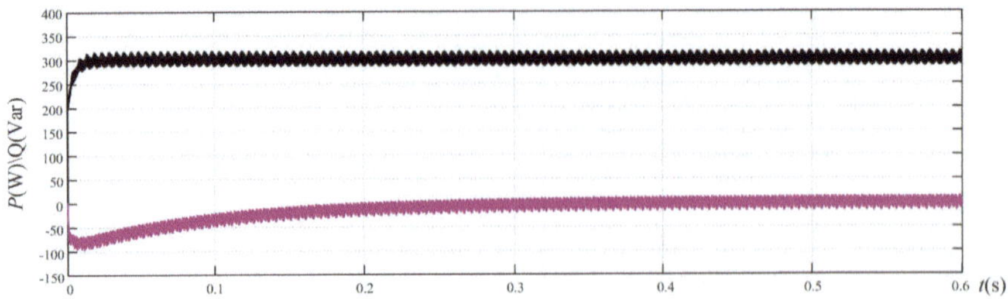

Fig. 7.12 Input active power and reactive power of the two-phase converter

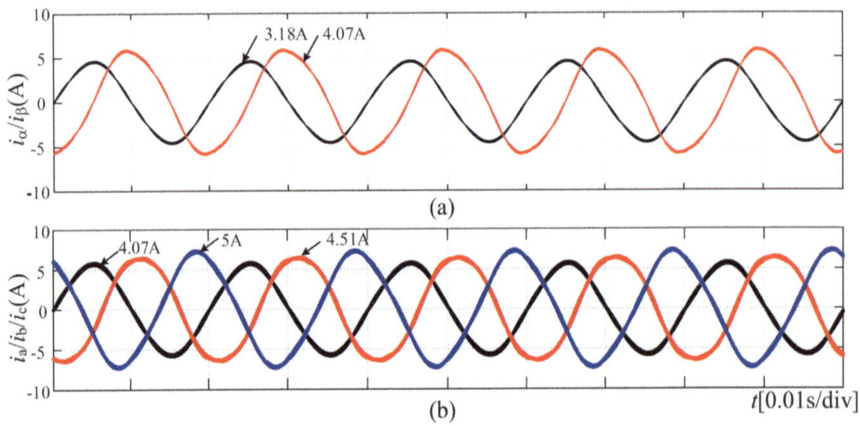

Fig. 7.13 Current waveforms of two-phase converter under independent control

Figure 7.12 shows the input active and reactive power waveforms under two-phase DQ decoupling control. After 0.3 s, the active power P stabilizes at 300 W, equal to the output power, while the reactive power Q remains at zero, demonstrating effective control of active and reactive components with a fast dynamic response.

In practice, line impedance inevitably arises due to line length and device errors. In a two-phase converter, the line impedances of the two single-phase converters rarely match exactly. When controlled independently, the impedance difference will cause differences in the working state of the two-phase converters. The simulation is based on differing line impedances between the DC-side capacitor and load of the two converters. When the two-phase converters adopt transient current control, respectively, Fig. 7.13 shows the input current waveforms (a) and the three-phase grid current waveforms (b). The figure shows that different line impedances result in effective two-phase input current values of 3.18 A and 4.07 A, causing current unbalance of the three-phase grid. The effective values of the three-phase grid currents are 4.07 A, 5 A, and 4.51 A, with an imbalance of approximately 10.6%. Additionally, each current waveform exhibits slight distortion, and the harmonic distortion rate increases.

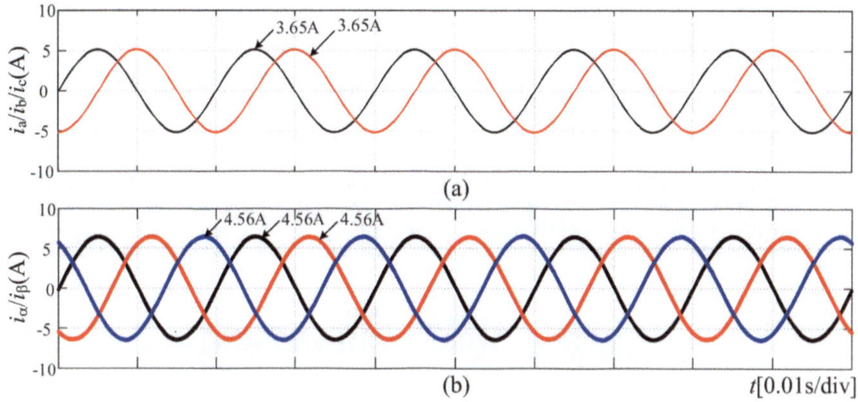

Fig. 7.14 Current waveforms of the two-phase converter under two-phase DQ decoupling

The imbalance calculation formula used in this study is

$$\text{Imbalance} = MAX\left(\left|I_p - I_{\text{ave}}\right|\right) / I_{\text{ave}} \tag{7.28}$$

where I_p is the phase current, I_{ave} is the average of the three-phase currents, and MAX(*) is a function of the maximum value. According to Fig. 7.13, the average I_{ave} is 4.52 A, and the "$I_p - I_{ave}$" is −0.45A, 0.48A, and −0.01 A for the three-phase currents, resulting in an imbalance of 10.6%.

Under the same conditions, Fig. 7.14a shows the two-phase converter current waveforms under two-phase DQ decoupling control, while Fig. 7.14b displays the current waveforms of the three-phase grid. As can be seen from the figure, since the two-phase DQ decoupling control treats the two single-phase converters as a unit, minor differences between them are unlikely to affect normal operation. The converter input currents have nearly identical effective values of approximately 3.65 A, the effective values of the three-phase currents are all 4.56 A with zero imbalance, and each current exhibits good sinusoidality.

This demonstrates that, compared to independent control of two converters, the two-phase converter under two-phase DQ decoupling control exhibits stronger resistance to system errors (such as line impedance) and offers improved system stability.

7.4 System Simulation Analysis Based on Two-Phase-to-Single-Phase Converters

To validate the topology, control, and modulation strategies of the two-phase three-level converter, as well as the feasibility and soundness of the continuous power supply system, this study employs MATLAB/Simulink for simulation and analysis.

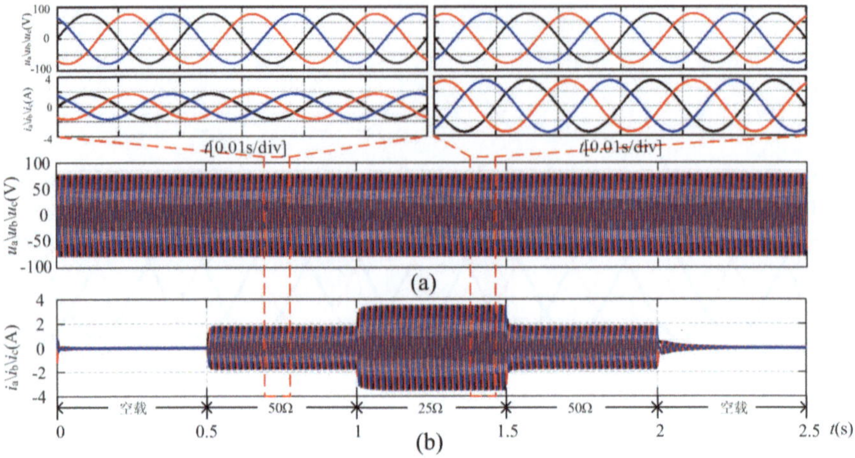

Fig. 7.15 Dynamic and static waveforms of three-phase voltages and currents

7.4.1 Two-Phase PWM Rectifier Simulation

Figure 7.15 shows the dynamic and static simulation waveforms of the two-phase converter. The two-phase converter starts with no load; a 50-Ω resistive load is connected at 0.5 s, another 50-Ω load is connected in parallel at 1 s (that is, the load of the two-phase converter is 25 Ω), one 50-Ω load is removed at 1.5 s, and no-load operation resumes at 2 s. Figure 7.15 presents the dynamic waveforms of the three-phase voltages and currents on the input side of the transformer. Subfigure (a) shows the waveforms of three-phase voltages u_a, u_b, and u_c, while subfigure (b) displays the waveforms of three-phase currents i_a, i_b, and i_c. At system startup, the three-phase currents experience a 1 A inrush, returning to 0 A after about 0.05 s. When the load is connected at 0.5 s, the three-phase currents increase to a stable value after 0.2 s, with an amplitude of about 1.7 A. During the dynamic response period, the maximum current amplitude does not exceed 1.8 A, with minimal overshoot.

A partial enlargement of the voltage and current waveforms, taken after the stabilization of the three-phase currents, indicates that the voltage and current are co-phase, the power factor is close to 1, the phase difference of the three-phase currents is 120°, the current amplitudes are nearly identical, and the currents exhibit good sinusoidality. When the load is connected at 1 s, the three-phase currents rapidly rise to approximately 3.4 A within 0.2 s. During steady state, the system still operates at unity power factor, with less harmonic content of currents and balanced three-phase system. After the load is removed at 1.5 s, the three-phase currents stabilize within 0.2 s. Following complete load removal at 2 s, the currents drop to 0 A in 0.2 s. The two-phase converter, utilizing two-phase DQ decoupling control and two-phase SVPWM modulation, demonstrates robust dynamic and static performance, ensures current balance on the three-phase side, and operates at unity power factor.

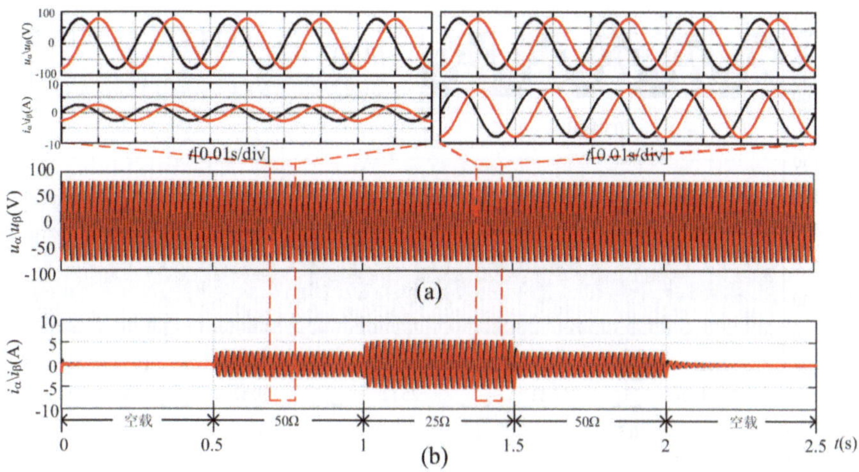

Fig. 7.16 Dynamic and static waveforms of voltages and currents on the AC side of the two-phase converter

Figure 7.16 presents the dynamic and static voltage and current waveforms at the AC side of the two-phase converter, with (a) showing the waveforms of voltages u_α and u_β and (b) displaying the waveforms of currents i_α and i_β. At system startup, the two-phase currents experience a 1 A inrush, returning to 0 A after about 0.1 s. Upon load connection at 0.5 s, the two-phase currents increase to a stable value of 1.77 A within 0.05 s. The currents are equal in magnitude, with a phase difference of 90°, and co-phase with their corresponding voltages, exhibiting good sinusoidality. After the load is connected at 1 s, the two-phase currents rise to 3.67 A in 0.02 s with minimal inrush. Following the load increase, the system continues to operate at unity power factor, ensuring good power quality. When half of the load is removed at 1.5 s, the two-phase currents reach steady state within 0.02 s. After all loads are removed at 2 s, the currents return to 0 A in 0.2 s. The two-phase converter, employing two-phase DQ decoupling control and two-phase SVPWM modulation, demonstrates strong dynamic and static performance, with the two-phase currents equal in amplitude and the system operating at unity power factor, effectively addressing potential negative sequence and reactive power issues on the three-phase side.

Figure 7.17 illustrates the dynamic and static waveforms of the AC-side port voltages u_{ab} and u_{cd} of the two-phase converter. As can be seen from the figure, the two port voltages fluctuate very little and settle rapidly during load switching, indicating good dynamic performance. In steady state, the port voltages exhibit five-level stepped waveforms with a phase difference of 90°. Notably, transitions between steps occur smoothly without jumps, consistent with the theoretical analysis of the two-phase SVPWM algorithm presented in this study.

Figure 7.18 illustrates the dynamic waveform of the DC-side voltage u_{dc} of the two-phase converter. At startup, the DC voltage stabilizes at the given value of 100 V in 0.15 s, with a minor overshoot of approximately 4.5 V. At 0.5 s, when the load

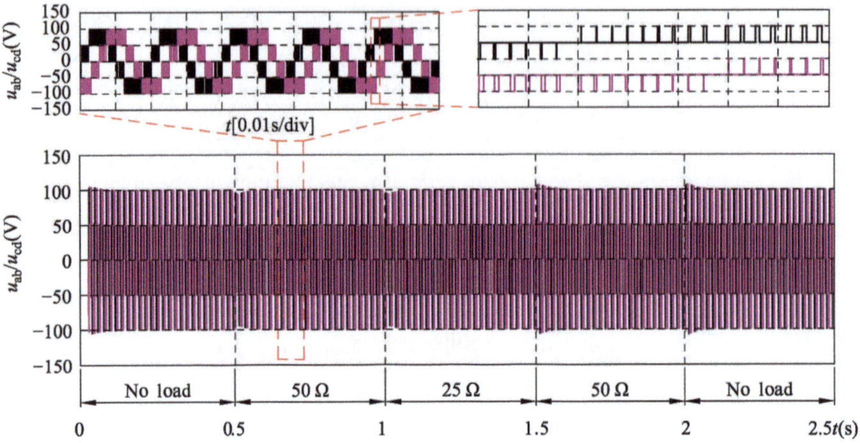

Fig. 7.17 Dynamic and static waveforms of AC-side port voltages of the two-phase converter

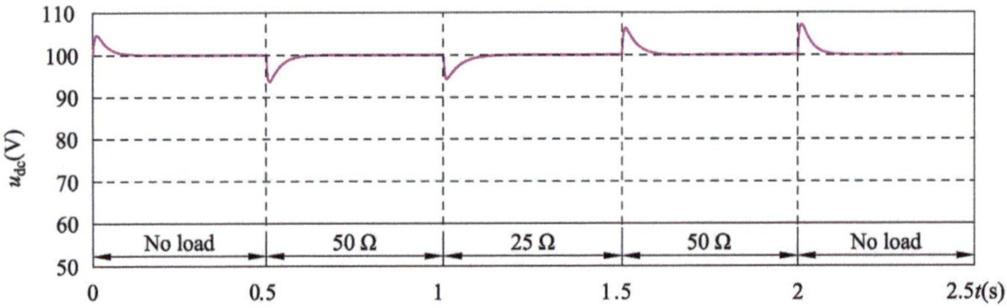

Fig. 7.18 Dynamic waveform of DC-side voltage of the two-phase converter

is connected, the DC voltage briefly drops to 94 V but returns to 100 V within about 0.15 s. As the load increases, the DC voltage again stabilizes at 100 V after a brief, small-amplitude drop at 1 s. Following load removal at 1.5 s and 2 s, the DC voltage rises slightly, with an overshoot of about 5 V, stabilizing within an adjustment period of 0.15 s. This demonstrates that the two-phase converter, employing two-phase DQ decoupling control and two-phase SVPWM modulation, effectively stabilizes the DC-side voltage.

The simulation results indicate that the two-phase DQ decoupling control and SVPWM modulation exhibit good dynamic and static performance, characterized by rapid dynamic response and minimal overshoot. The two-phase grid side operates at unity power factor, with equal current amplitudes and good sinusoidality. Consequently, the three-phase grid side also operates at unity power factor, ensuring balanced current and low harmonic content.

7.4.2 Simulation of Continuous Power Supply System Based on Two-Phase-to-Single-Phase PWM Converter

To achieve converter grid connection, traditional droop control and SPWM modulation are employed for single-phase converters, while two-phase DQ decoupling control and two-phase SVPWM modulation are used for two-phase converters to verify the correctness of the continuous power supply system based on the two-phase-to-single-phase converter.

When the traction power supply system is in the traction mode with a purely resistive load, the load voltage u_{out} is co-phase with the current i_{out}, as illustrated in Fig. 7.19a. At this time, the three-phase grid operates at unity power factor, characterized by low harmonic content, high sinusoidality, and balanced three-phase currents, as shown in Fig. 7.19 b and c. The input voltage and current of the two-phase converter are also co-phase, indicating balanced two-phase currents and operation at unity power factor, as shown in Fig. 7.19 d and e. The port levels of the two-phase converter exhibit five-level stepped waveforms with a phase difference of 90°, as shown in Fig. 7.19f. The simulation results demonstrate that when the load is resistive, the traction power supply system effectively transmits active power to the load, without power quality issues such as harmonics, low power factor, or imbalance on the three-phase grid side.

Figure 7.20 presents the dynamic waveforms of the traction power-supply system under various operating conditions: (a) voltage and current waveforms of the traction network; (b) voltage waveforms of the three-phase grid; (c) current waveforms of the three-phase grid; (d) input voltage waveforms of the two-phase converter; (e) input current waveforms of the two-phase converter; and (f) port waveforms of the two-phase converter. From 0 s to 1.5 s, the traction load is resistive; from 1.5 s to 3 s, it is resistive-capacitive; and from 3 s to 4.5 s, it is resistive-inductive. The figure illustrates that initially, the three-phase grid currents and the two-phase input currents experience significant inrush, accompanied by minor fluctuations in port voltages. After 1 s, the current waveforms stabilize. At the load-switching moment at 1.5 s, the three-phase grid currents and the two-phase input currents rise rapidly with a slight inrush, while the port voltages decrease slightly. The system stabilizes in approximately 0.5 s. At the load-switching moment at 3 s, the three-phase grid currents and the two-phase input currents exhibit a minor inrush, causing a slight increase in port voltages which quickly return to stability. Overall, the analysis indicates that the traction system demonstrates excellent dynamic performance, with minimal current inrush and short response time during load switching.

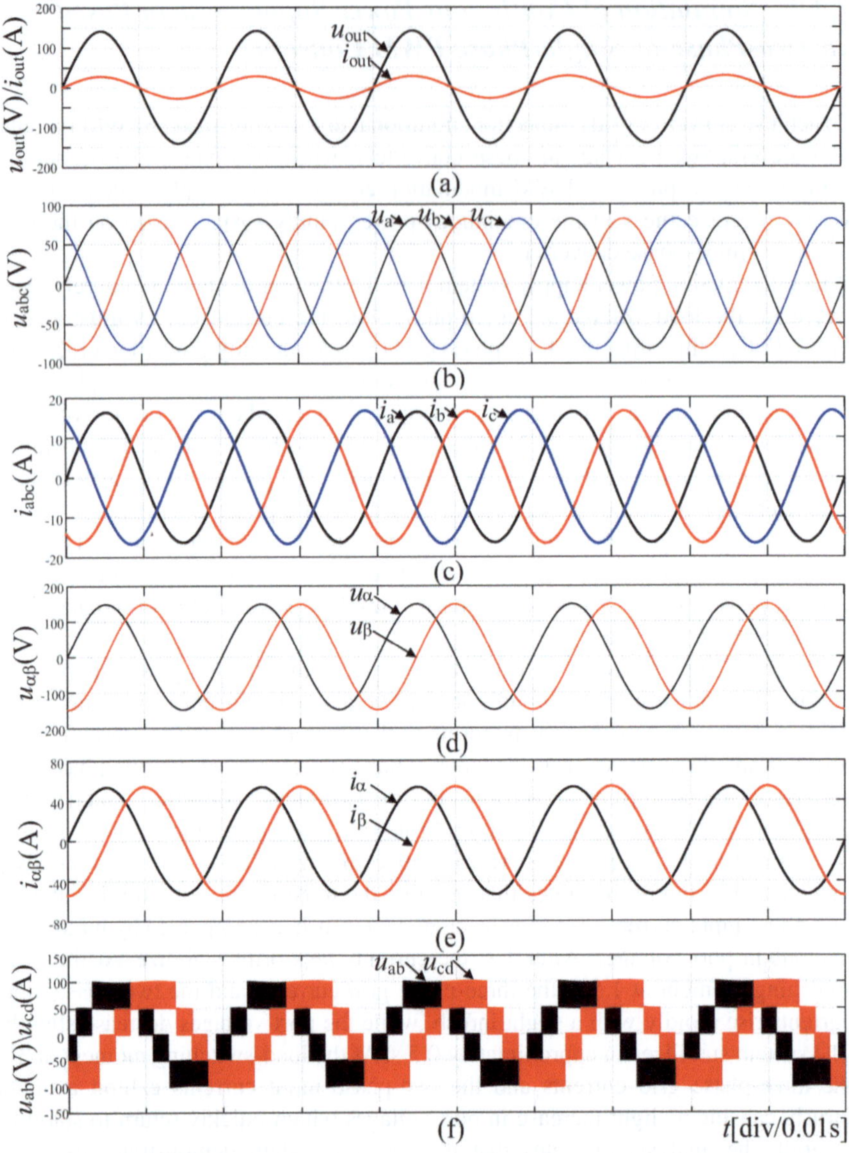

Fig. 7.19 Voltage and current waveforms of the traction power supply system with resistive load

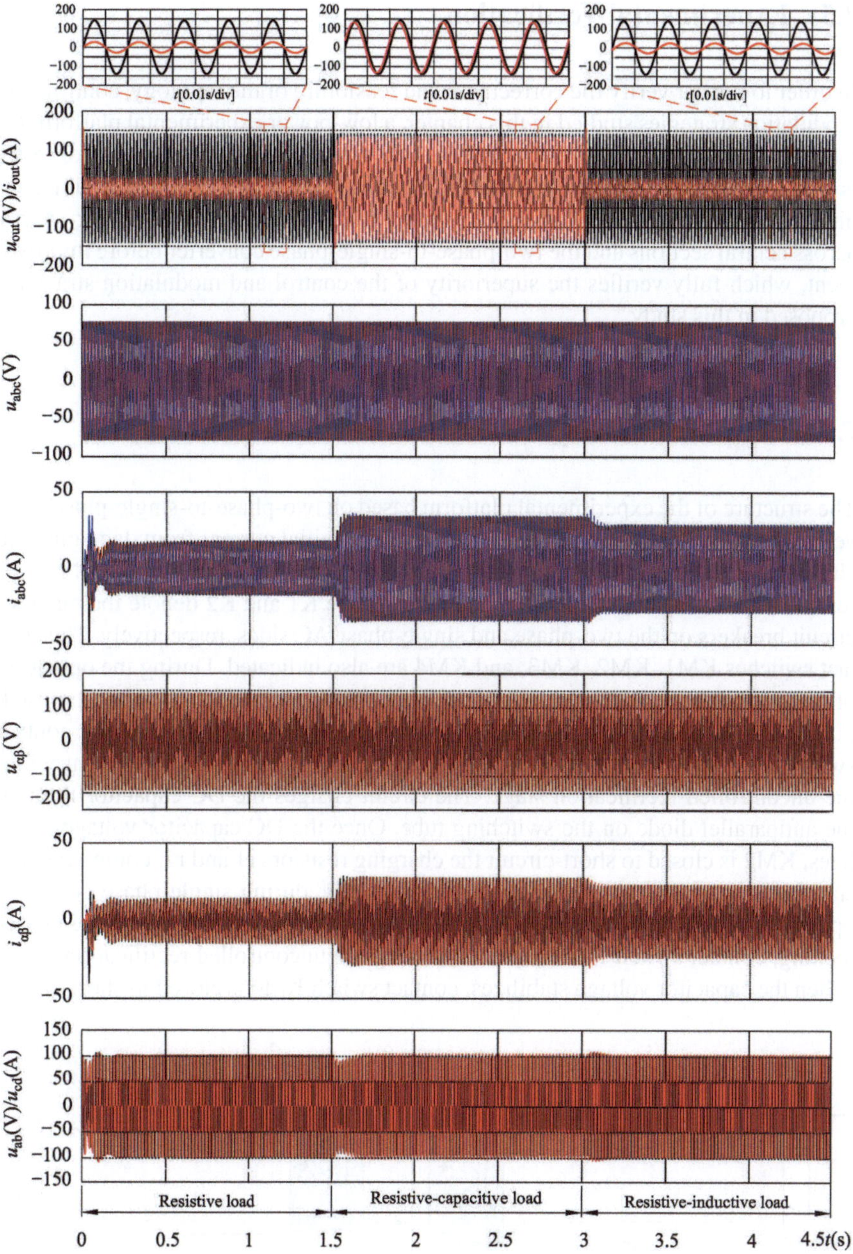

Fig. 7.20 Dynamic waveforms of the traction power supply system under different operating conditions

7.5 Experimental Verification

In order to further verify the correctness and feasibility of the topology, control, and modulation strategies studied in this chapter, a low-power experimental platform for the continuous power supply system based on two-phase-to-single-phase converter is built. Dynamic and static experiments are carried out under various working conditions, and a comparative analysis is conducted with traditional power supply across neutral sections and the two-phase-to-single-phase converter before improvement, which fully verifies the superiority of the control and modulation strategies proposed in this study.

7.5.1 Experimental System

The structure of the experimental platform based on two-phase-to-single-phase converter is shown in Fig. 7.21. To prevent excessive initial current from damaging the circuit devices, a pre-charging circuit is employed. In the figure, r1, r2, and r3 represent the charging resistors of the converter, while K1 and K2 denote the main air circuit breakers on the two-phase and single-phase AC sides, respectively. The contact switches KM1, KM2, KM3, and KM4 are also indicated. During the operation of the two-phase rectifier, the two-phase main air circuit breaker K1 is first closed. Following the completion of phase-locking and phase-sequence tracking, contact switch KM1 is closed, resulting in the blockage of switching signals and entry into the uncontrolled rectification stage. The circuit charges the DC capacitor through the antiparallel diode on the switching tube. Once the DC capacitor voltage stabilizes, KM2 is closed to short-circuit the charging resistors r1 and r2, completing the startup stage of the two-phase converter. Similarly, during single-phase converter operation, the main air circuit breaker K2 is first closed. After single-phase phase-locking, contact switch KM3 closes, initiating the uncontrolled rectification stage. When the capacitor voltage stabilizes, contact switch KM4 is closed to short-circuit

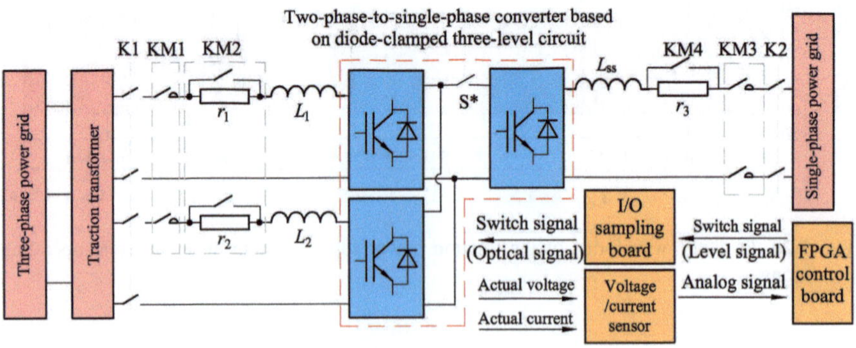

Fig. 7.21 Structure of the experimental platform based on two-phase-to-single-phase converter

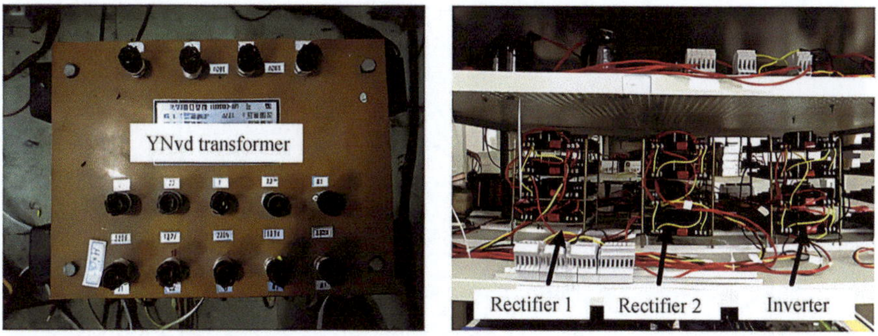

(a) (b)

Fig. 7.22 Experimental platform based on two-phase-to-single-phase converter

the charging resistor r3, completing the startup stage of the single-phase converter. During the experiment with the two-phase-to-single-phase converter, the single-phase converter is activated after the operation of the two-phase converter stabilizes, allowing the three-phase-to-single-phase converter to operate normally.

The low-power experimental platform described in this chapter is illustrated in Fig. 7.22. Figure 7.22a presents the YNvd traction transformer, while Fig. 7.22b displays the prototype of the two-phase-to-single-phase converter. The experimental parameters of the two-phase converter are listed in Table 7.1. In the experimental system, the *I/O* sampling board integrates the functions of signal conversion and sampling, outputting the IGBT driving signals via optical fibers. The voltage/current sensor transmits the collected analog voltage and current signals to the A/D conversion module on the main control board, which converts them into 12-bit digital signals for the FPGA control chip. The control algorithm is implemented by EP3C55F484C8, a third-generation Cyclone series chip from Altera.

7.5.2 Experimental Results and Analysis of YNvd Traction Transformer

Before the two-phase-to-single-phase converter is verified, the traction transformer is initially tested with a purely resistive load to confirm correct wiring and optimal performance, thereby preventing any interference with the converter's performance.

The output voltage of the traction transformer is 50 V, with a 25-Ω pure resistive load connected to each output terminal. Figure 7.23 illustrates the waveforms of the transformer's output voltage and current. Channel 1 represents the α-phase voltage uα, channel 2 the α-phase current i_α, channel 3 the β-phase voltage u_β, and channel 4 the β-phase current i_β. The figure shows that the output voltages are approximately equal with a 90° phase difference, and the output currents are also nearly equal.

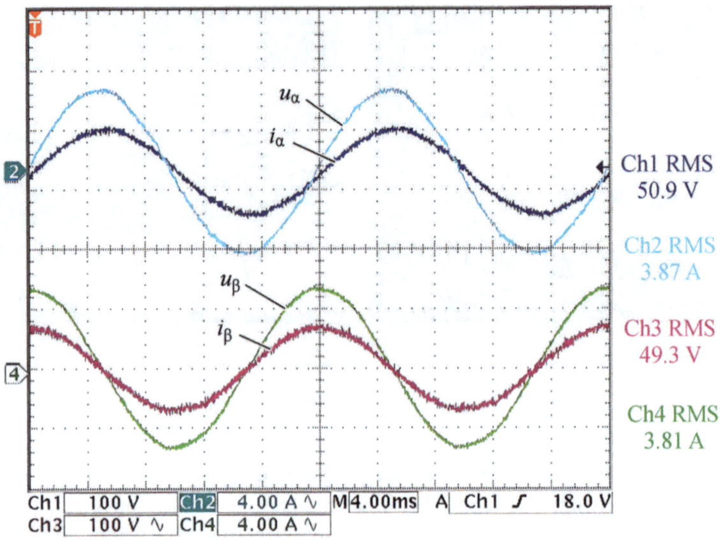

Fig. 7.23 Output voltage and current waveforms of the transformer

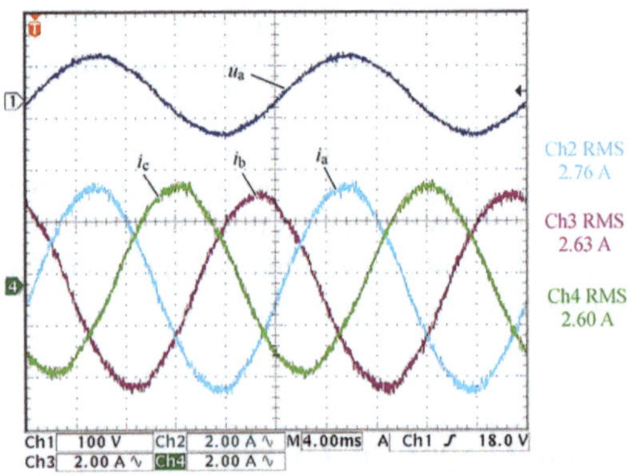

(a)

U	rms [V]	peak+[V]	peak-[V]	THD [%]
ch1	53.6	77.1	- 76.9	1.5
ch2	53.9	77.3	- 76.8	1.9
ch3	54.7	79.0	- 78.5	1.7

I	rms [A]	peak+[A]	peak-[A]	ITHD [%]
ch1	3.4	5.1	- 4.9	5.3
ch2	3.3	5.0	- 4.9	6.1
ch3	3.1	4.8	- 4.7	5.9
ch4	0.0	1.1	- 0.8	

	P [W]	S [VA]	Q [var]	PF
ch1	0.18k	0.18k	0.02k	0.995
ch2	0.18k	0.18k	0.01k	0.999
ch3	0.17k	0.17k	0.02k	0.995
sum	0.5k	0.5k	0.0k	0.997

Uave [V]	Iave [A]	Uunb [%]
54.1	3.3	0.8

(b)

Fig. 7.24 Input voltage and current of the transformer under pure resistive load

However, a phase difference exists between the α-phase voltage and current, indicating operation at a non-unity power factor.

The waveforms of the traction transformer's input voltage and current are presented in Fig. 7.24a. Channel 1 displays the a-phase voltage ua, while channels 2, 3, and 4 depict the three-phase currents i_a, i_b, and i_c, respectively. The figure illustrates an imbalance in the transformer's three-phase input currents. Figure 7.24b presents power quality data for the traction transformer's three-phase input, as measured by the harmonic analyzer. The three-phase system operates at unity power factor, with harmonic contents of 5.3%, 6.1%, and 5.9% and effective current values of 3.4 A,

3.3 A, and 3.1 A for the three-phase currents, resulting in an imbalance of approximately 5.1%.

The significant gap between the transformer's actual performance and the ideal requirements indicates that, even under purely resistive load conditions, the transformer operates at a non-unity power factor, exhibits harmonics in the three-phase input currents, and experiences high imbalance among the three-phase currents. Consequently, the performance of the two-phase-to-single-phase converter will be impacted.

7.5.3 Experimental Results and Analysis of Traction Power Supply Across Neutral Sections

To compare the power supply across neutral sections and the continuous power supply based on the two-phase-to-single-phase converter before and after improvement, only one α-phase output port of the traction transformer is first connected to the single-phase converter to simulate the power supply across neutral sections.
The load of the single-phase converter is 50 Ω. Figure 7.25 illustrates the waveforms of the three-phase voltage and current on the primary side of the traction transformer. Channel 1 represents the b-phase voltage u_b, while channels 2, 3, and 4 represent the three-phase currents i_b, i_a, and i_c, respectively. The figure indicates that in the mode of power supply across neutral sections, the reactive power output from the three-phase grid is substantial. The three-phase currents no longer maintain a

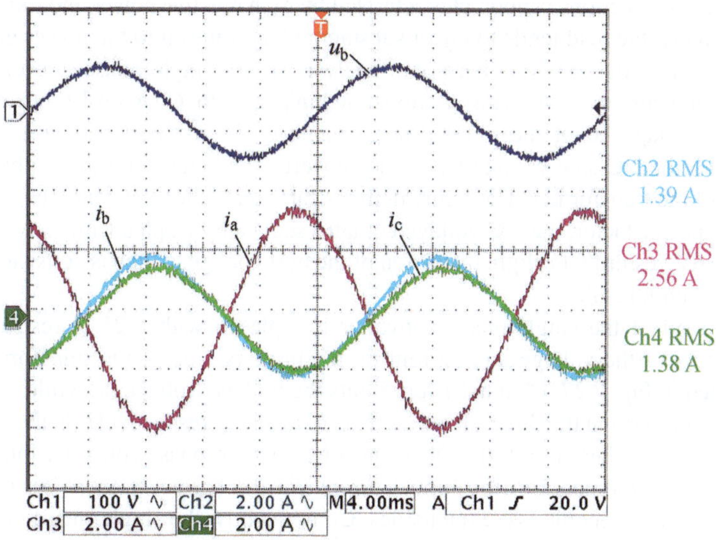

Fig. 7.25 Three-phase voltage and current waveforms of power supply across neutral sections under 50-Ω load

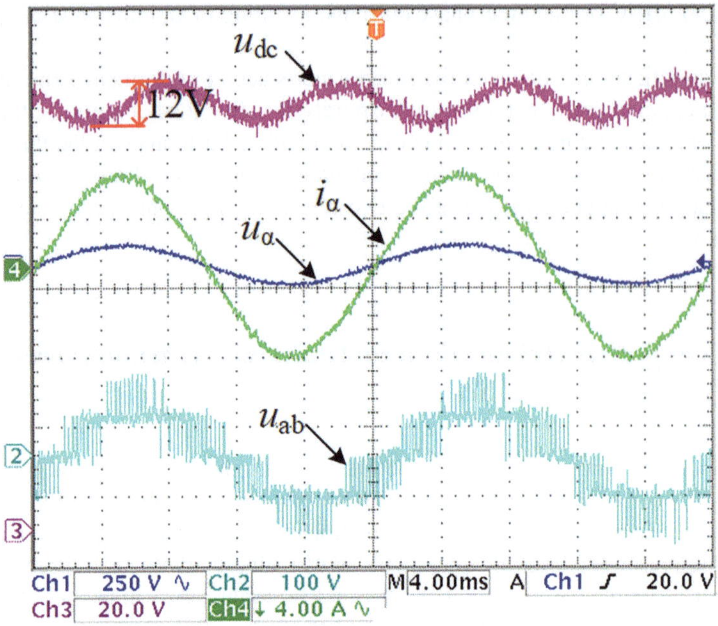

Fig. 7.26 Voltage and current waveforms of single-phase converter

120° phase difference, with the b-phase and c-phase currents nearly overlapping, resulting in a current imbalance of up to 43.8%. This suggests that without intervention, the power quality of the three-phase grid in the mode of traction power supply across neutral sections is poor, characterized by high reactive power and imbalance. Additionally, the load tends to inject substantial harmonics into the power grid.

To validate the previous harmonic analysis of the single-phase converter, the voltage and current waveforms are illustrated in Fig. 7.26. Channel 1 represents the α-phase voltage uα of the transformer, channel 2 the α-phase port voltage uαb, channel 3 the DC-side voltage u_{dc} of the converter, and channel 4 the α-phase current iα of the transformer. The figure demonstrates that when the converter operates at unity power factor, the DC voltage contains secondary ripples with an amplitude of up to 6V, resulting in slight distortion of the port voltage, which is consistent with the simulation results.

The load of the single-phase converter is replaced with a 25-Ω resistor. The waveforms of the voltage and current on the primary side of the transformer are illustrated in Fig. 7.27. Channel 1 represents the b-phase voltage u_b, while channels 2, 3, and 4 represent the three-phase currents i_b, i_a, and i_c, respectively. In the absence of a secondary filter circuit, the DC ripple of the single-phase converter introduces significant harmonics into the input current via the hardware loop. As the load increases, the harmonic issue intensifies. Consequently, the single-phase converter is used as a harmonic source to simulate locomotive load conditions. The figure shows that, with substantial harmonics in the traction transformer's load, the

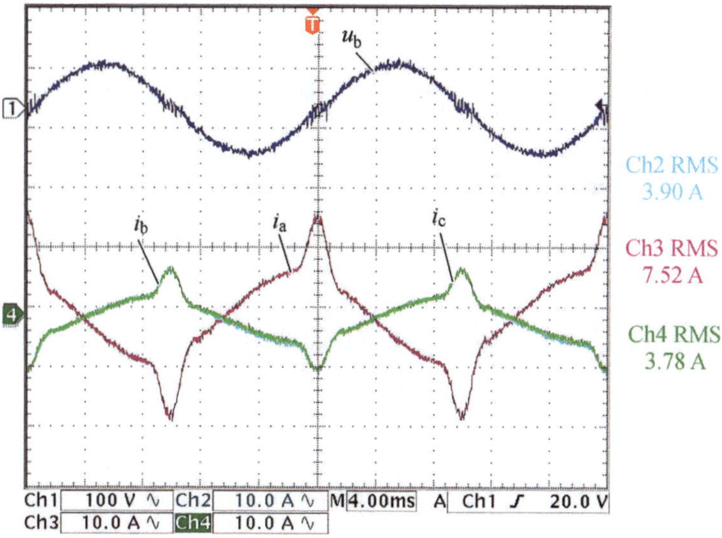

Fig. 7.27 Three-phase voltage and current waveforms of power supply across neutral sections under 25-Ω load

three-phase currents are severely distorted, affecting the three-phase voltages and leading to increased voltage distortion. Thus, the locomotive load significantly impacts the power system.

7.5.4 Experimental Results and Analysis of Continuous Power Supply Before Improvement

The two-phase converter serves as the transformer load for the continuous power supply experiment. The control and modulation of each converter operate independently, utilizing a transient current control strategy and a single-phase SPWM modulation strategy. Notably, the control and modulation parameters of both converters are completely consistent.

When the two-phase converter is under no-load condition, waveforms of the voltage and current on the transformer's secondary side are illustrated in Fig. 7.28. Channel 1 represents the transformer α-phase voltage u_α, channel 2 the transformer α-phase current i_α, channel 3 the transformer β-phase voltage u_β, and channel 4 the transformer β-phase current i_β. As can be seen from the figure, the α-phase voltage lags the β-phase voltage by 90°, the α-phase current is zero, and the β-phase current exhibits a high amplitude.

After the load of the two-phase converter is increased to 20 Ω, the voltage and current conditions of the converter are illustrated in Fig. 7.29. In Fig. 7.29a, channel 1 represents the transformer β-phase voltage u_β, channel 2 the β-phase current i_β,

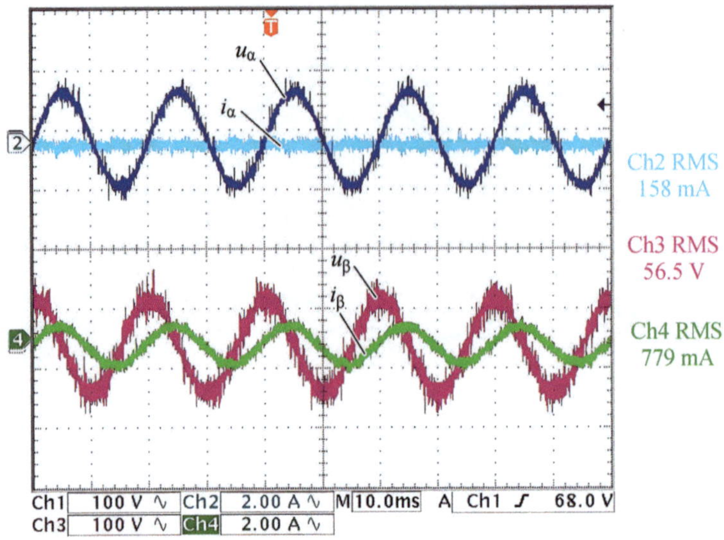

Fig. 7.28 No-load waveforms of the two-phase converter before improvement

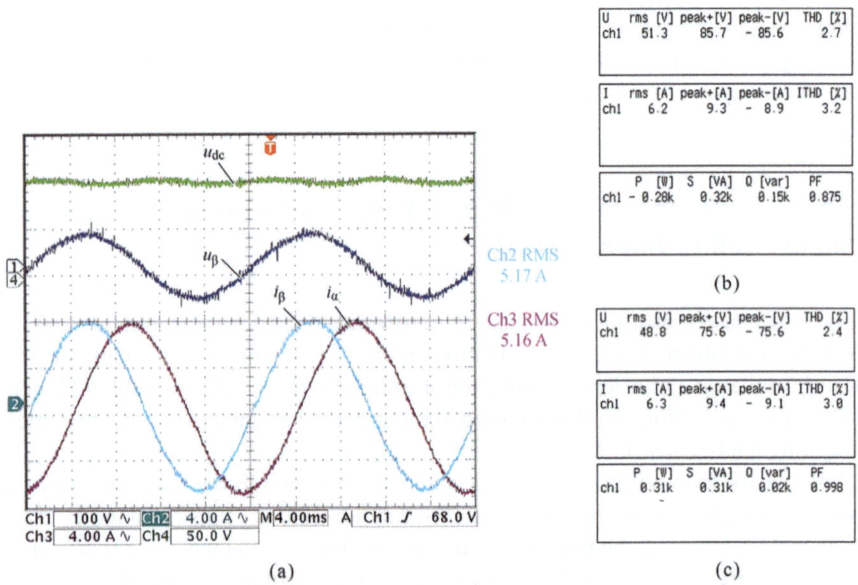

Fig. 7.29 Voltage and current of the two-phase converter under 20-Ω load

channel 3 the α-phase current i_α, and channel 4 the converter DC-side voltage u_{dc}. Figure 7.29b, c presents the power quality data for the α-phase and β-phase of the transformer, respectively. As can be seen from the figure, compared with a single-phase converter, when the two-phase input voltages of the two-phase converter have

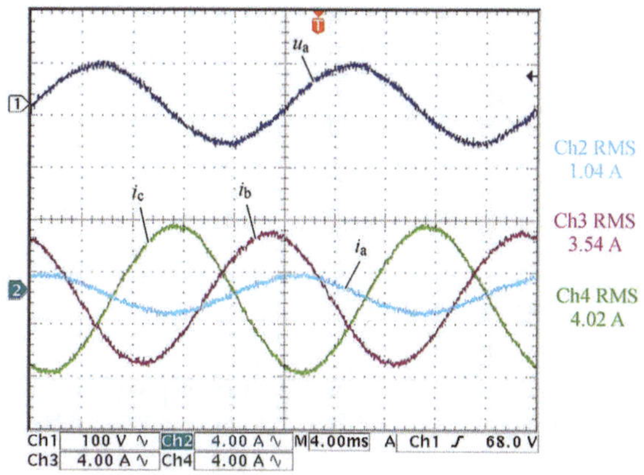

(a)

U	rms [V]	peak+[V]	peak-[V]	THD [%]
ch1	48.9	75.8	- 75.6	2.3
ch2	12.4	24.5	- 26.4	7.4
ch3	12.4	25.4	- 26.4	7.8

I	rms [A]	peak+[A]	peak-[A]	ITHD [%]
ch1	3.7	5.6	- 5.2	5.5
ch2	4.4	6.6	- 6.4	4.6
ch3	5.0	7.3	- 7.3	4.0
ch4	0.0	1.4	- 0.5	

	P [W]	S [VA]	Q [var]	PF
ch1	0.05k	0.18k	- 0.17k	-0.297
ch2	- 0.05k	0.05k	- 0.01k	-0.971
ch3	0.04k	0.06k	0.05k	0.642
sum	0.0k	0.3k	- 0.3k	-0.136

Uave [V]	Iave [A]	Uunb [%]
24.6	4.3	100.0

(b)

Fig. 7.30 Input voltage and current of the transformer under 20-Ω load

a 90° phase difference, the secondary ripple of the DC voltage is completely offset, the amplitudes of the two-phase currents are basically the same, and the harmonic content is low. However, the α-phase power factor is only 0.875, and the system operates at a non-unity power factor.

The waveforms of the transformer's three-phase input voltage and current are illustrated in Fig. 7.30a. Channel 1 represents the a-phase voltage u_a, while channels 2, 3, and 4 represent the three-phase currents i_a, i_b, and i_c, respectively. The figure illustrates an imbalance in the transformer's three-phase input currents. Figure 7.30b presents power quality data for the traction transformer's three-phase input, as measured by the harmonic analyzer. Channels 1, 2, and 3 correspond to phases a, b, and c, respectively. The reactive current is high, the phase angle relationships of the currents are irregular, and the effective values of the three-phase currents are 3.7 A, 4.4 A, and 5.0 A, with an imbalance of 15.33%.

The experimental results indicate that when two single-phase converters in a two-phase converter are controlled and modulated separately, one converter exhibits no-load reactive current. As the load increases, both the reactive current and current imbalance of the two-phase converter rise, ultimately resulting in significant reactive current and imbalance on the three-phase side.

7.5.5 Experimental Results and Analysis of Continuous Power Supply After Improvement

1. Experimental results and analysis of the two-phase converter

The two-phase converter examined in this study utilizes the two-phase DQ decoupling control and two-phase SVPWM modulation methods proposed herein. When the load is a 20-Ω resistor, the voltage and current of the two-phase con-

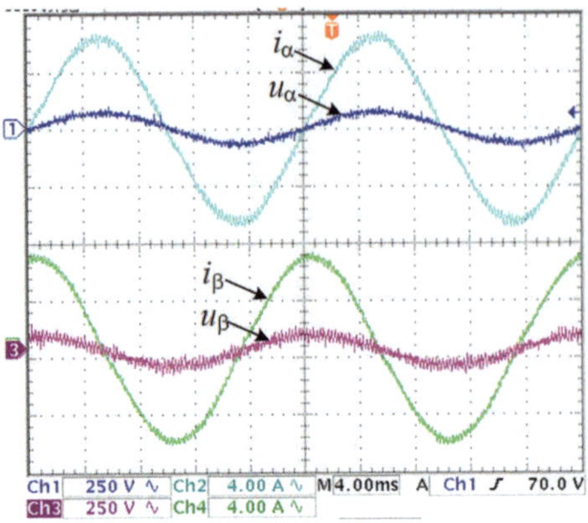

(a)

U	rms [V]	peak+[V]	peak-[V]	THD [%]
ch1	50.6	75.8	- 75.6	1.8

I	rms [A]	peak+[A]	peak-[A]	ITHD [%]
ch1	4.4	6.9	- 6.6	6.2

	P [W]	S [VA]	Q [var]	PF
ch1	0.22k	0.22k	0.03k	0.988

(b)

U	rms [V]	peak+[V]	peak-[V]	THD [%]
ch1	49.4	80.0	- 79.7	2.3

I	rms [A]	peak+[A]	peak-[A]	ITHD [%]
ch1	4.2	6.5	- 6.3	6.0

	P [W]	S [VA]	Q [var]	PF
ch1	0.20k	0.21k	0.06k	0.980

(c)

Fig. 7.31 Voltage and current of the two-phase converter under 20-Ω load

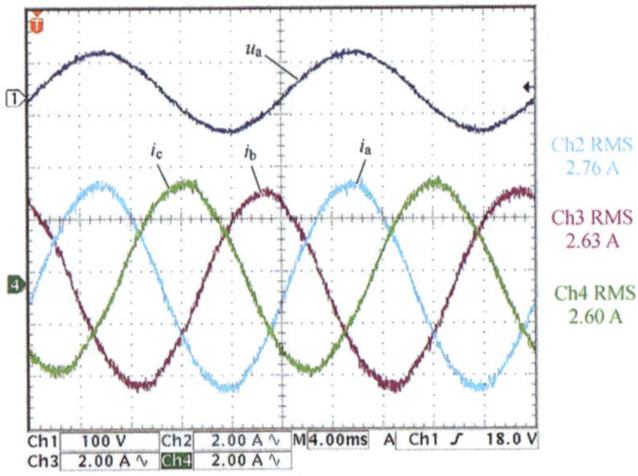

(a)

U	rms [V]	peak+[V]	peak-[V]	THD [%]
ch1	52.6	78.6	- 78.5	1.8
ch2	53.4	79.6	- 78.9	2.1
ch3	53.8	81.6	- 81.0	1.9

I	rms [A]	peak+[A]	peak-[A]	ITHD [%]
ch1	3.7	5.8	- 5.4	6.0
ch2	3.5	5.6	- 5.2	6.4
ch3	3.2	4.8	- 4.7	6.7
ch4	0.0	1.4	- 0.5	

	P [W]	S [VA]	Q [var]	PF
ch1	0.19k	0.19k	0.02k	0.993
ch2	0.19k	0.19k	- 0.02k	0.995
ch3	0.17k	0.17k	0.02k	0.995
sum	0.5k	0.6k	0.1k	0.994

Uave [V]	Iave [A]	Uunb [%]
53.2	3.5	1.1

(b)

Fig. 7.32 Input voltage and current of the transformer under 20-Ω load

verter are illustrated in Fig. 7.31. In Fig. 7.31a, channel 1 represents the transformer α-phase voltage u_α, channel 2 the α-phase current i_α, Channel 3 the β-phase voltage u_β, and Channel 4 the β-phase current i_β. Figure 7.31 b and c presents the power quality data for the α-phase and β-phase of the transformer, respectively. The figure indicates that the effective values of the transformer's two-phase output currents are 4.4 A and 4.2 A, with harmonic contents of 6.2% and 6%, and power factors of 0.988 and 0.980, respectively. It can be seen that the input currents of the two-phase converter are balanced within acceptable error margins and exhibit low harmonic content, and the system operates at unity power factor.

At this time, the voltage and current waveforms of the three-phase input of the transformer are illustrated in Fig. 7.32a, where channel 1 represents the a-phase

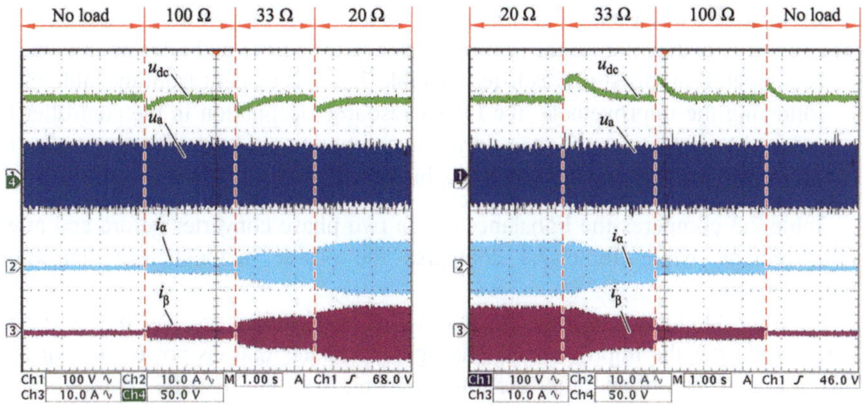

(a) (b)

Fig. 7.33 Dynamic waveforms of the two-phase converter

voltage u_a, and channels 2, 3, and 4 represent the three-phase currents i_a, i_b, and i_c, respectively. Figure 7.32b presents power quality data for the traction transformer's three-phase input, as measured by the harmonic analyzer. Channels 1, 2, and 3 correspond to the voltage, current, and power data of phases *a*, *b*, and *c*, respectively. The figure shows that the effective values of the three-phase currents are 3.7 A, 3.5 A, and 3.2 A, resulting in an imbalance of approximately 7.7%. Compared with that before improvement, the imbalance is reduced by about half. The three-phase power factors are 0.993, 0.995, and 0.995, indicating that the system operates at a near-unity power factor. Harmonic contents are low at 6.0%, 6.4%, and 6.7%, respectively.

Figure 7.33 presents the dynamic waveforms of the two-phase converter, where channel 1 represents the DC-side voltage u_{dc}, channel 2 the a-phase voltage waveform u_a, and channels 3 and 4 the input currents i_α and i_β of the two-phase converter, respectively. Figure 7.33a displays the dynamic waveforms for three load conditions, transitioning from no-load to 100 Ω, then to 33 Ω, and finally to 20 Ω. Figure 7.33b illustrates the dynamic waveforms for three load reductions, with the load transition sequence in reverse. It can be seen that the system reaches a steady state within 1.5 s after each load transition, with a stable DC voltage of 100 V and consistent amplitudes in the two-phase input currents. During load increases, the DC voltage drop is capped at 20 V, while during load decreases, the voltage overshoot remains below 30 V. Thus, the converter employing the two-phase DQ decoupling method demonstrates excellent dynamic and static performance.

The results indicate that the improved two-phase converter demonstrates good dynamic and static performance. Excluding the error of the YNvd traction transformer, the imbalance and harmonic content in the three-phase grid are minimal, and the system operates at unity power factor.

2. Comparative analysis

The resource occupation of the two-phase converter before and after improvement is compared in this study, as listed in Table 7.2. It is evident from the table that, following the improvement, the FPGA resource occupation in the controller is reduced by approximately one-third, the computational load of the controller is decreased, and the operating speed is increased.

Table 7.3 compares the imbalance of the two-phase converter before and after improvement, while Table 7.4 compares the power factor before and after improvement.

Figure 7.34 illustrates the comparison curves of imbalance and power factor before and after the improvement. The figure indicates that, as power and current increase, both before and after the improvement, the control system demonstrates

Table 7.2 Resource occupation of the two-phase converter before and after improvement

Logical resource	Logic unit	Logic memory	Storage unit	DSP unit
Before improvement	9431	8611	33,792	66
After improvement	6673	5167	33,792	42

Table 7.3 Comparison of imbalance before and after improvement

Power	100 W	200 W	300 W	400 W	500 W
Before improvement	100%	50%	50%	50%	15.33%
After improvement	/	16.67%	8.2%	7.4%	7.7%

Table 7.4 Comparison of power factors before and after improvement

Power	100 W	200 W	300 W	400 W	500 W
Before improvement	0.990	0.762	0.748	0.680	0.500
After improvement	1	0.976	0.990	0.993	0.950

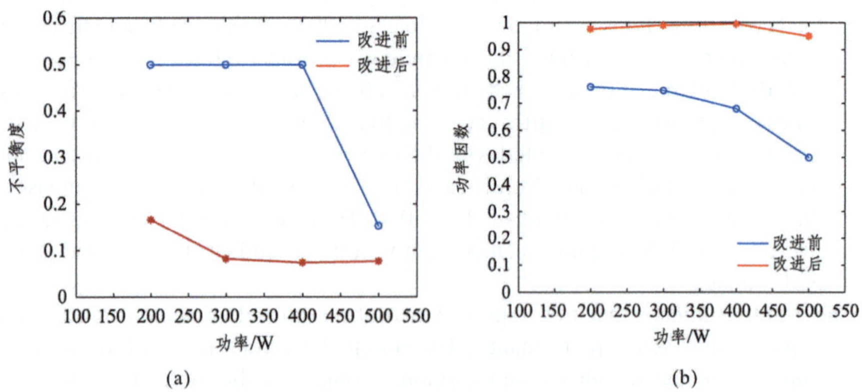

Fig. 7.34 Comparison of imbalance and power factor test results for the two-phase converter before and after improvement under varying power conditions

enhanced accuracy in current regulation, leading to a progressively smaller imbalance. Notably, the imbalance with the improved control method is significantly lower than that of the traditional method. Without the improved control method, the power factor of the two-phase converter gradually decreases with increasing power. In contrast, with the improved control method, the converter basically operates at unity power factor.

7.6 Summary

This chapter focuses on the advanced co-phase traction power supply system based on the diode-clamped three-level two-phase-to-single-phase converter. Comprehensive simulations are conducted on its mathematical model, control, and modulation strategies. Experiments validate the electric power quality of the output waveform and the dynamic and static performance of the system output. Additionally, comparative analyses are performed against the power supply across neutral sections and traditional three-phase control and modulation strategies. The two-phase-to-single-phase converter system exhibits the following characteristics:

1. The two-phase DQ decoupling control strategy treats two independent single-phase converters as a unified system, allowing for separate control of active power and reactive power. This approach ensures balanced two-phase input currents and operation at unity power factor. Additionally, the method features a low computational load, rapid response speed, no steady-state error, and a stable, reliable system.
2. The secondary ripples on the DC bus of the two-phase converter, characterized by a 90° phase difference between the input voltages, cancel each other out. This eliminates the need for filtering components in the hardware filter and control loop, significantly simplifying hardware and software design.
3. The traction network side of the continuous co-phase system needs to transmit active power and compensate for load reactive power. Consequently, a simulation and experimental platform for single-phase rectifiers is developed to validate that the single-phase rectifier employing the single-phase dq decoupling control method can effectively compensate for both inductive and capacitive reactive power in the system.
4. The feasibility of the co-phase power supply system based on three-level two-phase-to-single-phase converter and the validity of corresponding control strategy are verified through simulations and experiments. The system can transmit active energy bidirectionally and compensate for reactive energy. The three-phase grid consistently operates at unity or negative unity power factor.
5. Experimental comparisons reveal that the improved continuous power supply system experiences significant reactive power in the three-phase grid across neutral sections, resulting in high current imbalance and severe current distortion. Before improvement, the continuous co-phase power supply system experiences

no-load current and struggles to control reactive power to zero under load, resulting in unbalanced three-phase grid currents and elevated reactive power. After improvement, the continuous co-phase power supply system achieves better control of active power and reactive power in the two-phase converter, resulting in basically balanced three-phase grid currents, low harmonic content, and reduced reactive power.

6. The two-phase-to-single-phase converter system aligns with the two-phase output characteristics of existing traction transformers, enabling higher compatibility and lower transformation costs when existing substations for advanced co-phase traction power supply are upgraded.

References

Chen Yao, Tong Yibin, Jin Xinmin. A novel algorithm of SVPWM harmonic analysis based on PWM rectifier [J]. Proceedings of the CSEE, 2007;27(13): 76–80.

Lee TS, Liu JH. Modeling and control of a three-phase four-switch PWM voltage-source rectifier in dq synchronous frame [J]. IEEE Transactions on Power Electronics, 2011, 26(9): 2476-2489.

Liu Neng. Study on compensation optimization scheme for co-phase traction substation [D]. Master's Degree Dissertation of Southwest Jiaotong University, 2010.

Oettmeier M, Heising C, Bartelt R, et al. Flux-based multivariable control of a static converter feeding a 16.7-Hz single-phase load [M]. In: 2009 35th Annual Conference of IEEE Industrial Electronics. 2009, pp. 492-497.

Pi Junbo, Jiang Qirong, Wei Yingdong. Research on a scheme of co-phase power supply system for electrified railways based on cascade structure UPQC [J]. Electrical Equipment 2008 (10).

Wei Guang. Co-phase traction power supply system based on V connection [J]. Electric Power Automation Equipment 2010;30(12).

Zhang Rui. Research on protection solution of electric energy conversion equipment in continuous co-phase power supply system [D]. Master's Degree Dissertation of Southwest Jiaotong University, 2012a.

Zhang Rui. Research on protection solution of electric energy conversion equipment in continuous co-phase power supply system [D]. Master's Degree Dissertation of Southwest Jiaotong University, 2012b.

Zheng Zhongjiu. Research on control strategy and application of three-phase voltage type PWM rectifier [D]. Doctoral Dissertation of Dalian University of Technology, 2011.

Zhou Fulin, Li Qunzhan, Qiu Daqiang. A co-phase traction power supply system based on new AT power supply mode [J]. Journal of Hunan University (Natural Sciences), 2010;37(4):24–30.

MIX
Papier aus verantwortungsvollen Quellen
Paper from responsible sources
FSC® C105338

If you have any concerns about our products,
you can contact us on
ProductSafety@springernature.com

In case Publisher is established outside the EU,
the EU authorized representative is:
Springer Nature Customer Service Center GmbH
Europaplatz 3, 69115 Heidelberg, Germany

Printed by Libri Plureos GmbH
in Hamburg, Germany